The Emergence of
Professional Social Science

THOMAS L. HASKELL

The Emergence of Professional Social Science

THE AMERICAN SOCIAL SCIENCE ASSOCIATION
AND THE NINETEENTH-CENTURY
CRISIS OF AUTHORITY

THE JOHNS HOPKINS UNIVERSITY PRESS
Baltimore and London

For my parents,
MARTHA B. HASKELL
and A. PORTER HASKELL

Copyright © 1977 by the Board of Trustees of the University of Illinois
Copyright © 2000 by The Johns Hopkins University Press
All rights reserved
Printed in the United States of America on acid-free paper

Originally published by the University of Illinois Press, 1977
Johns Hopkins Paperbacks edition, 2000
9 8 7 6 5 4 3 2 1

The Johns Hopkins University Press
2715 North Charles Street
Baltimore, Maryland 21218-4363
www.press.jhu.edu

LIBRARY OF CONGRESS CATALOGING-IN-PUBLICATION DATA

Haskell, Thomas L., 1939–
　　The emergence of professional social science : the American Social Science
　Association and the nineteenth-century crisis of authority / Thomas L. Haskell.
　　　　p.　　cm.
　　Originally published: Urbana : University of Illinois Press, 1977. With new pref.
　Includes bibliographical references and index.
　　ISBN 0-8018-6573-5 (pbk. : alk. paper)
　　1. American Social Science Association. 2. Social sciences—United States—
　History. I. Title.

H11.A793 2000
300'.6'073—dc21   00-059092

A catalog record for this book is available from the British Library.

Contents

Preface to the Johns Hopkins Edition
page vii

Preface to the Original Edition
page xix

CHAPTER I
Introduction: What Happened in the 1890's?
page 1

CHAPTER II
Interdependence and the Rise of Professional Social Science
page 24

CHAPTER III
Frank Sanborn's Association
page 48

CHAPTER IV
The Antebellum Origins of the Movement to Establish Authority
page 63

CHAPTER V
The Founding and Formative Years of the ASSA, 1865–69
page 91

CHAPTER VI
The 1870's: Near Collapse
page 122

CHAPTER VII
The Proposed Merger of the ASSA and the Johns Hopkins University
page 144

CHAPTER VIII
The Founding of the Historical and Economics Associations, 1884–85
page 168

CHAPTER IX
From Social Science to Sociology: The 1880's and 1890's
page 190

CHAPTER X
Professionalism Unhinged
page 211

CHAPTER XI
Conclusion: Explanation and Causal Attribution in Modern Society
page 234

Appendix A
page 257

Appendix B
page 259

Bibliographical Note
page 285

Author-Title Index
page 287

Index
page 295

Preface to the Johns Hopkins Edition

The 1977 edition of *The Emergence of Professional Social Science* was born a stepchild. The university press that published it, though well meaning and eminently respectable, aspired to prominence in social history, not intellectual history. Chronic cash flow problems left over from the recession of 1973 set the publishing schedule back twelve months. The press run was kept so small that the book quickly went out of print without remainder, allowing it sometimes to slip through the net even of research libraries, which acquire academic monographs by blanket order. Meanwhile, the Library of Congress assigned the book a call number alongside the *Encyclopedia of Associations*—placing it aisles away from its natural home on library shelves devoted to social thought and professionalization.

In spite of inauspicious beginnings, *Emergence* garners a modest but respectable number of citations every year, many by scholars working in disciplines other than history. Given the book's marketing history, I am more than ordinarily gratified by the decision of Robert Brugger and the Johns Hopkins University Press to reissue it in this convenient and reasonably priced format, complete with appendixes that were deemed too costly for inclusion in the first edition. The first and most unconditional obligation of any scholarly monograph is to facilitate further research into its subject matter. The appendixes to this edition will do just that by providing researchers with a list of the chief officers of the American Social Science Association (ASSA), along with an index, by author, of all articles published in the association's principal publication, the *Journal of Social Science*, from its inception in June 1869, through November 1901.

Fresh archival sources, only recently opened up, make the appearance of a new edition especially timely. In 1969, when I began scouring the country for information, there was no single archival collection that embraced any substantial part of the ASSA's story, much less the whole of it. Lacking access to any central holding of ASSA papers, I and other scholars who were curious about the "mother of associations" had to piece the story together as best we could from widely scattered collections devoted to other subjects. I benefited immensely from the generosity of John W. Clarkson, who for many years had been collecting copies of the correspondence of Franklin Benjamin Sanborn, the main secretary and prime mover of the association from its founding in 1865 until decline set in around the turn of the century. But according to oral tradition still circulating in Sanborn's history-laden hometown of Concord, Massachusetts, upon his death Sanborn's sons sold what they could of his correspondence (including official ASSA records) and burned the rest. Since Sanborn and other officers of the association were in communication with nearly all the significant American reformers and social thinkers of the era, as well as many prominent politicians, the association's records were a potential gold mine—not for scholars alone but also for collectors who buy and sell the signatures of famous historical figures.

Having come to believe that many of the association's records had been lost to the world of scholarship, I was pleasantly surprised to learn last fall that a substantial part of Sanborn's files as ASSA secretary survived intact and are now available to scholars.[1] In 1989, twelve years after I published *Emergence,* the head librarian of Yale Law School, Morris L. Cohen, discovered in the basement of the law library an old trunk full of ASSA papers. The collection had been sitting there in the gloom since 1906, unexamined and uncataloged for nearly a century. Simeon E. Baldwin, Yale professor of law and president of the ASSA in 1898–99, about the time of Sanborn's resignation, probably had a hand in the acquisition.

A recent three-day tour of the new collection yielded mixed results. On the one hand, I am relieved to report that, although access to these records would of course have made my research easier and my history more complete, I did not uncover any heart-sinking surprises that unravel the interpretation set forth in *Emergence.* On the other hand, neither did I uncover any conclusive answers to what I regard as the outstanding puzzles and gaps in the history of the ASSA. But of course three days is not enough to scrutinize all parts of the collection, so the jury is still out on both counts.

Even a whirlwind visit was enough to convince me of the collection's importance, not only for scholars interested in the history of the social sciences, but for anyone who wants to speak authoritatively about late-Victorian culture in America or the antecedents of Progressive Era

---

[1] I am much obliged to Professor Dalia Tsuk, a recent graduate of the Law and History Program of Yale University, for bringing the collection to my attention.

and New Deal reform. There are, in all, thirteen boxes of correspondence, meticulously organized and fitted out with a splendid finding aid by the staff of the Manuscripts and Archives Division of Yale's Sterling Memorial Library.[2] Interpretive novelties are bound to emerge from a collection as rich as this one, but it is not for me to predict what they might be. There may be some value, however, in registering a few preliminary impressions about the scope and significance of the collection, especially insofar as it bears on issues of fact that have taken shape since the publication of *Emergence*.

Priceless though it undoubtedly is, the Yale collection is less complete than historians could wish. It contains only a scattering of items from the late 1870s and '80s and practically nothing from the '90s or later. There are important documents relating to the founding of the ASSA, including an interesting draft of bylaws proposed by Caroline Healey Dall but never adopted.[3] The collection is especially rich in information about the association's activities during the late 1860s and early '70s. This happened in part because Henry Villard, a far more orderly and systematic person than Frank Sanborn, kept good records during his term as secretary (1868–70) and partly because the executive committee maintained a remarkable set of minutes that runs continuously from 9 January 1867 to 25 October 1873.[4] In short, what we have is an important fragment of the official records of the association, rich in its coverage of the early years, but disappointingly thin on the 1880s and '90s, the pivotal decades when the rise of the modern American university put the old association on the road to extinction.

I hoped above all that the new collection would shed light on the merger that Harvard mathematician and astronomer Benjamin Peirce proposed between the ASSA and the new Johns Hopkins University in 1878—that, and the subsequent decision by Hopkins President Daniel Coit Gilman not to consummate the merger but instead to recruit the Hopkins faculty into the ambitious project of academic professionalization that eventually deprived the ASSA's brand of social science of the considerable authority it once enjoyed.[5] If ever there was a fateful decision in the development of academic social science in America, this was it. The collection does contain letters and reports written by Sanborn

---

[2] The finding aid, prepared by Diane E. Kaplan and Katherine M. Lewis, is titled "American Social Science Association Records, Manuscript Group Number 1603, New Haven, Connecticut, June 1991."

[3] "Some Hints on the Formation of a Society of Social Science," Box 1, folder 2.

[4] The book of minutes will be found in Box 7, folders 178 and 179. See also folder 182, "Records of Constitutional Revision."

[5] *Emergence*, chs. VII, VIII.

during 1878 and 1879 that fill in gaps in the correspondence available to me when I wrote *Emergence,* but they add nothing of dramatic import.[6] I found nothing new about what Peirce had in mind in proposing the merger; nothing that would help us understand how plausible or implausible the proposal appeared in Gilman's eyes, or exactly when and why he decided against it.

Although disappointing on this score, the collection opened my eyes on a closely related issue. It turns out that Gilman was not only a charter member of the ASSA and a member of the committee empowered to nominate its first slate of officers, as I reported in *Emergence.* He was also in steady correspondence with ASSA headquarters all through the 1860s and '70s. This surprised me. Whether writing from New Haven, where he seems to have organized a branch association that was more substantive than most, or from the University of California, where, as president, he aspired to organize yet another branch, he attached more importance to the work of the association than I had realized. This makes his decision against the merger all the more interesting. Gilman was deeply involved in the management of the association and had invested a great deal of his own time and energy in it when Peirce presented him with an opportunity to take it over and incorporate it into the structure of the Johns Hopkins University. Deciding against the merger could not have come easily; it was even less a foregone conclusion than I realized when I wrote *Emergence.*

One might also have hoped that the Yale collection would tell us more about the founding of the American Historical Association (AHA) and the American Economic Association (AEA). These two associations were organized in 1884 and 1885, respectively, at the annual meetings of the ASSA, by members of Gilman's faculty, acting with his encouragement and advice. Together with the Modern Language Association (1883), also organized by Hopkins faculty at Gilman's urging, the new associations established a model of disciplinary specialization and professionalization that would soon spread like wildfire through the entire academy. About these fateful developments, my quick survey of the Yale collection yielded one important finding: a new letter from the organizer of the AEA, Richard T. Ely, to Sanborn, defending the AEA statement of principles, which seemed to bar membership to any economist unwilling to repudiate laissez-faire doctrines. The handwriting is badly faded, but one can infer that in earlier correspondence Sanborn had urged Ely to meet with his critics and reconsider the controversial language of the statement. Ely resisted. Speaking out of both sides of his mouth, he acknowledged that "we are purposely exclusive," even as

---

[6] Box 7, folder 179; Box 8, folder 185.

he lamely tried to assure Sanborn that "we wish to encourage the largest freedom of discussion both within and without our colleges."[7]

The most pressing issues of fact that have surfaced since the publication of *Emergence* lie in another direction altogether. They concern the role of women in the association and the place of the association itself in the feminist movement. It is of course true that in the years 1969–75, when I was researching and writing *Emergence*, these questions were nowhere near the forefront of my consciousness. As anyone of my generation can testify, they did not assume their present politically charged status among members of the historical profession until the 1980s. Although similar tactics would become commonplace in the years ahead, I remember being astonished when a historian who was a complete stranger to me sent me a note archly announcing that *Emergence*, then newly published, had been found wanting by a prize committee on which she served because so little attention had been devoted to the ASSA's women members. How, I wondered at the time, could anyone not directly acquainted with the archival evidence claim so confidently to know how important a role women played in the ASSA?

But once political passions are aroused, familiarity with the evidence is easily trumped by more visceral sources of confidence. Today it is widely believed by historians that, however little I may have recognized it, the ASSA was a bastion of feminist activism. This notion may have originated in a passing comment made in the compendious *History of Woman Suffrage* published by Elizabeth Cady Stanton and her colleagues in 1886. In the chapter devoted to the state of Massachusetts, Harriet H. Robinson, an associate of Sanborn's, credited the ASSA and its Boston branch with being the "first large organizations in the country to admit women on an absolute equality with men."[8] When I first read those words, several years after publication of *Emergence*, I remember being surprised that Robinson thought the ASSA so distinctive. I knew that women had constituted a significant fraction of the ASSA membership (about 13% in 1867) and I knew that a few, Caroline Healey Dall prominent among them, had served on the executive committee.[9] But given that the association was a gathering of intellectually ambitious social reformers, mainly New Englanders, its receptivity to women did

---

[7] R. T. Ely to F. B. Sanborn, 24 August 1885, Box 4, folder 64.

[8] Elizabeth Cady Stanton, Susan B. Anthony, and Matilda Joslyn Gage, *History of Woman's Suffrage* (Rochester, N.Y.: N.p., 1886), 3:306. Eileen Yeo likewise credits the British National Association for the Promotion of Social Science, upon which the ASSA was modeled, with being the "first middle-class forum in Britain to welcome the public voice of women." Eileen Janes Yeo, *The Contest for Social Science: Relations and Representations of Gender and Class* (London: Rivers Orem Press, 1996), 129.

[9] See membership list in pamphlet titled "Samuel Eliot's Address at the 5th General Meeting, New York, 19 November 1867," Box 1, folder 9.

not seem especially noteworthy. I was also aware that Sanborn was an outspoken feminist and that the American Woman Suffrage Association was one among the multitude of organizations to which he devoted time and energy, but he would have been the last to claim that his views were typical of the association's members or officers. To be sure, in the course of reading thirty years of the *Journal of Social Science* and examining hundreds of letters in a dozen or more manuscript collections, I had come across praise of coeducation, pleas for expanding the franchise, and expressions of concern about women's wages. But when compared to the association's major preoccupations—prison reform, charity organization, the care of lunatics, the silver question, civil service reform, cooperative savings banks—my impression was that women's issues did not loom large. Any reader can put my impression to the test by scanning the list of *Journal of Social Science* articles that appears in Appendix B.

That feminism nonetheless tapped a deep reservoir of latent support in the association was evident from a single, thinly documented episode in 1870, in which Sanborn and other leaders successfully fought off an effort to exclude women from the executive board. I reported what little I knew of this defensive measure in *Emergence:*

> Reformers fell to quarreling [in the early 1870s] not only about national policies, but also about the internal affairs of the association. In 1870 E. L. Godkin of the *Nation* led an effort to raise the level of discussions in the ASSA by means of a restrictive classification of membership. Sanborn opposed this change, as he always opposed measures designed to exclude "cranks" and "eccentrics." He also vigorously opposed a movement in 1870 to remove women from the governing board of the association; he had insisted from the beginning on a prominent place for women in the affairs of the ASSA.[10]

But this did not suffice. In the militant atmosphere of the 1980s, allegations of male indifference to the historical achievements of women needed only to be voiced to be believed. The degree of my alleged culpability has varied from one commentator to another, but the presumption that I overlooked, ignored, downplayed, or suppressed the central part played by women in the ASSA has been by far the most common complaint about *Emergence,* repeated in many a footnote.[11]

---

[10] *Emergence,* 129.

[11] For recent and admirably mild-mannered examples that allege nothing worse than negligence, see the footnotes to essays collected by Helene Silverberg, ed., *Gender and American Social Science: The Formative Years* (Princeton: Princeton University Press, 1998). One essay in the volume even goes so far as to suggest that the ASSA may not have been quite as devoted to women's issues as its current reputation supposes: Mary G. Dietz and James Farr, "'Politics Would Undoubtedly Unwoman Her': Gender, Suffrage, and American Political Science," 66.

The complaint took on an aura of self-evident validity after the publication in 1980 of William Leach's *True Love and Perfect Union: The Feminist Reform of Sex and Society*. Here was a striking, full-blown portrait of the ASSA, depicting it not only as exceptionally receptive to women's interests but as the headquarters for a nascent science of society the principal aim of which was to transform relations between the sexes, thereby forging a unified ideology for a fractured bourgeoisie.

> The American Social Science Association was the queen of bourgeois reformist organizations. It spread its wings over lesser reform bodies and brought to life a myriad of new organizations. The ideas of the association were those of many, often superficially diverse societies and groups—reform spiritualists, moral educationists, Free Religionists, dress reformers, and others. Ideologically it expressed at the highest level the major themes of the most important feminist organizations and showed how thoroughly post–Civil War feminism was shaped within the context of an emerging new world view. . . . It was the [ideological] response of men and women eager to settle social and institutional arrangements within their own class and between the sexes.[12]

In Leach's eyes, it was not only individual members or factions, but "the association itself [that] showed a strong feminist bias almost from its inception" (300). And just as feminism was central to the mission of the association, so he believed social science was central to feminism: "Social science perspectives colored every facet of the feminist movement and tended to unify the vast array of feminist organizations that appeared after the Civil War" (292). Of 193 male ASSA members in 1874, Leach counted "forty at least [who] locked hands with the feminist movement" (313). But however numerous the feminist members of the ASSA and however deep their commitment to feminism, women's real stronghold, Leach implied, lay not in the national organization but in its local branches and affiliates. These he assumed to be numerous, active, closely coordinated with the parent association, and predominantly female. Taking at face value an 1880 report in the *Journal of Social Science,* he set forth a list of local branches so extensive as to suggest that the real heart of the association's work lay not at the national level, but in its local, predominantly feminine affiliates (318).

Could Leach be right? His is a stimulating interpretation that gives us much to think about. In its favor, Eileen Janes Yeo, in her discussion of

---

[12] William Leach, *True Love and Perfect Union: The Feminist Reform of Sex and Society* (New York: Basic Books, 1980), 323. Far from claiming an identity of feminist purpose across time, Leach was out to show that nineteenth-century feminism was tangled up with an array of diverse bourgeois interests, of which the only common denominator was the hope of bringing forth a revivified ruling class. Subtleties such as these were lost on the profession at large, however.

the ASSA's godparent, the British National Association for the Promotion of Social Science (NAPSS), reaches somewhat similar conclusions about the importance of gender in that organization—even though she acknowledges that women accounted for less than 7 percent of the NAPSS membership.[13] I remain exceedingly skeptical, but only further research will answer the question conclusively. I have spent more time in the ASSA archives than Leach has; he has spent more time in the archives of women's history than I have. To reconcile our interpretive differences, one would need at a minimum to know what, if anything, was really going on in all those local branches. Who belonged; how often they met; what they *did* when they met; how long each local branch endured; what relation, if any, it had to the ASSA. Careful readers will find few answers to questions like these in Leach's text. I take on faith his archivally based impression that lots of people, women especially, saw in the idea of "social science" a useful framework within which to think about gender conventions. But *social science* was a notoriously elastic term, which had many sources and could be used in a multitude of contradictory ways. That a group of people met periodically and talked about social science is not in itself evidence of affiliation with the ASSA, yet Leach hardly ever supplies any more specific evidence of linkage.

As for the supposed vitality of the branch associations, there could be no frailer reed on which to pin an interpretation. From the founding on, ASSA officers talked endlessly about the urgent necessity of rooting their enterprise in a network of local branches extending across the nation. Sanborn was as enthusiastic for local branches as anyone and worked assiduously to realize the dream. Yet in his farewell address in 1897 he conceded that the association had not lived up to its aspirations and blamed its inadequacies squarely on its failure to construct that elusive foundation of local societies.[14] It is not that local branches were mere figments of the imagination. Many were christened with all due fanfare, only to lead ephemeral existences, fading out of operation within a few years after their creation. The archival record of their functioning is too slight to carry such a heavy burden of interpretation. To the best of my knowledge the only significant exceptions were Philadelphia, New York, and Chicago, each of which developed branches that attracted substantial clienteles and managed to sustain a program of public meetings and occasional publications over periods of at least

---

[13] Yeo, *Contest for Social Science*, 153. Seven percent is the figure she gives for "women and aristocrats;" she does not explain how women and aristocrats happened to be lumped together.

[14] *Emergence*, 213.

several years. But the leadership of all three was decidedly masculine and as far as I know there is no evidence that the Chicago or New York branches existed after the mid 1870s. Even the pride of the litter, the Philadelphia branch—which in 1890 transformed itself into the remarkably prosperous American Academy of Political and Social Science (still alive today) and even considered the possibility of coming to the rescue of its own mother, the ASSA—always marched to the beat of a different drummer, never really functioning as a tributary of the ASSA.[15]

In the last analysis, anyone wishing to uphold the Leach thesis will have to come to terms with the low profile of women in the principal activities of the association. Of the 516 authors who published articles in the *Journal of Social Science* through 1901, the vast majority (468, or 91 percent) were men.[16] The finding aid for the new Yale collection provides an even more telling indication. In sorting through the ASSA correspondence, the Yale archivists established separate file folders for all correspondents represented in the records by five or more letters. This unbiased procedure resulted in ninety file folders, presumably bearing the names of the people who were most often in communication with ASSA headquarters during the association's heyday. Surely if local branches were as active as Leach's argument requires, their leaders would appear in this group. Yet of the ninety names, eighty-five are men.[17] Compared to the nation as a whole, the ASSA was undoubtedly ahead of the times in many ways, not least in its acceptance of women as members, as officers, and as both the objects and the practitioners of social inquiry. But that has never been the issue. The question is how to reconcile Leach's portrait of the ASSA as a vanguard of the feminist movement, deeply influenced by women and centrally concerned with woman-specific issues, with the fact that only about 6 percent of the people in frequent correspondence with ASSA headquarters were women. Until we know a lot more about the membership and activities of the local affiliates, it is wishful thinking to credit them with a feminine orientation that the national association itself plainly lacked.

Another of the unresolved puzzles about the ASSA's history may turn out to hold more interest for historians of feminism than the Leach thesis. I mentioned above that E. L. Godkin's effort in the early 1870s to elevate standards of discussion was part and parcel of an effort to displace women from governing positions in the association. The evidence available to me as I wrote *Emergence* shed little light on this struggle or on the larger crisis of which it was a part, which nearly cost

---

[15] *Ibid.*, 215.
[16] See Appendix B.
[17] The names appear in the finding aid of the Yale collection.

the association its life. Lacking hard evidence, I broadly attributed the whole affair to factional infighting over national politics, calling it a "crisis of Liberal Republicanism." This still seems to me the most plausible context in which to understand the turmoil that culminated in late October of 1872, when an "adjourned annual meeting" was held in Boston "to consider the expediency of continuing the Association."[18] The upshot of that meeting was, of course, a decision to keep the association alive. Not long afterward, the all-important position of secretary, left empty ever since Henry Villard had resigned in September 1870, was dropped into the reluctant lap of Frank Sanborn, whose influential place in the life of the association did not depend on his holding office.

Now the resources of the Yale collection, especially the executive committee minutes, hold out the possibility of a much fuller understanding of what was afoot. My fleeting visit to the archives was not enough to formulate any adequate interpretation of what was going on, but I did come away with a strong impression that Villard's two-year tenure as full-time secretary (1868–70) brought the association to the peak of its powers. Never before or after did it accomplish so much. Probably working closely with President Samuel Eliot, Villard got branch associations up and running in Philadelphia, New York, and Chicago. On the suggestion of E. L. Godkin, leader of the New York contingent, Villard (himself an immigrant from Germany) enthusiastically undertook the production of an *Emigrants' Handbook,* designed to encourage immigration to America by the "better classes of English working-men."[19] At $16,000, the funding for this project, supplied by railroad and shipping interests, considerably exceeded the association's entire annual budget for most years. Meanwhile, Villard also organized an illustrious series of Lowell lecturers in Boston, established close relations with Congressman Thomas Jenckes, and enlisted the ASSA in the fight for Civil Service Reform.

No sooner had this unusual burst of energy occurred than it began dissipating. Villard began complaining of illness and resigned after only two years in office; Samuel Eliot, president of the association, shocked the members of the executive committee by announcing that he would have to abandon the helm during an extended trip abroad. The minutes are too cryptic to permit anyone to speak with assurance, but there are hints of frayed relations all around. Then, on 5 October 1870, Eliot brought before the executive committee a proposed amendment to the constitution that he said he had invited Godkin to draft. No copy

---

[18] Memorandum of meeting dated 28 October 1872 in Box 1, folder 15.
[19] "Executive Committee Minutes," 24 April 1869, Box 7, folder 178.

appears in the minutes. Whatever the amendment's provisions, it was initially accepted, leading immediately to several resignations, including those of Abby May and Mary E. Parkman.[20] Caroline Healey Dall, the most influential woman on the committee, was uncharacteristically absent from the meeting, perhaps in anticipation of the vote. When the association was resuscitated and various structural reforms adopted in 1872, the most conspicuous innovations were the return of those women who had resigned and the expansion of the number of directors to twenty-five. Seven of the first slate of directors appointed were women— including May, Parkman, and Dall.[21] This represented an increase, though slight, in the proportion of women participating in the governance of the association.

Did Godkin's amendment explicitly exclude women from office? Or did it offend by calling more generally for an elevation of the level of discussion? Since he was spokesman for the New York contingent, Godkin's attempt to reshape the association's policy suggests that the New England group's receptivity to women may have been a liability in organizing branch associations. Or was New England's commitment to female equality up for grabs, as the initial vote in favor of Godkin's amendment could be taken to imply? Of course the turmoil within the association may have been more a matter of national politics, or even a result of clashing personalities. These and other questions about the ASSA have wide implications and would be well worth answering, but I am delighted to leave them to a younger generation of scholars.

---

[20] *Ibid.*, 5 October 1870.
[21] Minutes of meeting on constitutional revision dated 7 December 1872, in Box 7, folder 182.

Preface to the Original Edition

Only the modern taste for brevity prevents this book from being titled *"Preconditions for* the Emergence of Professional Social Science: *An Interpretation of* the American Social Science Association and the Nineteenth-Century Crisis of Authority." My principal concern is to investigate the rise of a distinctively modern perspective on human affairs, one that has been institutionalized and greatly elaborated in academic social science disciplines, but which pervades many other areas of modern thought. The perspective whose roots I seek to uncover is by no means the exclusive possession of social scientists; indeed, its adoption, in elementary and inarticulate form, by a substantial part of the educated lay public was a precondition and precipitant for the professionalization of social science. Without it, the public would have responded to claims of professional expertise in social science with smiles of incredulity and indifference.

Although this book is not intended to be a full history of the professionalization of the social sciences in the United States, it does make certain contributions to that subject. It is an interpretation of the last generation of "amateurs," a speculative inquiry into the reasons for their demise, and an assessment of the larger meaning and cultural significance of their displacement by the first generation of professional social scientists.

The book is addressed both to specialists in the history of the social sciences, who have tended to work in isolation from the mainstream of historical writing, and to general historians and students of history, who seldom have recognized the vital importance of this subject to their broader interests. In trying to bridge the gap between these two sets of interests, I have felt obliged to devote more space to conceptualization than historians usually do, and I have occasionally strained at the outer limits of the historian's customary universe of discourse. In defense of these transgressions against custom I can only plead the virtues of provocativeness in a field that has thus far lacked sharply defined issues.

The focal point of the study is the American Social Science Association (or ASSA). Founded in 1865, this pioneer effort to institutionalize social inquiry was sponsored by genteel New England intellectuals and reformers who wanted both to understand and to improve their rapidly changing society. For many years before its death in 1909, the ASSA's members struggled valiantly to accommodate social theory and social practice to the emerging realities of an urbanizing and industrializing society. Known as "mother of associations," the ASSA spawned such seemingly diverse activities as the civil service reform movement, the National Conference of Charities and Correction, the American Historical Association, and the American Economic Association. Although the ASSA had only modest influence on public affairs and did not include among its members any great social thinkers, it is, I argue, a key to understanding the urban-industrial transformation in this country.

The history of the Social Science Association illuminates three problems of general importance: first, the nineteenth-century crisis of professional authority; second, the growing persuasiveness in the late nineteenth and early twentieth centuries of the social scientific mentality and its professional practitioners; and third, the decisive reorientation of social thought that took place in or near the watershed decade of the 1890's. To reveal the intimate connections between these developments and the emergence of modern consciousness is one task of the pages that follow.

As for the first problem, the ASSA affords a uniquely favorable vantage point from which to examine the general crisis of authority that preoccupied the professional classes of this country throughout most of the nineteenth century. Scholars have misunderstood the meaning of professionalizing activities not only in social science but also in law, medicine, and many other fields because they have failed to recognize that the context of these activities was a broad movement to establish or reestablish authority in intellectual and moral matters. This pervasive movement, for which the ASSA served as something like a "headquarters," was a reaction against profound changes in the conditions of adequate explanation—changes that directly threatened the professional role, provoking professional men to search for new ways to institutionalize sound opinion. One surprisingly central outcome of their search was the development of specialized academic disciplines devoted to the science of society.

The ASSA is indispensable to an understanding of the transition from amateur to traditional social science. ASSA members represent the last generation of amateur social scientists. As such, they and their mode of inquiry were precisely what the first generation of fully professional social scientists were fighting against when they began to construct the modern disciplines of history, economics, political science, and sociology. Since the professional defines himself largely by the amateur practices he rejects, one cannot understand the professionalization of social science without first understanding the ASSA. Moreover, the full meaning and complexity of the transition from amateur to professional social scientist only becomes apparent when it is recognized that ASSA members, though amateur social scientists, often were professional men according to the prevailing division of labor that defined the professions as divinity, law, and medicine. By clarifying the role of the ASSA, this study suggests that the emergence of professional social science signified the breakdown of the classical division of professional labor, and the evolution within the professional class itself of a new division of labor suited to modern conditions.

The decline of the ASSA after the 1890's sheds light on a third problem, the profound reorientation of social thought and culture that occurred during that important decade. The ASSA died because its explanations of human affairs began to appear less credible than those of its new professional competitors. The Association's diminished credibility reflects not the diminution of its members' intelligence, however, but their stubborn adherence to a strategy of explanation which, even though it had deep roots in mankind's past, no longer seemed plausible to the serious thinkers of a younger generation accustomed to the highly interdependent conditions of a mature urban-industrial society. The reorientation of social thought that overtook the ASSA in the 1890's was essentially based upon the recognition of interdependence and the development of new strategies of causal attribution appropriate to that fact of modern life.

The ASSA was finally overwhelmed, then, by changing conditions of explanation triggered by the very wave of social change to which its founders had set out to accommodate themselves and their society. The reorientation of the 1890's confirmed the obsolescence of the ASSA and guaranteed the ascendancy of that most revealing symptom of modernity, the professional social scientist.

Since the reorientation of the 1890's has been the subject of important interpretations by Morton G. White, H. Stuart Hughes, and Talcott Parsons that are well known even to readers who do not ordinarily concern themselves with the history of the social sciences, I have departed from this study's mainly chronological format in order to address that problem in the first chapter. There also I try to clarify Thomas S. Kuhn's influence upon my thinking. The next three chapters carry out introductory functions: chapter II develops the major causal scheme of the study; chapter III is devoted to the career of

Frank Sanborn, the individual who most influenced the character of the ASSA; and chapter IV examines the movement to establish authority. The next six chapters trace the vicissitudes of the Association from its founding to its proposed merger with the John Hopkins University, and finally its death. The concluding chapter returns to the central question of how the development of modern society altered the way we explain human affairs and conceive of man.

My work has been supported at various stages by fellowships and grants from Stanford University, the Richard D. Irwin Foundation, and Rice University.

Many people have contributed to this book, often in ways I no longer can specify. Of the obligations that I can enumerate, one of the most important is to the late David M. Potter, who graciously endured many frustrating discussions as I struggled toward a clear formulation of the topic. I deeply regret that he lived to see only what is now the heart of chapter II. Albert Hastorf contributed more to my thinking than he had any reason to imagine at the time. Barton J. Bernstein gave much time and good advice on a subject far removed from his main interests, and George H. Knoles went far beyond the call of duty and saved me from missing a deadline by taking special care with the manuscript. An earlier advisor, John William Ward, introduced me to the field, suggested the ASSA as a research topic, and supplied both encouragement and a good critical reading as the revised manuscript neared completion.

Special thanks go to John W. Clarkson, who generously permitted me to see his collection of the correspondence of Franklin B. Sanborn. Without his labors and his openness this study would have been substantially diminished.

One pair of intellectual debts I incurred anonymously, without the knowledge of my creditors. In the spring of 1968 I had the good fortune to read Robert Wiebe's *Search for Order* and Leo Marx's *Machine in the Garden*. From that happy conjunction came the central notion on which my analysis rests.

Gale Stokes, Martin Wiener, and Carol Wiener gave me the close reading and stiff criticism that only the best of friends can give. I am grateful to them and to all of my colleagues at Rice University for their trust and patience.

For unusually thoughtful comments on the manuscript or portions of it I wish to thank Hugh Hawkins, John Higham, and Christopher Lasch. Many other friends and colleagues at Rice and at Yale's National Humanities Institute (1975–76) read parts of the manuscript or helped me by discussing related subjects. I regret that I cannot thank each reader for his or her special contribution, but must resort instead to a list of those whose comments were most valuable: Daniel Aaron, John Burnham, Hamilton Cravens, Chandler Davidson, Martin Diamond, Mary Furner, Charles Garside, Henry Glassie, Victor Greene, Sanford Higginbotham, Harold Hyman, Julie Jeffrey, Kirk Jeffrey, John Juricek, Allen Matusow, James Mohr, David Patterson, Lewis Perry, Kathryn Kish Sklar, and Gordon Wood.

After such a long list of benefactors, it is embarrassing to admit that I often have not taken their advice. Of course they are not responsible for my errors.

Dorothy, however, may as well help bear responsibility even for them, for she has lovingly shared in this work and my fate from the beginning. She was at first breadwinner, then mother, all the while acting as fellow researcher, critical editor, and infinitely patient and responsive audience. That after eight years of it she still likes both the book and me is my greatest source of confidence.

—T. L. H.

The Emergence of
Professional Social Science

CHAPTER I

Introduction: What Happened in the 1890's?

The immediate subject of this study, the birth, life, and death of the American Social Science Association, is not an end in itself but a point of entry into a larger subject: the rise to cultural dominance of the social sciences in the late nineteenth and early twentieth centuries. These years mark a decisive boundary in cultural history, a division between two different constructions of social reality, two quite different modes of understanding man's nature, his relations in society, and his place in the cosmos. Of course it is true that no matter how sudden any break with the past may seem, no major transformation can occur without long preparation. Indeed, if one looks for the origins of social science in the history of ideas alone—ignoring social context, looking always for antecedents, and ranking each thinker by the originality of his "contribution" to modern views—then the origins appear to lie further back, in the eighteenth-century Enlightenment or even earlier.

But origins must be distinguished sharply from the moment at which the social sciences achieved cultural dominion. One must consider not only the genealogy of ideas but also their institutional setting, their meaning and relevance for contemporaries, and their depth of penetration, both inward into individual consciousness and outward through society's many levels. It is my conviction that when the intellectual transformation of the late nineteenth century is understood in this way, it will appear no less profound than the economic and social changes of that century—the century of the urban-industrial transformation, undoubtedly the most profound and rapid alteration in the material conditions of life that human society has ever experienced.

The Social Science Association is of interest, then, primarily for what it can reveal about two central developments in the making of the modern mind: the crystallization of the social sciences in their present form as specialized academic disciplines, and the corresponding rise of the professional social scientist as paramount authority on the nature of man and society. At the outset it must be acknowledged that the ASSA serves only as a useful point of entry into these larger questions; it cannot supply complete answers to them. The contrast drawn in these pages is between concrete cases, ultimately unique, and not between modern and pre-modern, or professional and pre-professional per se. The confinement of this study to the experience of a single nation is one obvious limitation on the generality of the conclusions. Moreover, the history of the ASSA is more relevant to some social science specialties than to others. The Association was involved, in varying degrees, in the professional development of history, economics, political science, and sociology—the policy sciences, as they are sometimes called—but the institutional origins of psychology and anthropology lie elsewhere. Finally, not all important amateur social thinkers, even in the policy science field, were members of the ASSA, for it had a decided philosophical bias toward Emersonian and Hegelian idealism. To many readers this bias may seem at first to place the Association far outside the mainstream of late nineteenth century social theory. The popular vogue enjoyed by Herbert Spencer's evolutionary system, or certain fragments of it, has created among many historians the illusion that his bleak necessitarian positivism overwhelmed all competing views. But this is to confuse vitality with typicality. Spencerian positivism was one of the vital poles of late nineteenth century thought, not its mainstream. Most serious social thinkers in England and America read Spencer with mingled fascination and horror, clinging hopefully to a far more voluntaristic and spiritual view of human affairs.[1] Men of this temperament staffed the ASSA. Thinkers of a more naturalistic bent, including such notable figures as William Graham Sumner and Lester Ward, worked elsewhere or alone.

[1.] For balanced views of the role of positivism in late nineteenth century thought, see Philip Abrams, *The Origins of British Sociology, 1834–1914: An Essay with Selected Papers* (Chicago: University of Chicago Press, 1968), 67; J. D. Y. Peel, *Herbert Spencer: The Evolution of a Sociologist* (London: Heinemann, 1971), 238; R. Jackson Wilson, *In Quest of Community: Social Philosophy in the United States 1860–1920* (New York: John Wiley, 1963), 155; Maurice Mandelbaum, *History, Man and Reason: A Study in Nineteenth Century Thought* (Baltimore: Johns Hopkins University Press, 1971), 5; and Burton J. Bledstein, "Cultivation and Custom: The Idea of Liberal Culture in Post–Civil War America" (Ph.D. dissertation, Princeton University, 1967), 25, 71–81. See also Dorothy Ross, *G. Stanley Hall: The Psychologist as Prophet* (Chicago: University of Chicago Press, 1972). Positivism is given heavier stress by Richard Hofstadter, *Social Darwinism in American Thought*, rev. ed. (New York: Braziller, 1955); Stow Persons, *American Minds: A History of Ideas* (New York: Holt, 1958); and Paul F. Boller, Jr., *American Thought in Transition: The Impact of Evolutionary Naturalism, 1865–1900* (Chicago: Rand McNally, 1969). Hofstadter's influence is reflected in the first two chapters of Edward A. Purcell's interesting book, *The Crisis of Democratic Theory: Scientific Naturalism and the Problem of Value* (Lexington: University of Kentucky Press, 1973).

To say, then, that the ASSA had an idealistic bias is to locate it not far from the center of late nineteenth century American thought.²

The existence of a major cultural divide near the turn of the century has been recognized by many scholars, and, in characterizing the changes that took place, nearly all of them acknowledge the special importance of the new social sciences. Henry Steele Commager captured both the depth of the discontinuity and something of its essential nature in the observation that Jacob Riis's *How the Other Half Lives* falls chronologically halfway between Theodore Parker's sermons on the *Perishing and Dangerous Classes of Boston* and John Steinbeck's *Grapes of Wrath*, but intellectually Riis and Steinbeck are contemporaries, while Parker resides in another world. Similarly, another book of the 1890's, *The Theory of the Leisure Class*, is in spirit closer to *Middletown* than to *Walden*, despite being chronologically equidistant. "To the Americans of 1950," said Commager, "the political figures of the nineties seem almost contemporary; move back but a decade or two and they become quaint and archaic."³

The turning point seemed to Commager to lie in the 1890's, which decade he boldly labeled "the watershed of American history."⁴ To John Higham, also, "the reorientation of American culture in the 1890's" has been a subject of special interest.⁵ Richard Hofstadter regarded the nineties as a traumatic decade of psychic crisis for Americans, and he treated the years from 1890 to the outbreak of World War I as the decisive period in the accommodation of social thought to Darwinian evolutionary theory.⁶ The year 1889 serves as benchmark for Christoper Lasch's investigation of the "new radicalism" and the emergence of the intellectual as a social type.⁷ Edward A. Purcell, Jr., locates the intellectual origins of the "crisis of democratic theory" in the first decade of the twentieth century.⁸ Samuel P. Hays and Robert H. Wiebe portray the whole period from the end of Reconstruction through the Progressive era as years of massive social transformation, and both attach extraordinary significance to the few years either side

---
² Sumner served briefly as secretary and chairman of the Finance Department of the ASSA between 1874 and 1878, but he was not among the active regular leaders of the Association.

³· Henry Steele Commager, *The American Mind: An Interpretation of American Thought and Character Since the 1880's* (New Haven: Yale University Press, 1950), 53–54.

⁴· *Ibid.*, 41.

⁵· John Higham, "The Reorientation of American Culture in the 1890's," in his *Writing American History: Essays on Modern Scholarship* (Bloomington: Indiana University Press, 1970), 73–102.

⁶· Richard Hofstadter, "Manifest Destiny and the Philippines," in *America in Crisis*, ed. Daniel Aaron (New York: Knopf, 1952), and Hofstadter, *Social Darwinism*, ch. VIII.

⁷· Christopher Lasch, *The New Radicalism in America, 1889–1963: The Intellectual as a Social Type* (New York: Knopf, 1965).

⁸· Purcell, *Crisis of Democratic Theory*.

of 1900.⁹ The intellectual aspects of the transformation have been further developed in penetrating studies by David Noble, R. Jackson Wilson, and Jean B. Quandt.¹⁰ Henry F. May locates the "first years of our own time" somewhat later, 1912–17, perhaps because he is more concerned with high literary culture than with social thought or objective changes in social organization.¹¹

At the level of social thought, which is the prime concern of this study, the turn-of-the-century transformation has been investigated by three eminent scholars: Talcott Parsons, H. Stuart Hughes, and Morton White. Cited more often than read, but influential nonetheless, is Parsons's *Structure of Social Action* (1937). Impressed as any sociologist must be with the generation that produced Weber, Durkheim, Tönnies, and Pareto, among others, Parsons asked the central question that still guides research today. Referring to the achievements of the 1890's, he declared: "A revolution of such magnitude in the prevailing empirical interpretations of human society is hardly to be found occurring within the short space of a generation, unless one goes back to about the sixteenth century. What is to account for it?"¹²

H. Stuart Hughes adopted Parsons's question (and part of his answer as well) in his more readable book, *Consciousness and Society: The Reorientation of European Social Thought, 1890–1930* (1958). Many of the main themes of Parsons's analysis reappear in Hughes's work, but there is one distinct, if unannounced, point of conflict between the two accounts. The conflict concerns the respective roles of positivism and idealism.

Parsons's study consisted of an intensive analysis of the thought of four men, three of them drawn from the positivist tradition—Alfred Marshall, Vilfredo Pareto, and Émile Durkheim—and one whose background was idealist, Max Weber. The cardinal point of Parsons's analysis was that despite their apparent differences of method, approach, and concern, all four men arrived at a single set of essential

---

⁹· Samuel P. Hays, *The Response to Industrialism: 1885–1914* (Chicago: University of Chicago Press, 1957) and "The New Organizational Society," in *Building the Organizational Society: Essays on Associational Activities in Modern America*, ed. Jerry Israel (New York: Free Press, 1972); Robert H. Wiebe, *The Search for Order, 1877–1920* (New York· Hill and Wang, 1967).

¹⁰ David W. Noble, *The Paradox of Progressive Thought* (Minneapolis: University of Minnesota Press, 1958), and *The Progressive Mind, 1890–1917* (Chicago: Rand McNally, 1970); Wilson, *In Quest of Community;* and Jean B. Quandt, *From the Small Town to the Great Community: The Social Thought of Progressive Intellectuals* (New Brunswick: Rutgers University Press, 1970).

¹¹· Henry May, *The End of American Innocence: A Study of the First Years of Our Own Time, 1912–1917* (Chicago: Quadrangle, 1964). See also Paul F. Bourke, "The Social Critics and the End of American Innocence, 1907–1921," *American Studies*, 3 (July 1969), 57–72.

¹²· Talcott Parsons, *The Structure of Social Action: A Study in Social Theory with Special Reference to a Group of Recent European Writers* (New York: McGraw-Hill, 1937), 5.

theoretical assumptions about the basic structure of human behavior in society. This single system, which Parsons called the "voluntaristic theory of action," was in his eyes an empirically validated foundation for all further discussion of human affairs. For Parsons, the period around the turn of the century was indeed a dramatic moment in intellectual history. The two contending poles of nineteenth-century thought, idealism and positivism, had then converged in a single, coherent theory of social action. The convergence yielded a view of man and society that Parsons regarded as a permanent acquisition of human knowledge, clearly superior to either of its one-sided predecessors and a suitable point of departure for a cumulative social science.[13]

The essential feature of the voluntaristic theory of action, as Parsons presented it, is not entirely evident in the name he gave it. The virtue of the theory lay not simply in its voluntarism, but in its fine *balance* between the age-old alternative emphases of freedom and determination. As the name suggests, the new theory supported one of the main claims of nineteenth-century idealism by insisting on the authenticity of human freedom. In doing so it rejected the radical positivist tendency to reduce the conscious, willing mind to little more than an epiphenomenal reflex of heredity and environment. The new theory would not utterly assimilate man to nature as the positivists had tended to do.

But in contrast to the idealist tradition, the new theory was not hostile toward science, and it acknowledged a wide range of external constraints upon human freedom. The new theory gave close attention to law-like regularities in human experience and reached for the power to predict and control, even as it admitted the authenticity of choice. Man was free, but only to a degree and only within a circumscribed realm. Constraints upon freedom included not only the palpable factors of heredity and environment that positivists had stressed, but also "ideal" elements such as the press of cultural norms and common values—a kind of factor to which nineteenth-century theorists had been blind, or for which they had tried awkwardly to account with untenable theories of emanationism. The scope of all these constraints upon human freedom was admitted by the voluntaristic theory to be large; so large, in fact, that one might say that freedom pertained more to mankind collectively than to men individually. Certainly the new theory was methodologically far less individualistic than its predeces-

---

[13.] *Ibid.*, 11–15. That positivism and idealism did constitute the poles of nineteenth-century European thought is confirmed by Mandelbaum, *History, Man and Reason*, 5. Werner Stark adopts the theme of convergence in his *Fundamental Forms of Social Thought* (London: Routledge and Kegan Paul, 1962).

sors. The freedom it insisted on was the same collective freedom (of small comfort to the individual) that Vico had suggested in the insight that society made man, but was in turn made by man.[14]

Although Hughes relied heavily on Parsons's analysis, he arrived at a quite different understanding of the turn-of-the-century transition. What Parsons had described as a *convergence* of idealism and positivism, Hughes described as a *"revolt against positivism."*[15] Of course, to the extent that a revolt is successful—and Hughes thought this one was—it means not the convergence of two traditions, but the triumph of one over the other, or at least the invigoration of one at the expense of the other. Hughes ranged widely through the intellectual world, briskly surveying the thought not only of Pareto, Durkheim, and Weber, but also of Croce, Sorel, Bergson, Freud, Jung, James, Hesse, Proust, Mann, and many more. Everywhere Hughes looked he found a growing disenchantment with reason, a heightened concern for the subjectivity of experience, and a new awareness of the role of the unconscious. Only in the case of Max Weber did Hughes fully concur with Parsons's view of convergence: "Alone of his contemporaries, Weber was able to bridge the chasm between positivism and idealism."[16] According to Hughes, the common task of the transitional generation was not to transcend the two poles of nineteenth-century thought, but to rally round one while fleeing from, resisting, or aggressively assaulting the other.[17]

[14.] The points of similarity and dissimilarity between the voluntaristic, the positivistic, and the idealistic theories of action are compactly presented in schematic form by Parsons, *Structure of Social Action*, 77–82. Parsons treats utilitarianism as an early branch of positivism that preserved the all-important "rational schema"—*i.e.*, the voluntaristic image of man as an authentically free actor. However, it preserved this vital element only by making the untenable assumption that the ends which men pursue are random (51–60). The utilitarian position was inherently unstable because of this assumption and tended always toward a radical positivism in which rational purposes or ends are assimilated to the objective conditions of the situation (primarily heredity and environment)—thus erasing the element of voluntarism (60–74). See esp. his discussion of the "utilitarian dilemma" (64–67). Parsons summarizes the breakdown of the positivistic theory as manifested in the work of Marshall, Pareto, and Durkheim on 451–470. On the idealist contribution, see 473–487. A summary of his whole argument appears on 697–726. My two-paragraph summary is, I think, faithful to Parsons's argument, but the distinction between social constraint and natural causation (see 709) is far more important and more comforting to him than it is to me.

[15.] H. Stuart Hughes, *Consciousness and Society: The Reorientation of European Social Thought, 1890–1930* (New York: Knopf, 1958), ch. II, "The Decade of the 1890's: The Revolt Against Positivism." Hughes refers to Parsons's *Structure of Social Action* as "one of three extraordinary works of synthesis on which I have relied most heavily" (433).

[16.] *Ibid.*, 335.

[17.] Hughes never calls attention to his departure from Parsons's account of the convergence of positivism and idealism. He does call Parsons's reconciliation of Durkheim and Weber a *"tour de force. . . .* From the more general standpoint of intellectual history, the dissimilarities between the two have remained unresolved." But this cautionary statement is coupled with the apparently approving statement that Parsons's work is

Even if one rejects Parsons's claim that all major minds of the 1890's converged in a single theory of social action, one may still agree with him that social thought since the turn of the century owes at least as much to the positivist tradition as to the idealist. Surely in this sense, at least, he is correct in speaking of convergence, and Hughes is wrong in speaking only of rebellion. In the closing pages of his book, Hughes diluted his earlier imagery of "revolt," implicitly acknowledging the strength of Parsons's formulation. "It was," said Hughes, speaking of the new style of thought, "a tenuous synthesis. . . . It had originated in a revolt from positivism—explicit or implied—yet it retained from the positivist faith its trust in the procedures of exact science. It marked the revival of idealist thinking—yet it rejected the idealist belief in the eternal character of spiritual values."[18]

As descriptions of the intellectual transformation of the 1890's and early 1900's, Parsons's image of convergence and Hughes's one-directional image of revolt are logically incompatible. But it is easy to see how Hughes moved so casually from Parsons's formulation to his own, in spite of their differences. In part, Hughes may have succumbed to the above-mentioned illusion that positivism was not just a vital pole of nineteenth-century thought, but actually its mainstream. If Spencer had captivated the minds of the 1870's and '80's as fully as is sometimes supposed, then the 1890's, when his reputation was on the wane, would indeed have to be construed as a time of revolt against positivism and revitalization of idealism. That this view exaggerates and misunderstands the character of Spencer's influence, even in America, where he was most popular, is best shown by the career of the Social Science Association, which was the chief American forum for an anti-Spencerian form of social science during his supposed heyday.

But even if Hughes did not exaggerate the typicality of positivist views in the nineteenth century, his portrait of the transitional generation as rebels against positivism is readily understandable as a reflection of what they themselves thought they were doing. In this limited subjective sense, his portrait is accurate. From our own retrospective viewpoint it is essential to recognize that objectively they rebelled against (and drew upon) *both* traditions, idealist as well as positivist, and in something like equal measure. But in their own minds what mattered most was the rejection of positivism.

Positivism demanded active refutation precisely because it remained a live option even in the minds of those who opposed it most bitterly. In

---

"a dramatic illustration of the extent to which early twentieth century social thought was heading in a single direction, despite the most extraordinary personal and intellectual divergences among its creators" (*ibid.*, 287).

[18.] Hughes, *Consciousness and Society*, 429.

contrast, much of the idealist tradition could be lightly dismissed or ignored, for by the end of the century it simply had become implausible as an account of social action (though not as a philosophical position, of course). Many of the bolder formulations of idealism needed no elaborate repudiation in the 1890's and after. To serious social thinkers living in an interdependent urban-industrial society, non-causal schemes of explanation seemed inadequate and notions like the Absolute, World-Spirit, and emanationism seemed simply irrelevant. As the most volatile elements in idealism evaporated, positivism continued to flourish in the form of Marxist economic determinism (chastened but hardly demolished by Weber), the mechanistic physiology and psychology of Jacques Loeb, and the psychological behaviorism of Pavlov and John B. Watson.[19] Indeed, Watson's radical environmentalism, which went far beyond Darwin and Spencer to treat the contents of conscious mind as purely epiphenomenal and utterly beneath the interest of science, was very likely one of Parsons's chief polemic targets in the *Structure of Social Action*—perhaps accounting for the selection of the word "voluntaristic" as the distinguishing mark of the new synthesis.[20]

Hughes's thesis of a "revolt against positivism" is, then, an accurate rendering of the *polemic posture* of the generation of the 1890's, but it is incomplete and therefore misleading as a description of the *direction of movement* of intelligent social opinion in those critical years. The transition was not simply a movement away from positivism and toward idealism, but a more complex convergence involving the repudiation of parts of both traditions, and also the synthesis of certain key elements from each. Half of the convergence, the revolt against positivism, is readily visible. The other half that might be labeled "the revolt against idealism" has been lost to sight, historically speaking, because it was so successful—the enemy having been vanquished effortlessly—and because the transitional generation of social scientists, busily seeking their own legitimation as professionals, were eager to forget and even hide their linkages with the last, embarrassing practitioners of an extreme form of anti-positivistic social science. These practitioners in the United States worked in the American Social Science Association, and a proper understanding of the Association, which went into severe decline in the 1890's, will show that the turn of the century is best viewed as a time of convergence rather than of a revolt against positivism.

[19.] Jacques Loeb, *The Mechanistic Conception of Life* (Chicago: University of Chicago Press, 1912); John B. Watson, "Psychology as the Behaviorist Views It," *Psychological Review*, 20 (1913), 158–177.
[20.] Parsons never mentions Watson by name, but see *Structure of Social Action*, 115–121.

The analyses of Parsons and Hughes go far toward an explanation, or at least an illuminating characterization, of the turn-of-the-century transformation of social thought. In doing so they provide a suggestive point of departure for understanding the whole cultural transformation of these years. Fully acknowledging their achievement, it may nonetheless be useful to ask if the concepts of positivism and idealism have not been made to carry too heavy a burden of meaning. Is there some other frame of reference that will permit a further penetration into the character of the transition?

One alternative that has proved extremely influential among American historians was advanced by Morton G. White in his *Social Thought in America: The Revolt Against Formalism* (1949). White, writing a decade after Parsons and a decade before Hughes, was the first to adopt the imagery of revolt. Moreover, his dichotomy between formalism and anti-formalism establishes a frame of reference which parallels, to some extent, the positivist-idealist division employed by Parsons and Hughes. But White insisted upon the peculiarly American quality of his subject. His choice of anti-formalism as the central reference point followed naturally from his interest in John Dewey, who is in many respects the best exemplar of early twentieth century American thought. And although White regarded the turn of the century as a distinct watershed in American social thought, his concept of anti-formalism also serves implicitly to link Dewey's generation with a national heritage marked by recurring polar tensions between Arminian and antinomian, head and heart, form and spirit.

No one as familiar with John Dewey's career as Morton White could describe the turn-of-the-century transition simply as a revolt against positivism. Dewey's own path of intellectual development during the nineties, a critical period for him personally, was quite the reverse. For him it was a revolt against idealism, a rejection of his earlier heavy commitment to Hegel. Although Dewey moved away from idealism in the nineties, he did not abandon his idealist origins altogether. In White's earlier study, *The Origin of Dewey's Instrumentalism* (1943), he had stressed the continuing strength of certain idealist predispositions in Dewey's later years. Indeed, among these predispositions the most important was what White in his earlier book had already designated as a hostility to "formalism."

Anti-formalism, then, was not original with Dewey or new to the transition period. On the contrary, it was the chief element of continuity between Dewey's mature instrumentalism and his idealist past; it was a lesson learned at the feet of his mentors, George S. Morris and the British neo-Hegelians, who had long waged war against British empiricism, formal logic, dualism, atomism, and mechanism. This was

Dewey's first and most basic commitment. Only after he found in Darwinian evolutionary theory and in experimental psychology a new and sturdier support for his anti-formalist stance did he surrender his allegiance to idealism.[21]

Since the anti-formalist pole of White's interpretive framework is, at least in its origins, so closely identified with the idealist tradition, it is no surprise to find that the other pole, formalism itself, is associated with what Parsons and Hughes would have called the positivist tradition, or one segment of it. The formalist enemy against which Dewey and his generation rebelled was essentially that branch of early nineteenth century positivism known as utilitarianism. Jeremy Bentham is the central figure in the enemy camp; John Stuart Mill and John Austin are other main targets. White also mentions David Hume and Adam Smith.[22]

But White's scheme of analysis is not merely another version of positivism versus idealism. Although it overlaps that division at some points, its axis has a fundamentally different orientation. White's contrast is not between two schools, but between two styles of thought. As we have seen, the original basis for Dewey's anti-formalism was idealist, but in his later thought the same anti-formalist stance is supported by, and integrated with, very different Darwinian, positivistic, foundations. If compelled to define Dewey's mature position in terms of positivism and idealism, White presumably would have to speak of synthesis or convergence. But his preferred formulation abandons Parsons's framework altogether and establishes a new frame of reference which he defines in terms of two concepts, *historicism* and *cultural organicism*. These two concepts define the central tendency of Dewey's thought and link him with Oliver Wendell Holmes, Jr., Thorstein Veblen, James Harvey Robinson, Charles Beard, and other Progressive social thinkers in a general "revolt against formalism."

Historicism and cultural organicism are separable tendencies, says White, but they generally go together in the same minds. They are concretely manifested in Dewey's contempt for the sterility of formal logic; in Veblen's depreciation of the abstractive and deductive emphases of classical economics, with its artificial "economic man"; in Beard's battle against the formal, juridical approach to the constitution; in Robinson's insistence on the instrumental and interdisciplinary character of the "new history"; and in Holmes's famous slogan of legal realism, "The life of the law has not been logic: it has been experi-

---

[21.] Morton White, *The Origin of Dewey's Instrumentalism* (New York: Columbia University Press, 1943), 151, ch. II, 40–41, ch. VII, and 111.

[22.] Morton White, *Social Thought in America: The Revolt against Formalism*, expanded ed. (Boston: Beacon, 1957), 14, 22, 24.

ence."²³ At its worst, the assault on formalism led to an unhealthy distrust of all systematic thought and conceptual rigor; it encouraged an ad hoc expediency and a reluctance to come to grips with long-term goals and ultimate value commitments. But at its best it represented a grand flowering of the nineteenth-century philosophies of change and process, with their acute sense of growth, context, and function.²⁴

White defined the two concepts comprising anti-formalism as follows: "By 'historicism' I shall mean the attempt to explain facts by reference to earlier facts; by 'cultural organicism' I mean the attempt to find explanations and relevant material in social sciences other than the one which is primarily under investigation. The historicist reaches back in time in order to account for certain phenomena; the cultural organicist reaches into the entire social space around him."²⁵

White's division of anti-formalism into two aspects—historicism and cultural organicism—has a special significance of which he was surely aware, but which he did not discuss in detail. The two concepts correspond to the two most elementary alternative perspectives from which one can view human behavior; the "horizontal" and "vertical" dimensions, so to speak, which together constitute the whole universe of human affairs. Historicism pertains to the diachronic or developmental dimension; cultural organicism pertains to the synchronic dimension, or the cross-section, "perpendicular" to time. For example, to trace the birth, growth, and death of an institution is to view it diachronically, through time—"the attempt to explain facts by reference to earlier facts." A synchronic treatment of the same institution might characterize its internal principles of operation, or examine its functional role vis-à-vis other parts of society, irrespective of temporal development (e.g., "America in 1800," "the seventeenth-century mind," and "the function of the political boss" are synchronic treatments). This is what White meant by explanations that reach out into the "entire social space." Sometimes it is said that the historian's approach is diachronic and the sociologist's synchronic, but, as the best sociologists and historians have always recognized, a full account of human affairs requires a combination of the two approaches.²⁶

By singling out the concepts of historicism and cultural organicism White suggests the positive assumptions underlying the merely negative attitude of anti-formalism. He was primarily interested in (and disturbed by) the negative attitude itself, and so did not develop the

---

²³ *Ibid.*, 12–13.
²⁴ *Ibid.*, 236–246.
²⁵ *Ibid.*, 12.
²⁶ Robert F. Berkhofer, Jr., *A Behavioral Approach to Historical Analysis* (New York: Free Press, 1969), 235–237.

positive doctrine in detail. However, it can be derived from his categories and is compatible with his evidence.

The root assumption underlying anti-formalism is that the universe of human affairs is a seamless web, or nearly so; that human experience is a dynamic flux, virtually uninterrupted by intrinsic divisions and discrete boundaries. The essential interdependence of all phenomena in the universe of human affairs is manifested in both the diachronic and the synchronic dimensions. Historicism in White's sense means a dramatic stress on the interconnectedness of events in time, a stress so great that one speaks of *evolution* and *process,* rather than a mere *sequence* of discrete happenings. The historicist construes each moment in history genetically, as a growth out of some past and the seed from which a future will develop. White's concept of cultural organicism is to the synchronic dimension of analysis what historicism is to the diachronic. It merely repeats in the synchronic dimension a similar demand for dramatic stress on the interconnectedness of social and cultural phenomena at any given moment. Again, the stress is so great that a new vocabulary is invoked: one speaks of *roles* and *functions, adaptations* and *reflexes,* rather than autonomous *acts,* and of *systems of action* rather than discrete *actors.* Historicism and cultural organicism are the manifestations, each in its own dimension, of a single, vital insight—that the phenomena composing the universe of human affairs are intensely interdependent; that truly independent action is exceedingly rare.[27]

The cardinal intellectual sin committed by formalists, then, was their failure to recognize interdependence. Judging from Dewey's extreme hostility to formalism and abstraction, one can infer that he and his generation believed themselves to be living in a highly interdependent universe of human affairs, one whose multitudinous interactions and criss-crossing mutual dependencies were only beginning to be sorted out and identified. In such a universe, the ever-threatening source of error and misunderstanding is abstraction or isolation. For the anti-formalist no less than for the Romantic, to isolate is to kill. To abstract from context, to snatch a thing out of the temporal processes which bring it into existence, or to tear it away from the living web of synchronic factors interacting with it and maintaining it at each moment, is

---

[27.] Charles Beard carried the notion of human affairs as a seamless web to a logical, though indefensible, extreme in the 1930's by trying to do without the concept of "cause" altogether. In *The Discussion of Human Affairs* (New York: Macmillan, 1936) he said: "A search for the causes of America's entry into the [First World War] leads into the causes of the war, into all the history that lies beyond 1914, and into the very nature of the universe of which the history is a part; that is, unless we arbitrarily decide to cut the web and begin at some place that pleases us" (79).

to isolate the part from the whole which gives meaning. In such a universe the mistake most feared (because most easily made) is to confuse parts with wholes, to see as separable what is really integral, to construe as autonomous and self-sufficient what is actually the dependent and reflexive component of some larger system. Formalism springs from this error.

To anticipate a future stage of my argument and to establish the close parallel between it and White's conclusions, still more can be said about the positive assumptions implicit in the anti-formalist stance. In a highly interdependent social universe, prediction and explanation (or retrodiction) are critically important tasks, because change is pervasive and ceaseless: little can be taken for granted. But by the same token, prediction and explanation are exceedingly difficult tasks, because nothing moves independently. Where all is *inter*dependent, there can be no "*in*dependent variables." Without independent variables, causation can be attributed only after exhaustive empirical research and then only in the most tentative and probabilistic manner. Causal attribution, and the tasks of prediction and retrodiction that depend upon it, become in a highly interdependent universe the business of specialists, beyond the reach of lay opinion. But even the specialist is liable to attribute causation too close at hand; even he often fails to see the subtle causal links that make his subject dependent on another. No less than the radical positivists of the nineteenth century, the investigator in an interdependent universe must vigorously seek out remote causation, trying never to mistake secondary causes for primary ones or epiphenomenal reflex for underlying reality.

To insist on the interconnectedness of social phenomena in time and in social space is to insist on the improbability of autonomous action. This presumption against autonomy applied to all social phenomena, including especially the irreducible constituent atoms of the social universe, individual human beings. Self-reliance and interdependence are mutually exclusive. If the whole is interdependent, its parts can only be mutually dependent. Individuality, which nineteenth-century Americans had understood in terms of self-mastery and independence, either would be condemned by the intellectuals of Dewey's generation as "rugged individualism," a variety of selfishness; or the concept would be transformed into a measure of the *quality* of the inevitable mutual dependence between the individual and his social environment.[28] From this denial of individual autonomy there fol-

---

[28.] John William Ward, *Red, White and Blue: Men, Books and Ideas in American Culture* (New York: Oxford University Press, 1969), 213–266, "Mill, Marx, and Modern Individualism," and "The Ideal of Individualism and the Reality of Organization."

lowed momentous consequences that reached far beyond the esoteric realm of social thought. To the extent that the individual was viewed not as the autonomous creator, but the created product of society, nineteenth-century ethical views and the politics based on them would have to give way to a new welfare-oriented humanitarianism and a new politics.

Although the universe in which Dewey and his generation believed themselves to live was highly interdependent, it was not completely or rigidly so. Had its interdependence been perfect, its parts would have fused into one whole and it would have approached a state of solidity incompatible with either explanation or control. It was a highly interdependent universe, but it was not the solid, "block universe" that nineteenth-century thinkers, both positivist and idealist, had sometimes sensed about them, and whose icy touch had so terrified William James. It was a universe that permitted one to believe in a degree of human freedom—more meaningful collectively than individually—and which therefore supplied a basis for moral judgment, however fragile. By no means least important, it was a universe in which the professional social scientist had a vital role to play, for it was largely through his explanatory prowess that men might learn to understand their complex situation, and largely through his predictive ability that men might cooperatively control society's future and thus get the cash value of their small measure of freedom.

The concept of interdependence supplies the interpretive keystone of this study. Although the transformation in American social thought near the turn of the century can be regarded as a convergence of idealism and positivism, and although it can be construed even more exactly as a revolt against formalism in behalf of historicism and cultural organicism, I believe it is best understood in terms of the recognition and conceptualization of interdependence. This is the factor which most effectively distinguishes between the new and the old and thereby reveals not only the polemical orientation of the key thinkers but also the direction of change of intelligent opinion about the nature of man and society. Moreover, the concept of interdependence provides a natural link between the rarefied world of social thought and the everyday world in which social thinkers and everyone else actually lived. By doing so it offers some hope of achieving an integrated view of the turn-of-the-century transformation as a pervasive cultural event, an alteration in perspective experienced not only by a few social theorists but also by people in many other walks of life.

Dewey and his generation drew their notion of interdependence from their concrete social experience in an urbanizing, industrializing society.[29] Their sensitivity to interdependence and its implications was an intellectualization of the objective conditions of their society—or, more accurately, an intellectualization of the *direction* in which objective conditions were changing. For American society had not always been as interdependent as it was at the turn of the century. If Dewey and his generation had lived a century earlier, no amount of insight or contemporary evidence would have prompted them to place such heavy stress on interdependence. As long as life remained within the intimate confines of family, sect, and village, sensitivity to the interdependence of human relationships in time and social space was not a critical matter. It was plausible enough to construe each individual as self-reliant master of his own fate, and to construe society as an artificial and even ephemeral aggregate of individual wills.

But as society became increasingly interdependent, the conditions of adequate explanation and prediction changed. Dewey and his generation were the first to adopt the habits of mind appropriate to an interdependent society. Social thinkers who clung to older habits of mind, appropriate to a simpler society, became implausible and faded into obsolescence. The main point of cleavage between the rising professional theorists examined by Morton White and the declining amateur theorists studied in these pages is the recognition of, or the failure to recognize, the objective fact of social interdependence and its larger theoretical implications.

From this vantage point, the intellectual transformation of the 1890's can be seen to have deep roots in the complex social and economic changes that we too simplistically designate with terms like "urbanization," "industrialization," and "modernization." If this vantage point is valid, no history of ideas alone will suffice to explain what happened at the turn of the century. A major intellectual reorientation was produced—in the first instance, at least—not by an immanent progression of theory, as Talcott Parsons believed, but by a primitive recognition that the very constitution of the social universe had changed. And that primitive recognition was brought about not simply by detached scientific evaluation of an accumulating store of empirical evidence, but by a more elemental alteration of perspective, akin to the visual gestalt shifts that have so intrigued students of scientific discov-

---

[29.] Parsons acknowledged that the intellectual reorientation of the 1890's probably was "in considerable part simply an ideological reflection of certain basic social changes" (*Structure of Social Action*, 5), but he did not investigate those changes. White also felt obliged to apologize for the strictly intellectual focus of his study of Dewey, noting that a "Deweyan history of Dewey's philosophy" would require a systematic study of the extra-intellectual changes that took place in his environment (*Origin of Dewey's Instrumentalism*, 149).

ery.[30] The special sensitivity to interdependence possessed by Dewey and his generation was a product of their social experience—the whole range of that experience, not just the professional parts of it. They experienced society not only in the vicarious, deliberate, rational manner of the social researcher—presumably always master of his data, holding society at arm's length as a mere object of examination—but also in the immediate, haphazard, unaware manner of the voter, lover, consumer, and job-seeker who is himself an object and onrushing element of the social process. Both kinds of experience prompted their recognition of interdependence.

Interdependence was an objective condition of nineteenth-century society before social thinkers recognized it and articulated it as such. In the first decade or two after the founding of the Social Science Association in 1865, interdependence influenced thought and behavior in primitive ways, but was seldom given explicit attention. If acknowledged at all, interdependence was seen as an anomaly, not as the linchpin of theory. But once over the threshold of consciousness, the idea took on a life of its own. In the years around the turn of the century interdependence was not only recognized, but also detached from the objective conditions which it originally represented and put to use in a wide-ranging, paradigmatic manner to illuminate social phenomena at every level of analysis. In his study of the historicist and organicist emphases of Progressive social thought, Morton White documents the use of the idea of social interdependence as a guiding assumption, a controlling expectation, often implicit, that led men to distrust formalistic habits of thought even where those habits could not be shown conclusively to be erroneous.

The next chapter proposes a rigorous definition of interdependence, traces historically its emergence as a pronounced feature of society, and explores some of the ways in which the growth of interdependence must have altered perceptions of man and society, thus providing a basis for the claim of professional social science expertise.

[30.] Norwood Russell Hanson, *Patterns of Discovery: An Inquiry into the Conceptual Foundations of Science* (Cambridge: Cambridge University Press, 1965) and Thomas S. Kuhn, *The Structure of Scientific Revolutions*, 2nd ed. (Chicago: University of Chicago Press, 1970), esp. ch. X. Both authors draw upon visual gestalt shifts as at least an analogy to what happens in scientific discovery. Parsons's account of the breakdown of positivism, as manifested in the work of Pareto and Durkheim, and the rise of the voluntaristic theory, was couched in straightforward empirical terms: "In the thought of both [Pareto and Durkheim] the driving force of the change may be said to lie primarily in the realization of the empirical inadequacy of 'individualistic' theories as revealed by their own critical analyses of them and their own empirical investigations." And elsewhere: "In this breakdown [of the positivist theory of action] the sheer empirical evidence played a decisive role along with theoretical and methodological considerations." *Structure of Social Action*, 351–352, 470.

The remaining chapters interpret the American Social Science Association as an early and necessarily inadequate response to the changing conditions of explanation in a society growing ever more interdependent. By the turn of the century the ASSA for all practical purposes had been superseded by the emergence of intense professional communities of social inquiry, based in the university, that seemed far better suited to the task of explanation in an interdependent social universe.

To stress the growing interdependence of society is, of course, to reject or assign secondary importance to other characterizations of nineteenth-century social change. It is interdependence that occupies the central place in this study, not a changing mode of production, an alteration in class structure, the advent of machine technology, the extension of factory discipline, the fragmentation of the community, the growth of urban anonymity, the heightening of social mobility, the imposition of bureaucratic procedures by the nation-state, the rise of finance capitalism, the exhaustion of middle-class culture and the patriarchal family, or the separation of the worker from the tools of production. All of these and other factors have been cited in explanation of various aspects of modern consciousness, and each has its own area and degree of validity. Peter Berger has lent his exceptional imaginative powers to a recent book that squeezes what must be the very last drop of explanatory potential out of machine technology and bureaucracy as the two foundations of modern consciousness.[31] Certainly modernization is a multi-faceted process. It would be foolish to claim that the linkage between consciousness and social change in a modernizing society can be reduced to a single factor, even one as intrinsically multifarious as interdependence. But I do claim that interdependence can be seen as the common denominator of many previously proposed links between modern social structure and the mentality appropriate to it. Furthermore, the increasingly interdependent character of an urbanizing, industrializing society is critically important for intellectual history, because it is the dimension of social change that impinges directly on the way men attribute causation, and thus influences most profoundly the way they construe social reality. If the reader will accept the argument of the second chapter, and the conclusion—that many people living in an increasingly interdependent social environment are likely to change their customary habits of causal attribution—from that single change a great many other features of modern consciousness can be shown to follow quite naturally.

[31.] Peter Berger, Brigitte Berger, and Hansfried Kellner, *The Homeless Mind: Modernization and Consciousness* (New York: Random House, 1973).

The extent of my indebtedness to Thomas S. Kuhn's *Structure of Scientific Revolutions* (1962, 1970) will already be apparent to many readers. The scenario outlined here, of an amateur school of thought going into decline and a new school rising to authoritative professional status in its place, is a pattern of special significance within Kuhn's frame of reference. Although my account of the Social Science Association would be very different had I not read Kuhn, I have tried to avoid forcing the evidence into his mold, for I do not believe that it is entirely appropriate for historians of social science.[32]

Without qualification I adopt the same contextualist spirit that informs Kuhn's work. The "historiographic revolution" that he optimistically described in natural science must occur also in the history of social science, social thought, and intellectual history generally, where the need for it would seem even more obvious than in the history of science. If, as Kuhn says, discarded views of nature such as "Aristotelian dynamics, phlogistic chemistry, or caloric thermodynamics . . . were, as a whole, neither less scientific nor more the product of human idiosyncrasy than those current today," then surely historians must adopt an equally tolerant and respectful attitude toward discarded social theories, antiquated life styles, and abandoned frameworks of moral and ethical judgment.[33] The historian can be a moral critic, but he must not be merely a provincial booster of his own politics and morality. Men of the Mugwump class who dominated the ASSA are a good case in point, for their thought has been easy to dismiss from a non-contextualist point of view as merely quaint and archaic, if not vicious.[34]

My understanding of the meaning of professionalization in intellectual fields has been especially influenced by Kuhn. Throughout this study professionalization is understood to be a measure not of quality, but of community. A social thinker's work is professional depending on the degree to which it is oriented toward, and integrated with, the work of other inquirers in an ongoing community of inquiry. In the first

---

[32] David Hollinger, "T. S. Kuhn's Theory of Science and Its Implications for History," *American Historical Review*, 78 (Apr. 1973), 370–393, is a good guide to the literature spawned by Kuhn's own "paradigm" of historical scholarship. There are a number of useful essays, pro and con, including Kuhn's own "Reflections on My Critics," in Imre Lakatos and Alan Musgrave, eds., *Criticism and the Growth of Knowledge: Proceedings of the International Colloquium in the Philosophy of Science, London, 1965, vol. 4* (Cambridge: Cambridge University Press, 1970).

[33] Kuhn, *Structure of Scientific Revolutions*, 2–3.

[34] There is something oedipal in the condescending tone which almost all historians adopt when talking about their Victorian forbears. The hint of a new view is visible in David P. Thelen, "Rutherford B. Hayes and the Reform Tradition in the Gilded Age," *American Quarterly*, 22 (Summer 1970), 150–165. See also ch. V, below, and *American Quarterly*, 27 (Dec. 1975), a special issue devoted to "Victorian Culture in America."

instance, at least, the quality of his work is irrelevant. Amateurs in this sense range from stupid to brilliant, just as professionals do. Admittedly, I share the widespread impression that professionalization has raised the general level of social inquiry, but I think it would be exceedingly presumptuous and dangerous to fix upon this impression as a chief point of distinction between amateur and professional. As Alfred North Whitehead observed, "Each profession makes progress . . . but it is progress in its own groove. Now to be mentally in a groove is to live in contemplating a given set of abstractions. The groove prevents straying across country, and the abstraction abstracts from something to which no further attention is paid. . . ."[35] No profession can afford to believe that its groove is the whole of the reality it professes to understand.

Professions are collective enterprises, so their emergence and growth must be gauged in terms of collective, not individual, criteria. Professionalization in social science, then, may be defined as a three-part process by which a community of inquirers is established, distinguishes itself from other groups and from the society at large, and enhances communication among its members, organizing and disciplining them, and heightening their credibility in the eyes of the public. Any act which contributes to these functions, which strengthens the intellectual solidarity of this very special kind of community, is a step toward professionalization. Solidarity in this context does not mean harmony or an absence of conflict. The community of inquirers is knit together not by affection but by attention. Its members constitute a solidary community to the extent that they recognize each other as working in the same field and feel compelled to address themselves intellectually to each other. Conflict and even hatred are perfectly compatible with solidarity in this sense; solidarity is diminished only when inquirers ignore each others' work, or dismiss it as irrelevant.[36]

Of the many factors that can contribute to the solidarity of a community of inquirers, Kuhn was principally interested in one: the paradigm. By paradigm Kuhn originally meant a concrete scientific achievement on the order of Copernicus' heliocentric theory or Newton's *Principia* which serves "as a locus of professional commitment," a seed about which there crystallizes a community of researchers and a research tradition.[37] The paradigm is the achievement itself, which logically precedes, and is a good deal more mysterious in its functioning than any of the specific rules, concepts, laws, theories, or points of

---

[35.] Alfred North Whitehead, *Science and the Modern World: Lowell Lectures, 1925* (New York: Macmillan, 1939), 282–283.
[36.] Professionalization is further defined in the next chapter.
[37.] Kuhn, *Structure of Scientific Revolutions,* 11.

view that may be derived from it. It supplies the consensual basis for a community of researchers and guides their "normal" work by posing problems for solution and assuring them that a solution exists. But no paradigm can master the richness of reality. Normal scientific research undertaken in accordance with the paradigm necessarily generates anomalies, the accumulation of which leads to a sense of crisis and eventually perhaps to a "scientific revolution," which is the wholesale shift of the community from one paradigm to another. A new tradition of normal research then begins on the basis of the new paradigm, and practitioners who cling to the older paradigm are ignored or dismissed as cranks.

As Kuhn originally formulated the concept, a paradigm was a rare and exalted thing. Real scientists possessed them; social scientists probably did not, any more than poets, astrologers, or "electricians" before the discoveries of Ben Franklin.[38] Kuhn has since backed away from this rigorous formulation and now permits the word to be used loosely so that even immature or unscientific fields of inquiry, like philosophy and art, may be said to have paradigms. "What changes with the transition to maturity is not the presence of a paradigm, but rather its nature."[39] Moreover, he has proposed that paradigms be regarded as only one of several elements in the "disciplinary matrix" that provides the consensual basis for a community of inquiry.[40]

In the original exalted sense of the term, interdependence does not qualify as a paradigm. A sensitivity to interdependent relationships is no doubt one of the key elements in the disciplinary matrix of modern social science, and it certainly serves some paradigmatic functions, guiding thought in murky realms of inquiry where conclusive tests and proofs are not feasible. But to argue that it is a paradigm in the strict sense, one would have to show that the social sciences since the turn of the century have enjoyed that very special maturity which permits true science to *progress*, to advance in a cumulative and irreversible manner.[41] This has not happened.

In the looser sense of the concept, the whole intellectual reorientation of the 1890's and after can be fitted to Kuhn's model quite comfortably. The recognition of interdependence and the subsequent elaboration of a suitably non-formalistic social theory after the 1890's could be regarded as the acquisition of a new paradigm of a very elementary nature, basic to all the social sciences, and the initiation of a normal scientific tradition based upon it. The history of the social

[38.] *Ibid.*, 15.
[39.] *Ibid.*, 179.
[40.] *Ibid.*, 182.
[41.] *Ibid.*, 179, and ch. XIII, "Progress through Revolutions."

sciences in these years conforms to Kuhn's model in many ways. Certainly the recognition of interdependence was closely associated with the transition from a very loosely knit aggregation of social inquirers (the ASSA and other even less coherent groupings) to highly integrated professional communities of specialized academic social scientists.[42] The style of internal communication among the members of these groups changed as they moved toward integration in just the way Kuhn's model suggests.[43] The older inquirers of the ASSA did indeed suffer the fate of men who fail to shift their allegiance to a rising new paradigm.[44] The "revolution" itself, if such it was, has been "invisible" to its heirs, just as Kuhn predicts.[45] At the very least, Kuhn's model of paradigm shift is a suggestive analogy.

But if virtually all fields of knowledge possess paradigms, it does not make much difference whether interdependence is labeled a "paradigm" or not. Presumably we have always known that research and explanation are meaningful only within a framework of basic assumptions; that there are competitive "schools of thought" in most fields; and that both orthodoxy and innovation play a valid role in the advance of inquiry. Kuhn's originality consists not in subsuming these "frameworks," "schools," and "orthodoxies" under a new rubric, but in explaining how they or their essential elements come into existence, what sustains them, and how shifts occur from one to another. I have profited from Kuhn's suggestions in these respects, but I have not tried to portray the career of the American Social Science Association as a case study of paradigm shift. The questions "Is interdependence a real paradigm?" and "Was the intellectual reorientation of the 1890's a real social scientific revolution?" are not the foci of this study and are not, I think, profitable questions, save perhaps in a purely heuristic way.

Historians of social science can profit greatly from Kuhn's work, but they can also be misled by it. A too literal-minded reading of Kuhn could lead historians of social science to concern themselves exclusively with the internal dynamics of the community of inquiry, and to ignore the social and cultural context in which the community exists and on which it so heavily depends. The professional thought world of the social scientist, unlike that of the physicist, is not the product of a strictly intramural dialogue with his professional peers. Since some of the roots of social scientific paradigms may lie outside the professional community, in the everyday thought world shared by professional and layman alike, historians of social science will have to cast a far wider net

[42] *Ibid.*, 19.
[43] *Ibid.*, 20.
[44] *Ibid.*, 19.
[45] *Ibid.*, ch. XI, "The Invisibility of Revolutions."

than historians of natural science. Moreover, paradigms are not the only source of growth and solidarity even in natural scientific communities; as Kuhn recognized, they are not necessarily even the most important ones in fields like medicine, technology, and law, "of which the principal *raison d'être* is an external social need."[46] Surely this is true also of social science. The history of the ASSA shows that the process of professionalization in social science was propelled in its early stages not only by the acquisition of something like a paradigm, in the concept of interdependence, but also by non-scholarly needs and motives ranging from the desire to advance reforms by cloaking them in the mantle of science, to a hope of refurbishing the privileges of the genteel class by rebuilding its authority on a foundation of functional expertise.

What makes Kuhn's *Structure of Scientific Revolutions* so important to intellectual historians is not his concept of the paradigm or even his strategy of radical contextualism, but a still more primitive insight: namely, that mankind's most fundamental, paradigmatic assumptions are normally immune to the experiential evidence that might modify or falsify them. Intellectual history has often taught that the systems of belief by which men live possess a tenacity so powerful that assumptions shape experience far more often than they are shaped by it. The lesson is not new, but never before has it been taught so forcefully.

Kuhn's demonstration of the superior power that preconception and assumption ordinarily wield over the supposedly "brute facts" of experience carries special force because he applies it where it would seem least applicable: to scientists, those members of society who most deliberately and systematically submit their preconceptions to the tests of experience. There is no act more characteristic of the scientist than the experiment, the controlled experience deliberately contrived to bring the full weight of empirical evidence to bear upon a specific assumption. Yet during periods of "normal science," says Kuhn, the scientist relies totally on the operative paradigm in his field. From it he deduces hypotheses for empirical examination which, if found wanting, are freely modified or discarded; here theory dutifully submits to experience. But the overarching paradigm that generates theory and confers meaning on the scientist's experimental enterprise normally remains impervious to experience. Except in revolutionary moments, it is not tested. It controls experience rather than being controlled by it. It determines what the scientist will construe as falsifying evidence, but it is not itself susceptible to falsification.

No other member of society is as eager as the scientist to risk his assumptions against potentially falsifying experience. Even the scien-

[46] *Ibid.*, 19.

tist is open to reversals of opinion only in comparatively impersonal, technical matters. If even the scientist is normally prisoner of his most fundamental assumptions, then how much more so is the average person. The implications of Kuhn's thesis are in this respect profoundly deterministic.

Kuhn's insight implies certain cautions for intellectual historians. Historians ordinarily will be no more able than Kuhn was to posit a simple, straightforward connection between the gradual accumulation of anomalous and disconfirmatory evidence and the eventual, quite sudden, overthrow of an old system of assumptions. The question of why paradigmatic assumptions withstand contrary evidence for so long, and then so suddenly fall, remains as Kuhn left it, largely a mystery. We can be sure, however, that if fundamental assumptions are as resistant to disconfirmation as Kuhn suggests, historians of intellectual change will have to dig deep for the causes of important watersheds like the decade of the 1890's.

Most important, Kuhn warns us never to underestimate the ability of a basic assumption to survive untouched in the face of evidence that seems, in the light of the historian's own assumptions, to be plainly disconfirmatory. To take only one example from the following pages: the abandonment of the idea that pauperism resulted from a failure of character for which the impoverished individual himself was primarily at fault cannot be explained merely by pointing to the "brute facts" of proliferating poverty and unemployment in an industrializing society. The overarching assumption of individual potency that made the pauper seem a sufficient cause of his own poverty was supple enough to accommodate almost any increase in numbers. The poor probably were more visible and perhaps proportionally more numerous in preindustrial society, when the ethic of self-reliance was strongest, than later when it collapsed.

To explain the abandonment of the ethic of self-reliance by serious social thinkers near the turn of the century will therefore require a more elaborate explanation, one that reveals in many spheres of life a wide range of disconfirmatory experiences sufficiently vivid and personal to penetrate to the core assumption of individual potency. The concept of interdependence supplies the basis for such an explanation.

CHAPTER II

Interdependence and the Rise of Professional Social Science

When the American Social Science Association was founded in 1865, there were no specialized professional social scientists in this country or, strictly speaking, in any other. There were, of course, scholars and laymen engaged in various kinds of social inquiry, and it would be mere quibbling to deny that they can be regarded, retrospectively, as "social scientists" — though a less specific term such as "social inquirer" would be more accurate. In any case, these people were not *professional* social scientists. This is not to say that their work was unintelligent, but only to insist that it was not decisively oriented to any ongoing, disciplined community of inquiry. Whatever intellectual authority a nonprofessional social scientist possessed was a highly individualized and even idiosyncratic achievement that depended on particular feats of intellect and the approval of an audience whose members were difficult to identify, whose criteria of judgment were ill defined, and whose qualifications as judges were not intrinsically dependent upon the subject area of the scholarship they judged.

When the American Social Science Association died in 1909, the conditions of intellectual authority were different. By this time each of the specialized social sciences had formally declared its independence of all the others, and each had developed the familiar professional apparatus of journals and associations and more or less uniform training curricula at the university level. Intellectual authority now depended heavily upon membership and rank within a readily identifiable community of peers, whose members shared a similar training experience and hence a substantially similar set of evaluative criteria.

The Social Science Association died largely because its internal structure and mode of inquiry were not compatible with the new conditions of intellectual authority—conditions which, ironically, had been nurtured in infancy by the pioneering efforts of the Association and its individual members. The rise of professional academic social science, underway in the 1880's and in full swing by the 1890's, robbed the Association of its credibility as a fountainhead of sound social knowledge. What began in 1865 as a lively and often perceptive quest for social reality ended forty-five years later as a clubbish and stuffy enterprise, intellectually sterile.

To explain the profound changes in the conditions of intellectual authority that first gave birth to the Social Science Association and then rendered it obsolete, we need to know a great deal more than we do now about the professionalizing and institution-building activities of the first generation of professional social scientists. The modern disciplines of economics, history, sociology and political science are based upon disciplined communities of inquiry that were brought into existence by the mundane labors of men like Richard T. Ely, Herbert Baxter Adams, John Franklin Jameson, Edward A. Ross, Albion Small, and Jeremiah W. Jenks. Before such communities could exist, professional journals and other media of communication had to be created, professional associations had to be founded and periodic gatherings organized. Innumerable local intrigues and budgetary battles had to be waged on campuses across the country to reshape teaching institutions to the specialized, research-oriented needs of the new disciplines. Curricular reform, the introduction of new teaching methods such as the seminar, the establishment of a regular pattern of hurdles to be faced by those aspiring to professional status, the crystallization of a code of professional behavior whose outer limits were defined in the celebrated academic freedom cases of the period—all of these activities are a part of the process of professionalization that made the ASSA obsolete by the turn of the century. Historical research into most of these matters is sketchy and uncoordinated, and many archival materials have never been touched. Much remains to be done.[1]

[1.] What little work has been done in the field of academic professionalization is generally of high quality. In addition to biographies of the key figures and such general accounts as Richard Hofstadter and Walter P. Metzger, *The Development of Academic Freedom in the United States* (New York: Columbia University Press, 1955), and Joseph Dorfman, *The Economic Mind in American Civilization* (New York: Viking Press, 1949), there are a number of recent specialized studies. The essays contained in *The Social Sciences at Harvard: From Inculcation to the Open Mind*, ed. Paul Buck (Cambridge: Harvard University Press, 1965) are useful; see also Robert L. Church, "The Development of the Social Sciences as Academic Disciplines at Harvard University, 1869–1900" (Ph.D. dissertation, Harvard University, 1965). Only one other university is so well served: R. Gordon Hoxie *et al.*, *A History of the Faculty of Political Science, Columbia University* (New York: Columbia University Press, 1955). Several of Alfred W. Coats's articles are relevant, especially "The First Two Decades of the American Economic Association," *American Economic Review*, 50 (Sept. 1960), 555–572, and "The Political Economy Club: A Neglected Episode in American Economic Thought," *American Economic Review*, 51 (Sept. 1961), 624–637. In the same category is George W. Stocking, Jr., "Franz Boas and the Founding of the American Anthropological Association," *American Anthropologist*, 62

But no matter how much we learn about the concrete professionalizing activities of the first generation of professionals, that knowledge will not suffice to explain the rise to cultural dominance of the professional academic social scientist at the turn of the century. The professionalizers rode a wave of change over which they had little real control. A profession has many of the qualities of a self-perpetuating monopoly—in this case, a monopoly in authoritative opinion about the nature of man and society. But this kind of monopoly is neither created nor sustained simply by the autonomous efforts of those who exercise its privileges. The causes of professionalization do not lie in the specific labors of those who first construct the profession, or in their energy, dedication, or strength of mind—indispensable as these qualities are. Rather, they are related to the changing conditions of the larger society that encourage professionalizing activity and enable it to succeed.

To attribute the rise of professional social science simply to the labors of the founding generation would be to mistake instrumentalities for genuine causes. Professionalization is manifestly not a phenomenon unique to late nineteenth century American social scientists, but a process broadly characteristic of the modern world, touching many occupations in all advanced societies. It is, or accompanies, a stage of social development. Indeed, professionalization is one of those few critical processes by which we define the very character of the urban-industrial transformation of the nineteenth century. It is not only a response to or an effect of industrialization, but more accurately a part of it. As William J. Goode says, "An industrializing society *is* a professionalizing society."[2] Talcott Parsons asserts that "the development and

---

(Feb. 1960), 1–17. Mary Furner, *Advocacy and Objectivity: A Crisis in the Professionalization of American Social Science, 1865–1905* (Lexington: University of Kentucky Press, 1975), is especially strong on the founding of the American Economic Association and the elaboration of professional discipline in academic freedom cases. A broad view of university development is provided by Laurence R. Veysey, *The Emergence of the American University* (Chicago: University of Chicago Press, 1965), and other insights into professionalization can be found in Hamilton Cravens, "American Scientists and the Heredity-Environment Controversy, 1883–1940" (Ph.D. dissertation, University of Iowa, 1969); Jurgen Herbst, *The German Historical School in American Scholarship: A Study in the Transfer of Culture* (Ithaca: Cornell University Press, 1965); Bernard Crick, *The American Science of Politics: Its Origins and Conditions* (Berkeley: University of California Press, 1962), and Albert Somit and Joseph Tanenhaus, *The Development of Political Science: From Burgess to Behaviorism* (Boston: Allyn and Bacon, 1967). Recent work on the professionalization of social science in other countries includes Terry Nichols Clark, *Prophets and Patrons: The French University and the Emergence of the Social Sciences* (Cambridge: Harvard University Press, 1973), Anthony Oberschall, *Empirical Social Research in Germany, 1848–1914* (The Hague: Mouton, 1965), and Roger Lewis Geiger, "The Development of French Sociology, 1871–1905" (Ph.D. dissertation, University of Michigan, 1972).

[2.] William J. Goode, "Encroachment, Charlatanism, and the Emerging Profession: Psychology, Sociology and Medicine," *American Sociological Review*, 25 (Dec. 1960), 902. Emphasis added.

increasing strategic importance of the professions probably constitute the most important change that has occurred in the occupational system of modern societies."[3]

The rise of the professional social scientist, then, and the concomitant descent into obsolescence of the American Social Science Association, have deep roots in the professionalizing tendency inherent in the urban-industrial transformation. A proper explanation of the transit of authority from laymen and amateurs to professionals will seek out those aspects of human experience in an urbanizing, industrializing society that made some people receptive to expert advice about human affairs and gave others the confidence to offer such advice. Explanation lies not inside the professionalizing community, but in the linkage between it and the larger social context.

The professionalization of social inquiry should be seen as one among many facets of the urban-industrial transformation, but it also has a special claim to our attention. The emergence of the "social scientist" as a recognized occupational role is a pivotal event in the development of the mentality characteristic of an interdependent society. The linkages between the urban-industrial transformation and the rise of the social scientist therefore promise to be an especially fruitful subject of inquiry, because each term of the relationship potentially reveals the other—the effect illuminates its cause. In studying social scientists we study the most self-conscious and therefore potentially the most revealing symptoms of modernity itself.

Interdependence

The role played by the professional person illuminates the character of the society in which he acts. Professionals are persons who claim to possess esoteric knowledge which serves as the basis for advice or services rendered to the public for remuneration. The mode of remuneration is highly variable and need not be direct fee-for-service. The professional is never truly a lone wolf. He may enjoy a greater than ordinary degree of autonomy in his daily work (in the sense of freedom from routine supervision), but for his identity as a professional he is absolutely dependent upon admission to and reputation within a community of peers. By committing itself to selective admission, to the

---

[3.] Talcott Parsons, "Professions," in David L. Sills, ed., *The International Encyclopedia of the Social Sciences* (New York: Macmillan and Free Press, 1968), XII, 536–546. Everett C. Hughes speaks of the professions and the "professional attitude or mood" as "components of the professional trend, a phenomenon of all the highly industrial and urban societies; a trend that apparently accompanies industrialization and urbanization irrespective of political ideologies and systems." Hughes notes also that "the professional trend is closely connected with the bureaucratic." See Hughes, "Professions," in Kenneth Lynn, ed., *The Professions in America* (Boston: Houghton Mifflin, 1965), 1.

expulsion of charlatans, and to other kinds of internal policing, the professional community provides each of its members with a warrant to practice. The warrant is essential because laymen, in the very act of seeking professional help, confess their ignorance of the matters in question and therefore imply their inability to judge adequately the quality of services rendered. The credibility of the individual professional man in the eyes of laymen is not his own doing, but is to a great extent a product of his membership and good standing in the professional community.[4]

What is it about modern society that causes men to rely increasingly on professional advice? Under what circumstances do men come to believe that their own judgment, based on common sense and the customary knowledge of the community, is not adequate? It is true, but not very helpful, to answer that modern society is complex and that professionals thrive on complexity. "Complexity" is a uniquely uninformative word, little more than a mirror-image of confusion. What we have meant by complexity in this context, I think, is social interdependence. Virtually every student of the nineteenth-century transformation of society uses the word "interdependence" in an informal, untechnical manner; it appears in the literature only slightly less often than "urbanization" and "industrialization." It is an abstract term and is commonly used in a vague way, but it can be made precise. Properly understood, the concept goes far toward explaining the social and especially the intellectual implications of the urban-industrial transformation.

For America and the advanced nations of Europe, the nineteenth century was a period of growing interdependence among all the components of society, individual as well as institutional. By the term "growing interdependence" I mean to refer to something quite exact: that tendency of social integration and consolidation whereby action in one part of society is transmitted in the form of direct or indirect

[4.] Definitions of professionalism differ more often in emphasis than in substance. The most elegant definition to my knowledge is by Everett C. Hughes: "A profession delivers esoteric services—advice or action—. . . to the public [or some part of it]. . . . The nature of the knowledge, substantive or theoretical on which advice and action are based is not always clear; it is often a mixture of several kinds of practical and theoretical knowledge. But it is part of the professional complex, and of the professional claim, that the practice should rest upon some branch of knowledge to which the professionals are privy by virtue of long study and by initiation and apprenticeship under masters already members of the profession. . . . Professionals *profess*. They profess to know better than others the nature of certain matters, and to know better than their clients what ails them or their affairs. This is the essence of the professional idea and the professional claim. . . . The client is not the true judge of the value of the services he receives; furthermore the problems and affairs of men are such that the best of professional advice and action will not always solve them." Hughes, "Professions," 1–3; italics in the original. For a formal definition, see Wilbert E. Moore, *The Professions: Roles and Rules* (New York: Russell Sage Foundation, 1970), ch. I.

consequences to other parts of society with accelerating rapidity, widening scope, and increasing intensity. A society is interdependent to the extent that its component members or parts influence and are influenced by each other. A society is *inter*dependent to the extent that the activity of each of its members *depends* (consciously or not) upon the activity of its other members; conversely, it lacks the quality of interdependence to the extent that its members are able to act *in*dependently, without taking each other's acts into account. Interdependence is an objective condition of life which exists apart from anyone's perception of it. It can, however, be intensified by a growth of mutual awareness that prompts men to respond more deliberately and sensitively to their dependencies.

Of course it is true that all forms of human organization involve a degree of mutual dependence. The simplest, most isolated peasant village or primitive tribe may involve intense mutual dependence among its few members and within its narrow borders. What I mean by the term "interdependence" is involvement in a network of intense dependencies that is regional or global in scope, and which includes vast numbers of people, most of them strangers who will never encounter each other on a face-to-face basis.[5] The word is admittedly deficient in that it does not self-evidently distinguish between intimate relationships in primary groups and the vast webs of intense but impersonal interrelationships characteristic of modern society.[6]

Defining the term in this manner, one can hardly doubt that American society and other advanced societies became vastly more interde-

---

[5.] The typological dichotomy established here between autonomy and interdependence obviously overlaps and parallels to some extent the classic characterizations of Tönnies (*Gemeinschaft-Gesellschaft*), Weber (traditional-rational), Durkheim (mechanical-organic solidarity), and Cooley (primary-secondary groups). On the overlap between these and other typologies, see C. Wright Mills, *The Sociological Imagination* (New York: Oxford University Press, 1959), 152–153; Edward Shils, "The Contemplation of Society in America," in *Paths of American Thought*, ed. A. M. Schlesinger, Jr., and M. White (Boston: Houghton Mifflin, 1963), 398–399; and Gladys Bryson, *Man and Society: The Scottish Inquiry of the Eighteenth Century* (Princeton: Princeton University Press, 1945), 171–172. Despite the similarities, the concept of interdependence seems to me a preferable formulation if one is concerned with the intellectual consequences of modernization—and especially the rise of the social sciences—because it immediately reveals the impact of social change upon causal attribution, and, hence, upon explanation.

[6.] To talk about the growing mutuality of dependence among men is not to deny the importance of power and status. Some men occupied strategic positions in the growing web of interdependence: their actions influenced thousands, while the average man, with little access to the instruments that permitted influence-at-a-distance, was merely swept along. This study concentrates on interdependence at the expense of power because the former contains the key to the explanation of the social science mentality. The imbalance is justified, I think, by the fact that historians in general have neglected the factor of interdependence, for at least two reasons. First, it has often been assumed that power is synonymous with independence—that powerful men, by virtue of their power, cannot be regarded as dependent upon others in any interesting way. On the

pendent, internally and internationally, over the course of the nineteenth century. To document the process we need not start from scratch, but can simply discuss briefly three familiar areas of scholarly inquiry. First, the expansion of the capitalist marketplace and the intensification of its forces, culminating in triumphant industrial capitalism at the end of the century; second, the transportation-communication revolution, which should be understood to include not only technological advances but any innovation that enhanced the possibilities of acting and being acted upon across great distances; and third, the process of the division of labor, with its unending proliferation of ever narrower occupational specialities. Although these areas have been treated as separate fields by their respective investigators, they should be regarded as intimately related facets of a single process, the decisive consequence of which for our purposes is the growing interdependence of society.[7]

Sources of Interdependence: Market, Transportation, and Specialization

The earliest men whom one might plausibly call "social scientists," the moral philosophers and classical economists of the late eighteenth and early nineteenth centuries, made the dynamics of social interdependence one of their central concerns. Adam Smith's great achievement in *The Wealth of Nations* (1776) was to single out as an analytical unit the market, the most intensely interdependent sector of

---

contrary: dependence on other men is generally a vital condition of power and advantage. Second, as humane men, scholars have tended to approach all social questions with the balance scale of blindfolded Justice in hand. The peculiarity of the balance scale is that it registers *only* imbalance, and gives no indication of changes occurring simultaneously on both sides of the scale. This is the kind of change implied in the concept of heightening interdependence: both weak and strong men lost a degree of autonomy in a more or less simultaneous process. For discussions of parallel questions at the level of high theory, see Alvin W. Gouldner, "Reciprocity and Autonomy in Functional Theory," in *Symposium on Sociological Theory*, ed. L. Gross (Evanston, Ill.: Row, Peterson, 1959), 264–265, and Wsevolod W. Isajiw, *Causation and Functionalism in Sociology* (London: Routledge & Kegan Paul, 1968), 45–48.

[7.] A fourth area of scholarship, "mass society theory," covers much the same ground as the three I have named. Daniel Bell describes it as "probably the most influential social theory in the world today," apart from Marxism. He summarizes it as follows: "The revolutions in transport and communications have brought men into closer contact and bound them in new ways; the division of labor has made them more interdependent; tremors in one part of society affect all others. Despite this greater interdependence, however, individuals have grown more estranged from one another. The old primary group ties of family and local community have been shattered; ancient parochial faiths are questioned; few unifying values have taken their place. Most important, the critical standards of an educated elite no longer shape opinion or taste. As a result, mores and morals are in constant flux, relations between individuals are tangential or compartmentalized rather than organic. At the same time greater mobility, spatial and social, intensifies concern over status. Instead of a fixed or known status symbolized by dress or title,

society, and delineate its main principles of operation. Unlike later vulgarizers of his doctrine, he fully recognized that civilization means mutual dependence. "In civilized society," wrote Smith, "he [man] stands at all times in need of the co-operation and assistance of great multitudes, while his whole life is scarce sufficient to gain the friendship of a few persons. In almost every other race of animals each individual, when it is grown up to maturity, is entirely independent, and in its natural state has occasion for the assistance of no other living creature."[8] Smith's most striking discovery was that this extensive network of mutual dependence, without which man would be no more than an animal, was sustained unintentionally, not by benevolence but by self-interest: "It is not from the benevolence of the butcher, the brewer, or the baker, that we expect our dinner, but from their regard to their own self-interest."[9]

each person assumes a multiplicity of roles and constantly has to prove himself in a succession of new situations. Because of all this, the individual loses a coherent sense of self. His anxieties increase. There ensues a search for new faiths. The stage is thus set for the charismatic leader, the secular messiah, who, by bestowing upon each person the semblance of necessary grace, and of fullness of personality, supplies a substitute for the older unifying belief that the mass society has destroyed." Bell, "The Theory of Mass Society: A Critique," *Commentary*, 22 (July 1956), 75–83. See also William Kornhauser, *The Politics of Mass Society* (Glencoe, Ill.: Free Press, 1959), Maurice R. Stein, *The Eclipse of Community: An Interpretation of American Studies* (Princeton: Princeton University Press, 1960), and Arthur J. Vidich and Joseph Bensman, *Small Town in Mass Society: Class, Power and Religion in a Rural Community* (Princeton: Princeton University Press, 1968). This general line of argument has been exploited most effectively in the historical profession by Robert Wiebe, whose work is discussed below.

My argument is closely related to this theory, but I wish to avoid certain of its emphases—especially its preoccupation with aristocratic values, its obsessive search for harbingers of fascism, and its implicit faith that a revitalization of group life might rescue modern man from alienation. In my view, the groups (family, church, village, etc.) which once mediated between the individual and the larger society thrived precisely because the society was not interdependent, and they were destroyed or weakened by the loss of autonomy that followed necessarily from their absorption into an expanding network of mutual dependencies. Mediating groups, in short, were destroyed by a condition—interdependence—to which each contributed, but which each experienced as external to itself. The same condition exists today. Although the reconstruction of such groups might provide the individual with some temporary relief, it would at best only change the units of interdependence (to groups rather than individuals) and do nothing about the condition itself. Under conditions of interdependence, increasing the power of each group does not make it autonomous; it merely amplifies the influences that each transmits to the others, thus further reducing autonomy. In such a society self-determination can exist only as an illusion. Given existing cultural norms of behavior and the existing technology of influence-at-a-distance, it is difficult to see how there can be any escape from the condition of interdependence.

[8.] Adam Smith, *An Inquiry into the Nature and Causes of the Wealth of Nations*, ed. Edwin Cannan (New York: Putnams, 1904), I, 16.

[9.] *Ibid.* On the emergence of the problem of unintended consequences among the moral philosophers, see Louis Schneider, ed., *The Scottish Moralists on Human Nature and Society* (Chicago: University of Chicago Press, 1967), xxix-xlvii. On its fundamental role in modernسociology, see *The Idea of Culture in the Social Sciences*, ed. L. Schneider and C. Bonjean (Cambridge: Cambridge University Press, 1973), 140–141.

Beginning with the assumptions that men act rationally to seek pleasure and avoid pain, and that they have a natural propensity to barter and exchange, Smith showed that the separate economic activities of widely dispersed individuals were intricately linked through the laws of supply and demand so as automatically to allocate resources, distribute goods, and assign tasks in a manner more efficient than any human magistrate could hope to devise. This intricate linkage, this web of unintended interdependencies that manifests itself as the law of supply and demand, stands at the very heart of Smith's doctrine. Smith was the first to discuss systematically a new, invisible form of social discipline that for many years had been growing unseen in the cracks and crevices of traditional society. It was an impersonal discipline of the carrot rather than the stick. Men preferred it to the overt, personal discipline of a lord and master because it permitted one to feel free even as he performed the task demanded by society. In the hands of the classical liberals of the early nineteenth century, Smith's conception became a rationale for dismantling the last major vestiges of traditional society, with its hierarchy and its relationships of command and obedience and organic connection, all of which were obstructive to the natural laws of the market. Where there was little to dismantle, as in the United States, society approached the atomistic ideal of liberal capitalist theory.

The network of unintended mutual dependencies that knit society together was orderly and beneficent in Smith's view. Indeed, the self-regulatory feature of the system was cause for awe and wonderment—seemingly an empirical evidence that God was in his place and all was well with the world. Smith's conception merged easily with much older imagery of society as an organic whole, each of whose members, however lowly, was indispensable to the rest. As late as 1840 Alonzo Potter was typical of American moral philosophers in his belief that "all society is, in fact, one closely-woven web of mutual dependence, in which every individual fiber gains in strength and utility from its entwinement with the rest." In the same vein, Francis Bowen of Harvard taught that "we are all servants of one another without wishing it, and even without knowing it. . . ."[10]

The idea that men were bound in a web of mutual dependence was a cheering thought, merely a pious proof of human brotherhood devoid of analytical implications, as long as the discipline of the market was mild. When Smith wrote, on the eve of the Industrial Revolution, most men lived on the land and entered the market only occasionally. In 1790, 95 percent of the U.S. population lived on scattered farms or in

---

[10.] Both quotations from Donald Harvey Meyer, "The American Moralists: Academic Moral Philosophy in the United States, 1835–1880" (Ph.D. dissertation, University of California at Berkeley, 1967), 277, 278.

settlements of fewer than 2,500 persons. A semi-subsistence farmer, the average American was vulnerable to the hazards of nature and the whims of his neighbors, but his life seldom depended on people or events outside his own locality. The only people fully exposed to the harsh, complicated ways of the market were merchants—generally well rewarded for their exposure—who were clustered in seaboard cities, close to the only efficient means of transportation.

The chief significance of the early nineteenth century transportation revolution is that it carried the discipline of the market and the other consequences of interdependence into the lives of people who previously had been protected by a moat of time and space. The whole pattern of human relationships in society was torn apart and made over again by the multitudinous influences thrust into men's lives by canal, steamboat, and railroad. The complex life of the merchant was imposed on all men. Here, in the expanding, intensifying influence of market forces, lies the most important development of the Industrial Revolution.[11]

For the individual, the advance of communication and transportation technology meant opportunity—but also the obverse, competition. For every producer who gained by access to remote markets, there was another who found his customary local market invaded by more efficient outsiders. What one person experienced as opportunity was necessarily experienced by others as competition. The local community, which had once seemed to contain almost all the people and events that one needed to take into account in managing his life, was now penetrated by lines of influence that originated among strangers

[11.] The argument that the establishment of the market economy is the centrally important development of the Industrial Revolution is presented by Karl Polanyi, *The Great Transformation* (Boston: Beacon, 1957), 40. Polanyi does not stress the role of communication-transportation facilities as much as I have. On the latter subject the classic treatment, limited to the United States and very modest in its conclusions, is George Rogers Taylor, *The Transportation Revolution* (New York: Rinehart, 1951). See also the splendid research of Allen Pred, *Urban Growth and the Circulation of Information: The United States System of Cities, 1790–1840* (Cambridge: Harvard University Press, 1973), Richard Brown, "Knowledge Is Power: Communications and the Structure of Authority in the Early National Period, 1780–1840," paper delivered at 1974 American Historical Association convention, and Douglass C. North, *Growth and Welfare in the American Past: A New Economic History* (Englewood Cliffs, N.J.: Prentice-Hall, 1966), 111. The effects of the transportation revolution have been a special concern of Lee Benson; see his *Merchants, Farmers and Railroads: Railroad Regulation and New York Politics, 1850–1887* (Cambridge: Harvard University Press, 1955), 249, and *Turner and Beard: American Historical Writing Reconsidered* (Glencoe, Ill.: Free Press, 1960), 42–52. Martin A. Knapp, chairman of the Interstate Commerce Commission, wrote perceptively about the broader social consequences of transportation technology: see "Transportation and Combination," *Science*, n.s. 23 (Feb. 1906), 178–184. Essentially the same article appeared as "Social Effects of Transportation," *Annals*, 20 (July 1902), 1–15. The sociologist Charles Horton Cooley, son of an earlier chairman of the ICC, was also particularly concerned with the broad consequences of transportation technology: see Quandt, *From Small Town to Great Community*, 51–66.

THE TRANSPORTATION REVOLUTION

The pace and magnitude of the transformation wrought by new transportation technology in the nineteenth century is reflected in Figures 1, 2, and 3. Douglass North's graph of inland freight rates from 1780 to 1900 shows that, even before the advent of the railroad, transportation costs fell precipitously to a small fraction of their former level. Assuming even the roughest correlation between transportation costs and the volume of market transactions, Figure 1 implies a staggering intensification and multiplication of interdependent relationships. Similarly, Allen Pred's two maps showing the time it took for news of public events to reach New York in 1794 and in 1841 suggest how rapidly ancient barriers to communication collapsed. By way of contrast, Fernand Braudel analyzed the speed of diplomatic correspondence to Venice, 1500–1765, and observed no significant change: "Goods, boats, and people travelled as fast, or as slowly, in the days of the Avignon Popes . . . as they did in the age of Louis XIV." Braudel, *The Mediterranean and the Mediterranean World in the Age of Philip II* (New York: Harper and Row, 1972), I, 365–369.

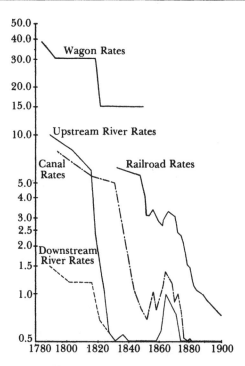

FIGURE 1 General Pattern of Inland Freight Rates (cents per ton-mile), 1780–1900.

SOURCE: Douglass C. North, "The Role of Transportation in the Development of North America," in *Les Grandes Voies Maritimes dans le Monde, xve-xixe Siècles* (Paris: S.E.V.P.E.N., 1965), 222, (c) 1965 by Ecole Pratique des Hautes Etudes.

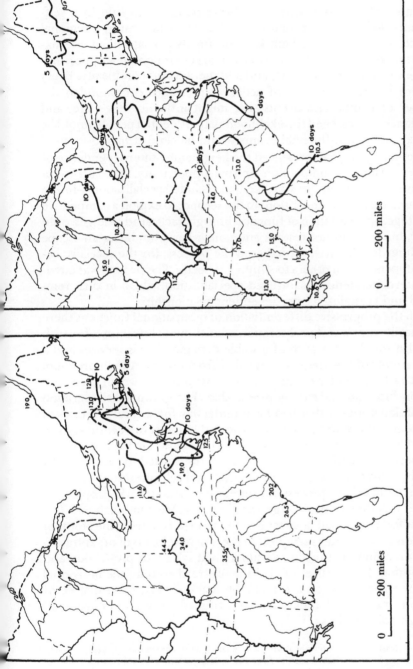

FIGURE 2 Mean Public Information Time Lags for New York in 1794 (Based on *American Minerva* and *New York Evening Advertiser*)

FIGURE 3 Mean Public Information Time Lags for New York in 1841 (Based on *New York Daily Tribune*)

SOURCE: Allen Pred, *Urban Growth and the Circulation of Information: The United States System of Cities, 1790–1840* (Cambridge: Harvard University Press, 1973), maps 2.5 and 2.9. (c) 1973 by the President and Fellows of Harvard College.

in remote places. Men who had always bought and sold in a self-contained local market found themselves working first in a regional market, then in a national market, and finally perhaps in an international market. Every expansion of the market brought new opportunities, new competitive dangers, and a higher level of complexity. Every advance in the technology of action-at-a-distance compounded the number of people and events that might influence one's life, and simultaneously lessened the likelihood that influential people would be familiar and that influential events would be close at hand.

Whether a person experienced the heightening interdependence of society as ruinous competition or ripe opportunity, his path to an improved position in life was likely to include specialization—in the first case, to escape competition; in the second, to exploit a competitive advantage. Specialization of function brought tactical advantages, even power, but it could not restore autonomy. Far from being an escape from interdependence, it furthered the process, for the specialist became dependent on others to supply his needs. The process was circular: interdependence encouraged specialization, which in turn reinforced and crystallized interdependent relationships.

With the progressive differentiation of occupational functions there necessarily came a vast proliferation of institutional structure, for each specialist sought assurance of a stable demand for his services and a stable supply of the services he needed from others. A high premium was placed on regularity and stability, anything that would crystallize relationships and make them predictable. The gradual disintegration of traditional society that had been underway for several centuries in European culture was reversed in the early nineteenth century. Even in the United States, where society had come closest to the liberal ideal of a mere aggregation of autonomous atoms, related to each other by only the most tenuous bonds of contract and self-interest, it was clear by midcentury that a process of reintegration was at work; people were coming together in new, larger and more stable combinations. The market forces that had once tended to pry men free from the bonds of traditional society now reconnected them in new forms of organization based on economic function within a great system of production and distribution. Like a sheet of chain mail stretched taut, the individual links of what had been a remarkably slack society lost a degree of freedom and found themselves constrained by forces transmitted through adjacent links.[12]

[12.] The classic study of specialization is Emile Durkheim, *The Division of Labor in Society*, trans. George Simpson (Glencoe, Ill.: Free Press, 1960). Contrary to the assumptions of this study, Durkheim viewed the advance of mutual dependence favorably, as the potential basis for a more extensive solidarity—an "organic" form of solidarity, as

The rise of new forms of organization based on economic function eroded older forms of organization based most often on geographical proximity. The archetypal experience of the nineteenth century, in matters intellectual and cultural as well as economic, is symbolized by the image of a small, familiar, local market being absorbed into progressively larger market systems. For every market that lost its identity in a larger system, a village was made to appear provincial in the eyes of its residents; local leaders were stripped of the aura of power and authority that community isolation had bestowed upon them; and individuals were deprived of the sense of boundless self-sufficiency characteristic of Jacksonian America and were encouraged to construe their own and other people's lives as a product of external circumstance. The most general result of the growing interdependence of American society was a sense of drift and disorder. As Robert Wiebe points out, near the heart of the matter was the erosion of the customary local units of society:

> The great casualty of America's turmoil late in the century was the island community. Although a majority of Americans would still reside in relatively small, personal centers for several decades more, the society that had been premised upon the community's effective sovereignty, upon its capacity to manage affairs within its boundaries, was no longer functional. The precipitant of the crisis was a widespread loss of confidence in the powers of the community. In a manner that eludes precise explanation, countless citizens in towns and cities across the land sensed that something fundamental was happening to their lives, something they had not willed and did not want, and they responded by striking out at whatever enemies their view of the world allowed them to see. They fought, in other words,

---

opposed to the merely "mechanical" solidarity, based on likeness, that existed in premodern "segmented" societies (406, 364, 398, and 62–63). The atomistic, non-institutional character of early nineteenth century America is a basic assumption of Stanley Elkins, *Slavery: A Problem in American Institutional and Intellectual Life*, 2nd ed. (Chicago: University of Chicago Press, 1968), and Marvin Meyers, *The Jacksonian Persuasion: Politics and Belief* (Stanford: Stanford University Press, 1960). It is reflected also in Eric Foner, *Free Soil, Free Labor, Free Men: The Ideology of the Republican Party before the Civil War* (New York: Oxford University Press, 1970). Several of John William Ward's essays are pertinent, especially "Jacksonian Democratic Thought: 'A Natural Charter of Privilege,'" in *The Development of an American Culture*, ed. Stanley Coben and Lorman Ratner (Englewood Cliffs, N.J.: Prentice-Hall, 1970), 44–63; and "Mill, Marx, and Modern Individualism" and "The Ideal of Individualism and the Reality of Organization," in Ward's own *Red, White and Blue*, 213–266. For the reintegration of American society, see Rowland Bertoff, "The American Social Order: A Conservative Hypothesis," *American Historical Review*, 65 (Apr. 1960), 495–514, and *An Unsettled People: Social Order and Disorder in American History* (New York: Harper and Row, 1971). See also Samuel Hays's introduction to *Building the Organizational Society*, ed. Jerry Israel, and Hays's *Response to Industrialism;* Wiebe, *Search for Order;* and Louis Galambos, "The Emerging Organizational Synthesis in Modern American History," *Business History Review*, 64 (Autumn 1970), 279–290.

to preserve the society that had given their lives meaning. But it had already slipped beyond their grasp.[13]

Everything that Wiebe formulated negatively in the concept of the declining sovereignty of island communities, I have meant to formulate positively in the concept of heightening interdependence. For the loss of confidence in the sovereignty of the community does not elude precise explanation if it is understood that the interdependence of the whole necessitates the dependence of the parts. Each element of society is part of the social environment of the other elements. As each element of nineteenth-century society exercised its own power and made use of the advancing technology of action-at-a-distance, it necessarily eroded the autonomy of other elements. Each person, community, business enterprise, political interest group, or other societal component experienced as external to itself a condition to which it in fact had contributed, simply by going about its routine business with new instruments of communication, transportation, and organization. In a highly interdependent system, the component parts must become dependent; independence becomes an illusion or an ideal, at odds with reality.[14]

It is inconceivable that people could have lived through such a transformation of society without profound changes in their habits of

[13.] Wiebe, *Search for Order*, 44. Although Wiebe may have drawn more heavily on Tönnies than Durkheim, his phrase "island community" neatly captures the essential character of what Durkheim called a "segmented society"—a society compartmented so that most events begin and end within a local community without generating consequences throughout the society. "When society is made up of segments, whatever is produced in one of the segments has as little chance of re-echoing in the others as the segmental organization is strong. The cellular system naturally lends itself to the localization of social events and their consequents." In contrast, a functionally integrated "organic" society offers few barriers to the spread of consequences. "According to their mutual dependence, what strikes one [component] strikes the others, and thus every change, even slightly significant, takes on a general interest." Durkheim expressly associated segments with markets: "In so far as the segmental type is strongly marked, there are nearly as many economic markets as there are different segments" (*Division of Labor*, 223, 369).

[14.] It is not my purpose here to present a general analysis of the conditions of freedom from the nineteenth century to the present. If it were, I would insist on the high priority of the trend which I have designated as heightening interdependence, but I would also acknowledge certain countercurrents running in the direction of greater freedom—a growing technological power over nature, for example, and a widening range of options of all kinds (many of them trivial). In civil liberties the trend unquestionably has been toward expansion, in terms of what is culturally tolerated and what is legally protected (though a pessimistic interpretation emerges if one focuses on the growing discrepancy between libertarian values and the collectivist realities of social structure). Dependence upon the will of others was taken for granted by pre-industrial man, of course, and modern man has gained real comforts by exchanging the specific and overt dependency relationships of traditional society for the multitudinous, but usually painless and anonymous dependencies of modern society. But none of these qualifications should blind us to the reality of the web of dependencies that today shapes each individual's behavior at every moment. That is all my argument requires.

mind, their modes of organizing experience, their very manner of perceiving human affairs. This was no ordinary transition from one generation to another, but a movement from one social universe to another governed by strikingly different conditions of action and explanation. To be born in one world and to have to live in the other—the typical fate of nineteenth-century men—was an intellectually wrenching experience, blinding to most but stimulating to a creative few. The fruits of long and painful reappraisal were gathered near the end of the century, as the modern view of society crystallized in the work of Weber, Durkheim, Dewey and other members of that unusual generation. To reconstruct the experience of those who lived through the critical decades leading up to the 1890's is risky and necessarily speculative, but the effort is indispensable if we are to understand the emergence of professional social science.

The vital nexus, I shall argue, lies in the influence of social interdependence on the way people attribute causation in human affairs. Causal attribution is the heart of all of man's efforts to explain the world around him, and it serves also as the basis for the imputation of important personal qualities such as moral responsibility, authority, and spontaneity. Any change in habits of causal attribution must have vast cultural ramifications. By hypothesizing that a people's habits of causal attribution are related to the degree of interdependence they experience in their society, we can bridge the chasm that still separates social from intellectual history, and show the rise of social science to be an integral and especially revealing facet of the urban-industrial transformation.[15]

Recession of Causation

As society became increasingly interdependent, we may speculate that the effect, as experienced by sensitive persons within the society,

[15.] I take it for granted that causation is not given in experience, but is *attributed* to people, events, and conditions in accordance with theories, paradigms, conventions, habits, expectations, wishful thinking and the like, as well as the particular facts of the case. I mean to define "cause" in a manner congruent with lay usage, and not in the far narrower sense employed, for example, by Mario Bunge, *Causality: The Place of the Causal Principle in Modern Science* (Cambridge: Harvard University Press, 1959). Bunge reserves the word "cause" for a particular category of determination sharply distinct from statistical, mechanical, dialectical, or structural determination (17–19). All categories of "determination," as he defines it, correspond with what are called "causes" in everyday affairs. Any factor that might be designated in response to the question, "Why did this happen? Of what is this the effect?" fits the notion of causation used here. In this context no rigid distinction can be drawn between causes and necessary conditions. Also relevant are Fritz Heider, "Social Perception and Phenomenal Causality," *Psychological Review*, 51 (Nov. 1944), 358–374; A. Michotte, *The Perception of Causality* (London: Methuen, 1963); R. M. MacIver, *Social Causation* (New York: Harper and Row, 1969); and much of the continuing work of Jean Piaget.

must have been a tendency for efficient causation to recede from the observer. Reflective individuals experienced a general recession in the perceived location of the important events, people, and institutions that influenced their lives. As society's components (individuals and institutions) became more interdependent, as they began to influence one another more frequently, more intensely, and in more varied ways, the effective cause of any event or condition in society, from the vantage point of any single component, became more contingent and more difficult to trace. The "distance" (understanding the term to involve more than linear extension) between the observer and those things which he regarded as having genuine causal significance in his own life, became greater and greater over the course of the century.

As causation receded, one's immediate social environment was drained of vitality. Things near at hand that had once seemed autonomous and therefore suitable for causal attribution were now seen as reflexes of more remote causes. Those factors in one's immediate environment that had always been regarded as self-acting, spontaneous entities—causes: things in which explanations can be rooted—now began to be seen as merely the final links in long chains of causation that stretched off into a murky distance. One's familiar milieu and its institutions were drained of causal potency and made to appear merely secondary and proximate in their influence on one's life.

Unambiguous causality was difficult to attribute confidently anywhere, but it seemed increasingly likely to be a property of remote and unfamiliar objects rather than of one's own environs. Small but growing numbers of people found it implausible or unproductive to attribute genuine causal power to those elements of society with which they were intimately and concretely familiar. One's personal milieu, which is the source of the average man's entire conception of the larger society, was drained of vitality and made transparent, so to speak, to the play of influences originating far beyond the individual's range of vision. Illusions of local and individual autonomy, challenged only sporadically in the past, now fell permanently from the eyes of many people.

Not only were the island community and other complex groupings of personal milieu drained of causal potency and devalued by interdependence; perhaps more importantly, so was the individual, the irreducible atomic constituent of society. The growing interdependence of society necessarily created at all levels of analysis a shortage of "independent variables"—a shortage of things, events, or people to which genuine causal potency could be attributed and which could serve, therefore, as anchors for explanation and pressure points for purposeful action. Nothing varies independently in an interdependent system, not even one's self.

Since today we are accustomed to this shortage of independent variables and have firsthand experience of no other perspective, we may say that, prior to the intense interdependence of the late nineteenth century, men lived in a universe of human affairs that had a *superabundance* of independent variables. Their world was unpredictable but richly explicable, compared to our own, for it was full of independent comings and goings whose explanation lay in the autonomous will of the mover and nowhere else.

It is of course true that in all ages, present as well as past, the limited and subjective evidence of untutored common sense tells men that they are free, that their mental experience of choice and decision is truly and irreducibly the "cause" of their behavior. Common sense always tends to view human affairs as a product of unconditioned individual intentions, complex in its mixture, perhaps, but simple in its elements. Only upon reflection and analysis does the experience of free choice come to seem partly or wholly illusory. Even today, in an interdependent society, common sense often ignores the remote determinants of behavior.

But today, unlike in the past, common sense is challenged and among educated persons is virtually overwhelmed by the institutionalized authority of social science. We are taught that human behavior flows not simply from the autonomous conscious will of the individual, but from a host of determinants external to the conscious mind, such as historical tradition, economic interest, social conditioning, the press of culture, or instinctual drives inaccessible to consciousness. These determinants limit individual autonomy and diminish man. In comparison with past images of man, the twentieth-century image is frail and even degraded. Even the most pessimistic of premodern social thinkers, Thomas Hobbes, conceived of man as a causally potent being—vicious, perhaps; but master of his own fate, if he could only overcome the jealousy of his neighbors and the hazards of nature. But, from the modern social scientific viewpoint appropriate to a highly interdependent society, no man can be viewed as autonomous master of his own fate. Even if he conquers his neighbors and triumphs over nature, he is still a puppet on the strings of instinct and environment.

Conversely, intelligent men in all ages have occasionally transcended common sense and acknowledged the existence of determinants of human behavior other than the conscious will of the individual. But this is eminently a matter of degree. Never before the watershed of the 1890's did so many thinkers so consistently and so authoritatively locate causation outside conscious mind. Before the nineteenth century, if a thinker acknowledged any remote, supra-individual or supra-local level of causation, he was likely to go all the way to the most remote cause of all, the First Cause, God. A providential interpretation of human affairs locates causation even more remotely than a social scientific interpretation.

But according to almost all orthodox theologies, God endowed His children with free will, which meant for all temporal purposes that causation would be attributed to conscious human intentions after all. As Perry Miller showed in a classic essay, even such a deterministic worldview as seventeenth-century Calvinism was able to sustain its original insistence on the remoteness of causation only by a logic so strained as to verge sometimes on hypocrisy. The commonsense view that men are causally potent—so potent, in fact, that they might by their good behavior influence even Calvin's God to save them—was too strong to be resisted. Few heeded Jonathan Edwards's lonely protest against American Calvinism's steady drift away from Calvin's remote and omnipotent God and toward Arminianism.[16] Not until the last half of the nineteenth century was the commonsense assumption of free will again seriously challenged, this time by an exaggerated secular determinism in the form of Spencerian positivism. And not until the turn of the century did the commonsense view finally meet its match in the less radical but far more authoritative determinisms of institutionalized social science.

By encouraging habits of remote causal attribution and thereby undermining a noble image of man and a semi-enchanted view of human affairs, the objective conditions of life in an interdependent society challenged not only naive common sense—always content to locate causation superficially—but also the accumulated wisdom of the ages. Only with the utmost reluctance would even the most sensitive and intelligent men give up their age-old predisposition to attribute temporal causation to events close at hand, to familiar people and things, and, above all, to the conscious will of individual human beings.

Consequences

The growing interdependence of society's component parts brought into existence some of the most distinctive features of modern culture. Not all are relevant to the emergence of professional social science; of those that are, three are particularly important. Interdependence made human affairs appear to be a far more inviting field for scientific inquiry than ever before; it supplied a quasi-paradigm or guiding

---

[16.] Perry Miller, "The Marrow of Puritan Divinity," in *Errand into the Wilderness* (New York: Harper and Row, 1964); also Miller's *Jonathan Edwards* (New York: Dell, 1967). For a brilliant discussion of the individualistic assumptions at the very heart of post-medieval political thought, see C. B. Macpherson, *The Political Theory of Possessive Individualism, Hobbes to Locke* (London: Oxford University Press, 1962).

assumption to focus research; and it provided a market for expert advice in human affairs by discrediting traditional systems of belief.

The recession of causation and the consequent devitalization of island communities, individuals, and personal milieux gave a new concreteness and uniquely modern salience to the very idea of "society" (or the "polity," or the "economy") as an entity apart from particular people and concrete institutions. By doing so it created a viable field for scientific inquiry. For men struggling to comprehend the changing texture of human affairs in the nineteenth century, "society" was that increasingly important realm of causation located in an intermediate position between two more familiar realms of causal attribution: it stood "behind" personal milieu, now increasingly drained of causal potency, but "in front of " (less remote than) Nature and God, hitherto almost the only plausible loci of remote causal influence.

The ever-denser interweaving of cause and effect within this intermediate realm of causation invited a uniformitarian perception of human affairs without which professional social science is hardly imaginable. As long as the human actors in the social universe were perceived as autonomous beings, endowed with a spark of divine potency by their Creator, no naturalistic model of explanation could be plausible. As long as most significant events in a man's life appeared to begin and end within the confines of personal milieu or island community, the notion of "society" or "the social" was irrelevant and incapable of serving as a field of professional expertise. Only when individual behavior was perceived as largely a product of external circumstance, when personal milieu became partly transparent to the play of remote influences, could there be a science of "society."

The growing interdependence of nineteenth-century society not only supplied a uniformitarian social universe that invited the application of scientific methods, but it also ultimately provided the generation of the 1890's with something like a paradigm about which a research tradition could coalesce. As discussed above, interdependence at first affected perception in primitive and inarticulate ways—the recession of causation, the devitalization of milieux. But near the end of the century interdependence was recognized and articulated as an objective condition of life, and then finally it was detached from its original objective referents and put to use paradigmatically as a presumption against autonomy, against the formalistic habits of thought appropriate to a simpler, less interdependent universe of human affairs.

In addition to providing an inviting field of inquiry and some of the conceptual tools with which to inquire, the growing interdependence of nineteenth-century society also contributed to the emergence of professional social science by creating a market for the results of inquiry, a demand for expert advice. This was its first and perhaps its most critical contribution. Professional social science could not exist as long as human affairs seemed readily and self-evidently explicable. As long as causation can confidently be attributed to familiar people and events, as long as the world can be understood as simply an extension of local community, as long as common sense and customary knowledge seem sufficient to master the tasks of explaining and managing one's life, students of social phenomena must remain lonely exotics, unable to command the attention or the deference requisite to professional status.

But when the more articulate members of the public experience a recession of causation, when they begin to sense that their familiar milieu has become merely proximate in its influence—or indeed, when causation threatens to evaporate altogether—they look upon the inquirer as a man of extraordinary importance. For his task is to pursue causation, to track it down. Virtually all inquiry is causal analysis, or a fact-gathering preparation for such analysis. To engage in inquiry is to search for genuine causation, to shear away merely secondary influences and necessary conditions so as to isolate those factors which, within a given frame of reference, can be regarded as self-acting, causal entities—"independent variables." As causes recede and as growing interdependence introduces more and more contingency into each chain of causation, the realm of inquiry must expand and the conditions of satisfying explanation must change. Common sense fails and the claim of expertise gains plausibility. Explanation itself becomes a matter of special significance, because the explainer promises to put his audience back in touch with the most vital elements of a receding and increasingly elusive reality.

It is no coincidence that the social sciences rose to cultural dominance at a time of rampant doubt and uncertainty among the better-educated members of Western society.[17] The late nineteenth century is commonly understood to be a period of spiritual crisis when the Christian cosmology, weakened by the rude shock of Darwin's evolutionary theory, first lost its grip on substantial parts of the intellectual class. But the exhaustion of faith and the growing popularity of naturalistic and uniformitarian views may themselves have been induced by subtle changes in the pattern of social experience. Certainly the inability of

---

[17] Nathan Glazer, "The Rise of Social Research in Europe," in *The Human Meaning of the Social Sciences,* ed. Daniel Lerner (New York: Meridian, 1959), 46–66; J. W. Burrow, *Evolution and Society: A Study in Victorian Social Theory* (Cambridge: Cambridge University Press, 1966), 264; and Paul A. Carter, *The Spiritual Crisis of the Gilded Age* (DeKalb: Northern Illinois University Press, 1971).

sensitive minds to attribute causation unambiguously in everyday affairs must have compounded the crisis of confident belief, even if the crisis has other independent origins.

One thinks of the psychic maladies and seizures of fear and trembling to which the intellectuals of the nineteenth century were notoriously prone. In the context of the present analysis, is it surprising that Herbert Spencer, who was taught as a boy to ask "What is the cause?" of every new thing encountered, as an adult took to his bed and wore earmuffs to prevent overstimulation of his senses? And think of the others: William James, intellectually absorbed by the question of free will, who was troubled throughout life by spells of nervous collapse; or his father, the elder Henry James, who was once struck dumb with terror for no apparent reason—in his own words, "reduced from a state of firm, vigorous joyful manhood to one of almost helpless infancy." There is the unforgettable image of Max Weber, sitting by a window dully picking at his fingernails rather than endure another moment of intellectual work. The list could be extended: the great minds of the *fin de siècle* were troubled minds.[18]

To be sure, the mood of uncertainty did not always reach such extremes. The early leaders of the Social Science Association, men like Samuel Gridley Howe, Louis Agassiz, and Frank Sanborn, seem not to have been afflicted with dark despair, nor certainly were they relativists. They did not toy with the idea that truth was beyond reach or that it was plural. Rather, they thought they possessed the truth, in fundamentals, and that all alternatives to their version were spurious. But they felt the sting of competition more deeply than earlier generations; the values and practices of their society had lost their accustomed quality of opaque givenness. Alternative interpretations of man and society were perceived as live, albeit repulsive, options, and it was necessary to defend the truth against them. The defense could not be *ex cathedra*. It would require search and even research, for the common sense of the individual and the customary knowledge of the community were admittedly inadequate grounds for belief.

The defensive posture of much early social science in America, and the undercurrent of doubt that provoked it, is epitomized in the career of William Graham Sumner. As the Episcopal minister of Morristown, New Jersey, the young Sumner was so impressed with Spencer's serialized essay, "The Study of Sociology," that in 1872 he abandoned the pulpit in order to become one of the first college teachers of social science in America. There can be no doubt that Sumner left the house

---

[18.] Peel, *Herbert Spencer*, 21–23; Burrow, *Evolution and Society*, 184; Gay Wilson Allen, *William James* (New York: Viking, 1967), 17–18; *From Max Weber: Essays in Sociology*, ed. and trans. H. H. Gerth and C. W. Mills (New York: Oxford University Press, 1958), 12.

of God only because he thought the church's work would be done more effectively in another sphere of action. In his final sermon before going to Yale as a professor of political science, he lay bare his motives before his congregation:

> Our modern society is deeply infected by philosophical skepticism and philosophical indifference. The number of men who are falling into a doubting position with regard to traditional religious doctrines is increasing every day.... A conflict is impending between the traditional dogmas and modern speculation and science in which it is possible that all religion may be lost. A chaos threatens the world of mind and spirit. Not two or three, not a score, but a hundred claimants wrangle for the faith of men. It is to be apprehended that utter ruin may come upon all universal ideas and spiritual faiths. The great question is whether there is an historical revelation of spiritual and universal truths which has authority for man, or whether each man and each generation must reason out the whole problem afresh, or rest contented with such knowledge of the shell of things, as he can win through the senses. I say that this conflict is impending, and I believe and have always believed that it was the duty of the pulpit to prepare for it, to enter into it, and to win a victory in it. It is certain that every tradition and every inherited faith of mankind is to be stated on a reexamination of authority and evidence within the next fifty years.[19]

The ironic outcome is well known. As Perry Miller said, Sumner "put his religious beliefs in a drawer and turned the lock... upon unlocking it, he found the drawer empty."[20]

Many of the earliest teachers of social science in America were motivated, like Sumner, by profoundly conservative impulses. They wished to defend cherished traditional values against modern skepticism, but they themselves were far more skeptical than their elders. Sumner soon found himself embroiled in public controversy with Yale president Noah Porter over the use of Spencer's books in an undergraduate course. John W. Burgess, who began teaching history and political science at Amherst in the 1870's, found that his senior colleagues "regarded the college as a place for discipline, not as a place for research. To them the truth had already been found. It was contained in the Bible, and it was the business of the college to give the preliminary training for acquiring and disseminating it. Research implied doubt. It implied that there was a great deal of truth still to be found, and it implied that the truth thought to have been already found was approximate and in continual need of revision and readjustment. Still

---

[19.] Quoted by Harris E. Starr, *William Graham Sumner* (New York: Henry Holt, 1925), 167–168. See also Hofstadter, *Social Darwinism*, 55.

[20.] Perry Miller attributes the statement to Sumner without citation: *American Thought: Civil War to World War I* (New York: Holt, Rinehart and Winston, 1954), xxvi.

more briefly expressed, they regarded research as more or less heretical."[21]

From the vantage point of a still later generation, Albion Small thought that "unsatisfied bewilderment was the original state of the American sociologists.... At the start, American sociology, always excepting Ward's [Comtean] system ... amounted in effect to little more than an assertion that all the traditional ways of interpreting human experience were futile."[22]

The point of departure for the professionalization of social science was a pervasive mood of doubt and uncertainty, triggered certainly by Darwin and historical criticism of the Bible, but rooted also in the intellectual quicksand of an increasingly interdependent social universe. By depriving sensitive thinkers of the opportunity for easy, unambiguous causal attribution that had enabled earlier generations to be serenely confident in their judgments of men and social affairs, the growing interdependence of society contributed to an erosion of confidence that made men receptive to claims of social science expertise.

The collapse of confident belief gave rise to a movement of cultural reform which intended to construct safe institutional havens for sound opinion. The American Social Science Association at first played a central role in the movement, but ultimately the logic of the movement passed by the ASSA and led, among other things, to the formation of its heirs, the modern social science disciplines. The complex of motives that inspired the movement may best be introduced by examining the life of a single man, the prime mover of the ASSA, Franklin Benjamin Sanborn.

[21.] John W. Burgess, *Reminiscences of an American Scholar: The Beginnings of Columbia University* (Morningside Heights, N.Y.: Columbia University Press, 1934), 147–148.

[22.] Albion Small, *Origins of Sociology* (Chicago: University of Chicago Press, 1924), 344, 335.

CHAPTER III

Frank Sanborn's Association

Franklin Benjamin Sanborn once described the *Journal of Social Science,* the annual publication of the American Social Science Association, as "my serial novel."[1] The phrase rings true in both its possessive tone and its equation of social science with an art form. The *Journal* (and sometimes, it seemed, even the Association itself) really did belong to Frank Sanborn, in spirit if not in law. And for Sanborn social science was a convenient rubric for a kind of inquiry and reform activity in which scientists had no necessary edge over novelists. From the founding of the Association in 1865 to Sanborn's resignation as secretary in 1898, no other individual gave it half as much time, energy, or character. Sanborn's preeminent place in the ASSA means that the fate of institutionalized social inquiry in the United States was shaped in its early years by one of the most cantankerous intellectual figures of the Gilded Age. To understand the meaning and significance of the Social Science Association, one must begin by understanding Frank Sanborn.

Sanborn led a fascinating life, but somehow it has slipped through the fingers of historians. Not the least of the reasons for his present obscurity is the fact that when he died his unappreciative sons burned nearly all of his papers.[2] He wrote his two volume memoirs at an advanced age, and they are predictably rambling and anecdotal. They say very little about his activities in the Social Science Association. A third volume, never published, probably was to have discussed these matters.[3] The scarce historical evidence about Sanborn's life is entirely out of proportion to his significance. Although not a figure of first impor-

[1] Sanborn to John Albee, Concord, Mass., 16 [26?] Feb. 1896, Clarkson Collection (original in Concord Free Public Library).

[2] Benjamin Blakely Hickok, "The Political and Literary Careers of F. B. Sanborn" (Ph.D. dissertation, Michigan State College, 1953), xviii. As the title indicates, Hickok's valuable study does not cover Sanborn's activities as reformer and social scientist.

[3] F. B. Sanborn, *Recollections of Seventy Years: By F. B. Sanborn of Concord* (Boston: Richard G. Badger, 1909). The manuscript of the third volume may yet turn up. Sanborn intended to finish it in 1910 and was still working on it in 1916; see Hickok, "Political and Literary Careers of F. B. Sanborn," xii.

tance, Sanborn was on the fringes, and sometimes close to the heart of a great deal that was vital in late nineteenth century America. Yet the record is pathetically thin. It is typical that Arthur Mann, in his study of Boston reformers from 1880 to 1900, recognized Sanborn as "the leading social worker of his day," but then found very little else to say about him.[4]

The sprawl and incoherence of Sanborn's life is partly a product of his own abundant idiosyncrasies, but it also reflects the kind of society that took shape in his adult years and which we take for granted today. Frank Sanborn defied the prime imperative of the late nineteenth century: *specialize*. He was a generalist in a society that demanded specialization, a man of diffuse mental habits and unfocused career aspirations in a society that paid ever higher rewards for narrow discipline. He was a teacher, poet, journalist, radical firebrand, polite reformer, government bureaucrat, philosopher, classicist, propagandist, and philanthropist. He was a free-lance intellectual of a sort not uncommon in the nineteenth century, but rare indeed in our own time.

If Sanborn is remembered at all today, it is in his role as one of the "Secret Six" who helped finance John Brown's raid on Harpers Ferry. Involved with Sanborn in the venture were Samuel Gridley Howe, Thomas Wentworth Higginson, Gerrit Smith, Theodore Parker, and Frank Luther Stearns. All of them were administrators of the Massachusetts State Kansas Committee. Sanborn, secretary of the committee, was the youngest; twenty-five years old and fresh out of Harvard, he cast John Brown in the role of a romantic hero from their first meeting in January 1857. What is more, Sanborn promptly set himself the task of persuading others to see Brown as a hero. He pledged to the veteran of Pottawatomie and Black Jack that he would support his ventures, protect his family, and defend his memory against any who might attack it. "I thank you for remembering me," wrote the impressionable young Sanborn in April 1857, "and I shall prize anything from you as a momento [sic] of the bravest and most earnest man it has been my fortune to meet. You need not fear that you will be reckoned an unprofitable servant."[5]

It was through the good offices of this young admirer that John Brown met Ralph Waldo Emerson and Henry David Thoreau. More important, it was Sanborn who introduced Brown to the other members of the "Secret Six," excepting Gerrit Smith, whom Brown had known earlier. Sanborn helped Brown gain the support of the Mas-

---

[4.] Arthur Mann, *Yankee Reformers in the Urban Age* (Cambridge: Harvard University Press, 1954), 17.

[5.] James C. Malin, *John Brown and the Legend of Fifty-six: Memoirs of the American Philosophical Society*, XVII (Philadelphia, 1942), 345.

sachusetts Kansas Committee in January 1857 and then accompanied him to New York, where Brown won additional aid from the National Kansas Committee. A year later Sanborn was among the first to hear Brown announce his more ambitious plan, not only to defend Kansas, but to foment a slave insurrection in the South. In March 1858 the six conspirators gathered with Brown in the American House in Boston. Without knowing where he intended to strike, they agreed to advise the old warrior and raise a thousand dollars for his "experiment."[6]

When news of the raid at Harpers Ferry reached Sanborn, he first fled to Canada, then returned to his home in Concord on the advice of Emerson, Howe, Higginson, Stearns, and other friends. Sanborn became a public figure in the spring of 1860 when the Senate voted his arrest after he refused to testify in Washington. In a nighttime drama ideally suited to his sense of the romantic, law officers who tried to arrest him in his home were set upon by his neighbors and sent flying back to Boston. Newspapers told how he bravely resisted arrest. As his captors tried to carry him out, he delayed them by bracing his feet first against the sides of the front door of his house, then against the posts of the veranda, then against the stone pillars at the front gate, and finally against the sides of the carriage, which he kicked in. Meanwhile his sister roused the neighbors and ran back to yank the beards of his tormentors. When William James brought his younger brothers, Wilkinson and Robertson, to attend Sanborn's Concord School in 1860, he wrote home about "the famous Mr. Sanborn."[7]

After the war, Sanborn kept his promise to Brown. For the remainder of his long life he did everything in his power to preserve an honorable place for John Brown in the patriotic mythology of the nation. He engaged in controversy with Brown's detractors; he wrote influential articles on Brown for the *Atlantic Monthly;* and he published a major biography of Brown in 1885. He did not pretend to be neutral. He withheld facts and bent them when necessary to further a goal he thought higher than objectivity. He remained convinced that "Brown was an instrument in the hands of Providence to uproot and destroy an evil institution."[8]

[6.] Stephen B. Oates, *To Purge This Land with Blood: A Biography of John Brown* (New York: Harper and Row, 1970), chs. 14 and 16, and esp. p. 238. The literature on Brown and the conspiracy fills a shelf. I have relied on Oates, Malin, and Sanborn, *Recollections,* I, ch. 3.

[7.] Ralph Barton Perry, *The Thought and Character of William James: Briefer Version* (Cambridge: Harvard University Press, 1948), 20. See Sanborn, *Recollections,* I, ch. 7 and pp. 208–211.

[8.] Sanborn, *Recollections,* I, 148. James C. Malin has shown in detail the major part played by Sanborn in fostering the legend of John Brown: see *John Brown and the Legend of Fifty-six,* ch. 13 and *passim.* See also Hickok, "Political and Literary Careers of F. B. Sanborn," 193–242, and Sanborn, *Recollections,* I, ch. 9, in which Sanborn explains some

The Genteel Tradition

Frank Sanborn was born in 1831 and died in 1917. His eighty-six years bridged that awesome time-span from Andrew Jackson to Woodrow Wilson, from a sparsely settled agrarian republic to an international colossus of cities and factories. His lifetime coincides with the years of most intensive change in the urban-industrial transformation.

Sanborn spent his youth in a rural environment strongly infused with latter-day Puritan intellectuality. He was the fifth of seven children in a family of moderate distinction in Hampton Falls, New Hampshire, where Aaron Sanborn, his father, was town clerk. The family was Unitarian, but as a boy Frank Sanborn owed more to the church library than to the religious functions of the institution. Hampton Falls had become a provincial center of learning and erudition no later than 1780, when ex-President Langdon of Harvard took up residence as parson. Before the age of eight Sanborn had read Plutarch's *Lives* in translation; before he even thought of going to Harvard he had begun the study of Latin, Greek, French, and German, all on the resources of the parsonage library.[9] Abolitionist John Gibson Hoyt of nearby Phillips Academy took Sanborn under his wing, preparing him for Harvard and introducing him to notable figures in state politics.[10]

Politics was an important part of the family culture. Frank Sanborn's father and grandfather were early supporters of Andrew Jackson and the Democratic Party. His brother Charles, older by ten years, was a local leader of the anti-slavery wing of the party. When the state party divided on the issue of Texas annexation in 1844, Frank, only thirteen years old, joined his brother in opposition to their father and grandfather, both of whom remained Democratic regulars. Since his brother subscribed to the *Congressional Globe,* the boy was well versed in congressional debates. His political attitudes were further shaped by Greeley's *Tribune* and the *National Era,* to which Whittier and Harriet Beecher Stowe were regular contributors.[11]

Neither shy nor lacking in well-placed relatives and friends, Sanborn entered Harvard as a sophomore in 1852 and soon found his way into an illustrious circle of literary and political figures. While taking his college examinations in July, he found time to call on Harriet Beecher Stowe, who was then at the peak of her fame as author of *Uncle Tom's*

of the reasons for his shifting account of Brown and the conspiracy. One hesitates even to comment on a question so long debated and so obviously political in nature, but perhaps it is worth noting that Malin's harsh criticism of Sanborn measures him by a standard of sober moderation and academic objectivity to which he never laid claim.

[9] Sanborn, *Recollections,* I, 16-20.
[10] *Ibid.,* I, 28; II, 284-287; Hickok, "Political and Literary Careers of F. B. Sanborn," 45-49.
[11] Sanborn, *Recollections,* I, 20-24.

Cabin. Within the year he also made the acquaintance of the Alcott family and became a friend of Theodore Parker. During his first two years at Harvard he became acquainted with Samuel Gridley Howe, Thomas Wentworth Higginson, John Greenleaf Whittier, and Ralph Waldo Emerson. Before graduating he became a member of an influential Massachusetts political group, led by Francis W. Bird and called the "Bird Club"; Sanborn became president of the group upon Bird's death in 1894. Among the political luminaries with whom he associated in the Bird Club were Charles Sumner, Henry Wilson, James Freeman Clarke, John M. Forbes, Anson Burlingame, John A. Andrew, and William S. Robinson. From within this circle of friends it was but a few short steps after Harvard to the secretaryship first of the Concord branch of the State Kansas Committee, then the secretaryship of the Middlesex County branch, and finally the secretaryship of the Massachusetts state committee itself.[12]

Despite his early success in reform politics, Sanborn's chief aspirations as a youth were literary rather than political. His childhood tastes were framed by the twin pillars of the "Genteel Tradition," as Santayana would later call it. On the one hand there was Romanticism (Byron's poems, Scott's novels, *Don Quixote*) and on the other the classics of Scottish and English Puritanism (*Pilgrim's Progress,* Doddridge's *Life of Colonel Gardiner,* and *The Scot's Worthies,* for example). By his own recollection it was in his fourteenth year that he began reading Hawthorne, Carlyle, and Emerson. He felt that he had a "natural affinity" for the school of thought which Emerson represented: "Something akin to his intuitions in my own way of viewing personal and social aspects, really brought me into relation with him before I ever saw him, or heard that thrilling voice...."[13]

Sanborn first met Emerson, to whom he bore a striking physical resemblance, at the end of his first year at Harvard. While walking from Cambridge to a friend's home in Sudbury, he stopped in front of the Old Manse in Concord, screwed up his courage, and stepped inside to pay the ten-minute call that protocol allowed. That autumn he and several classmates resolved to "consult the oracle within its own grove" and, with encouragement from Emerson, their visits soon became quite regular.[14] Sanborn's later career as eulogist and self-appointed historian of transcendentalism was set in late 1854, when Emerson invited him to take charge of the Concord school in the following spring. Sanborn accepted, and for the next eight years he supervised the schooling of Concord's younger generation. His pupils included

[12.] *Ibid.,* I, 50–51.
[13.] *Ibid.,* II, 256–261.
[14.] *Ibid.,* II, 435–440. For the resemblance of Sanborn to Emerson, see Alexander Johnson, "An Appreciation of Frank B. Sanborn," *Survey,* 37 (10 Mar. 1917), 657.

the children of Emerson, three sons of Horace Mann, the children of Judge Ebenezer Hoar, two daughters of John Brown, the two youngest sons of Henry James, Sr., and Frank Stearns and Julian Hawthorne. The school was run on principles as exotic as the community it served. Ellery Channing took an active interest and was enthusiastic about Sanborn's administration. Bronson Alcott was superintendent of schools in Concord, and Sanborn, as secretary of the school committee, worked with him to compile huge annual reports on the operation of the Concord schools.[15]

In his later years Sanborn made a career of editing the letters and writing biographical accounts of his famous friends in Concord and elsewhere. His efforts have not survived the critical temper of the twentieth century. His prose is notoriously flaccid; he had no sense of narrative development, and as an editor he took unconscionable liberties with his texts. His work epitomizes all that was lax and irresponsible about nineteenth-century editorship. Walter Harding has shown that his edition of Thoreau's letters, long the standard work, is rife with unacknowledged revisions that strip Thoreau of his natural strengths—such as the substitution of the word "perspiration" for "sweat."[16]

The same cloying delicacy that led genteel Victorians (even men of radical temperament like Sanborn) to avoid words like "sweat" is evident also in the story of Sanborn's "tragic love," without which no account of his life would be complete. Ariana Smith Walker was a bright and spirited girl with whom Sanborn fell in love when both were in their teens. To her he attributed his decision to apply himself intellectually and to seek admission to Harvard. She suffered from a painful lameness and was subject to nervous attacks that seemed to threaten her life.[17] Her attacks intensified in 1853 and 1854, so she and Frank decided to marry even though she was literally on her deathbed. She died eight days after the ceremony. Sanborn remarried in 1862, but his second wife, a cousin who assisted in teaching at the Concord school, occupied an embarrassingly small place in his life. In reverence

---

[15] *Ibid.*, II, 441–443; Hickok, "Political and Literary Careers of F. B. Sanborn," 98–107.

[16] Walter Harding, "Franklin B. Sanborn and Thoreau's Letters," *The Boston Public Library Quarterly*, 3 (Oct. 1951), 288–293. See also Hickok, "Political and Literary Careers of F. B. Sanborn," 314–315, 242. Sanborn's editorial and biographical work includes the following: *A. Bronson Alcott: His Life and Philosophy*, co-author William Torrey Harris (Boston: Roberts Bros., 1893); *Dr. S. G. Howe, the Philanthropist* (New York: Funk and Wagnalls, 1891); *Hawthorne and His Friends: Reminiscence and Tribute* (Cedar Rapids, Ia.: Torch Press, 1908); *Henry D. Thoreau* (Boston: Houghton Mifflin, 1882); *The Life and Letters of John Brown, Liberator of Kansas and Martyr of Virginia* (Boston: Roberts Bros., 1885); *Memoirs of Pliny Earle, M.D.* (Boston: Damrell and Upham, 1898); *Ralph Waldo Emerson* (Boston: Small, Maynard, 1901).

[17] Sanborn, *Recollections*, II, 291.

to Ariana's memory, Sanborn had their love letters sealed in the masonry of the fireplace in the home in which he and his second wife lived. He regarded his youthful romance as a story of classical dimensions. In his *Recollections* he devoted more than a chapter to it; of his second wife, virtually nothing is said.

There is of course genuine pathos in Ariana's premature death, but for Sanborn the matter took on an air of maudlin sentimentality. The parallel between Sanborn's Ariana and Emerson's Ellen Tucker, who died of tuberculosis two years after marriage, does not work in Sanborn's favor. Whatever one's personal reaction to it, the episode exemplifies the exaggerated and perhaps affected level of sensibility that many of Sanborn's contemporaries associated with transcendentalism. It was an aspect of transcendentalism and the romantic temperament generally that deeply repelled many notable intellectuals in the postwar years—men such as the nation's most eminent economist, Francis A. Walker, who always remained on the periphery of the Social Science Association and eventually presided over the founding of one of the ASSA's successors, the American Economic Association.[18]

After his brief moment of glory as a conspirator against the slave power, Sanborn receded from public view and never again rose above a middling level of prominence. Artistic and intellectual limitations aside, he was barred from success by a prickly personality ill suited to a polite age. "His normal attitude," complained one acquaintance, "was that of opposition." Contemporaries recoiled from his sarcastic and denunciatory ways, a personal style that was "not favorable to the acquisition and retention of intimate friends."[19]

Although never again a public figure, he continued to be active in political and literary affairs. As a member of the Bird Club he was involved in Massachusetts politics and active nationally in the Liberal Republican and Mugwump movements. In addition to writing biographies and editing the works of Concord's sages, he contributed articles and poems to the nation's leading magazines and occasionally toured the popular lecture circuit, telling tales of John Brown. Aside from the select but limited audience which he reached through his editorship of the *Journal of Social Science,* his only other steady avenue of influence over public opinion was through his connection with Samuel Bowles's Springfield *Republican,* one of the finest newspapers in the country. Sanborn became a correspondent for the paper in 1856, just after he left Harvard, and served as resident editor from 1868 to 1872. Even after leaving the editor's desk, he continued for most of his

---

[18.] See George M. Fredrickson, *The Inner Civil War: Northern Intellectuals and the Crisis of the Union* (New York: Harper and Row, 1965), 223.

[19.] Edward Stanwood, "Memoir of Franklin Benjamin Sanborn," *Proceedings of the Massachusetts Historical Society,* 51 (1918), 310–311.

life to contribute weekly columns to the *Republican:* "Our Boston Literary Letter" and "Our Weekly Boston Letter," on politics.[20]

Most of his mature years Sanborn spent as an unsung pioneer in social work and social science, two fields which in his mind were inseparably related. When Massachusetts created the first Board of State Charities in the country in 1863, Frank Sanborn was among the original members. Two years later, when the Board sponsored the formation of the American Social Science Association, Sanborn wrote the invitation to the charter, meeting, served on the committee of arrangements, and became a secretary of the new organization. Nine years after that, when the ASSA sponsored the creation of the National Conference of Charities and Correction, a national umbrella organization for state charity boards and the charity organization movement, Sanborn again was the organizer. Working without fanfare and for little or no remuneration, he was the prime mover in both the ASSA and the NCCC for the remainder of the century.

On top of his labors for these two institutions, his literary work, and his journalism, Sanborn also found time to serve as a public official with the Massachusetts Board of State Charities and related agencies. He appears to have been one of the key figures in the expansion and modernization of the state's welfare complex, working not only as administrator (secretary of the Board of Charities and general inspector of charities) but also as inside lobbyist and drafter of legislation.[21] His approach to the problems of poverty and dependency was comparatively humane and generous in an age that tended to hold the individual responsible for his own place in life. Like his mentor in matters of humanitarian reform, Samuel Gridley Howe, Sanborn generally opposed large-scale, impersonal institutions and advocated neighborhood solutions to the problems of dependency whenever feasible. His policy of boarding out the insane provoked a public controversy that terminated his career as a public official in 1888. When Sanborn died in 1917, Massachusetts legislators thought highly enough of his services to the commonwealth that they ordered the State House flag lowered to half-mast for three days in his honor.[22]

The Abolitionist as Transcendentalist

Except for the few years 1868–72, when Sanborn was away from Boston, editing the Springfield *Republican,* the day-to-day work of the

---

[20] Sanborn, *Recollections,* I, 80; Hickok, "Political and Literary Careers of F. B. Sanborn," 253.

[21] Johnson, "Appreciation of Frank B. Sanborn," 656–657. See also the *Annual Reports of the Board of State Charities of Massachusetts.*

[22] *Ibid.,* and David J. Rothman, *The Discovery of the Asylum: Social Order and Disorder in the New Republic* (Boston: Little, Brown, 1971), 251, 289.

Social Science Association fell almost exclusively on his shoulders. The correspondence of the leading members of the Association makes it clear that Sanborn and a shifting group of perhaps a dozen other men and women constituted the living substance of the Association; all else was largely show and institutional image. The group which exercised controlling influence was ordinarily made up of the department secretaries, some of the department chairmen, a few of the directors and vice-presidents of the Association, together with Sanborn and the current president. Many of the officers were figureheads or potential patrons who had no real effect on the Association's policy or character. The rank-and-file members for the most part were there to supply an audience at meetings and a subscription fund for the *Journal.* The thread of institutional continuity and coherence was Frank Sanborn.

Because Sanborn's influence on the Association was so great, it is important to ascertain its exact nature. Obviously his leadership in the Association establishes a strong line of continuity between it and the romantic reform tradition of the early nineteenth century. As a colleague of Theodore Parker and Samuel Gridley Howe, Sanborn carried on into the postwar years their abolitionist sentiments as well as a concern with prison reform, the care of the insane, and the humane treatment of deviancy and dependency in general—interests typical of the romantic era. Since Sanborn was at the same time an intimate member of Emerson's literary circle at Concord, one could hardly ask for a more suitable exemplar of "the transcendentalist as abolitionist," or "the abolitionist as transcendentalist."

As abolitionist and transcendentalist, one might expect Sanborn to display the anti-institutional attitudes and peculiarly abstract frame of mind that Stanley Elkins attributed to the romantic reformers of antebellum America in his brilliant work, *Slavery: A Problem in American Institutional and Intellectual Life* (1959, 1968). This expectation is borne out in some respects, but not in others. On close examination Sanborn turns out to be not only a link to the past, but also a bridge to the future. His ambivalent admiration both for the romantic individualism of the prewar era and for the community-oriented, institution-building activities that prevailed after the war suited him ideally to lead the Social Science Association, because it too occupied a transitional position in that great swing of cultural moods that makes the early nineteenth century so different from its end.

Elkins's thesis about the infantilization of the slave has excited so much critical attention that scholars have been distracted from the real core of his book, which was a penetrating conservative critique of the whole antinomian and individualist strain in American culture, especially as it flowered in the social ferment of the Jacksonian era. This critique is potentially no less seminal than his analysis of slave treatment, which after more than a decade of criticism requires considerable modification.[23]

Elkins drew a sharp contrast between the fluid, atomized quality of antebellum American society and the more traditional societies of Latin America, where democracy and the spirit of capitalism had made fewer inroads and self-interest was still restrained by a web of corporate institutional influences. The lack of institutional "backbone" in North American society complicated the task of abolishing slavery, suggested Elkins, because it produced in the Jacksonian era a class of rootless intellectuals — typified by Concord transcendentalists — who were cut loose from society and thus blinded to its real nature and mode of functioning. Their disengagement from effective institutions created among them an illusion of individual autonomy that discouraged any realistic perception of social process.

Impatient even with the few weak institutions available to them and eager to push still farther out on a course of perfectionist individualism, the men of intellect who might have approached slavery as a specific and solvable social problem approached it instead as an abstract evil crying out for blood atonement. Their exclusively moralistic view precluded any sustained concern with the mundane cause-and-effect relationships that produced the problem and determined its character. Driven by an overwhelming sense of personal guilt, but ignorant of the uses of power and deprived in any case of the institutional channels through which power might have been exerted, the "transcendentalist as abolitionist" finally led the nation into a needless war.[24]

Elkins had no reason to concern himself with the postwar period, but several historians, building explicitly on his analysis of antebellum society, have shown that a dramatic shift of cultural mood took place in midcentury. The radical egoism and anti-institutionalism of the first half of the century gradually gave way to a consolidating, organizing,

[23.] Ann J. Lane, ed., *The Debate over "Slavery": Stanley Elkins and His Critics* (Urbana: University of Illinois Press, 1971); Laura Foner and Eugene D. Genovese, eds., *Slavery in the New World: A Reader in Comparative History* (Englewood Cliffs, N.J.: Prentice-Hall, 1969); Allen Weinstein and Frank Otto Gatell, eds., *American Negro Slavery: A Modern Reader* (New York: Oxford University Press, 1968). See also Carl N. Degler, *Neither Black nor White: Slavery and Race Relations in Brazil and the United States* (New York: Macmillan, 1971); John W. Blassingame, *The Slave Community: Plantation Life in the Antebellum South* (New York: Oxford University Press, 1972); Robert B. Toplin, *The Abolition of Slavery in Brazil* (New York: Atheneum, 1972); and Robert Conrad, *The Destruction of Brazilian Slavery* (Berkeley: University of California Press, 1973).

[24.] Elkins, *Slavery*, 140–222. See also Bertram Wyatt-Brown, "Stanley Elkins' *Slavery*: The Anti-Slavery Interpretation Reexamined," *American Quarterly*, 25 (May 1973), 154–176; and the contributions of Aileen Kraditor and Elkins himself to Lane, *The Debate over "Slavery."*

institution-building thrust of major proportions. R. Jackson Wilson traced the emergence among social philosophers of a quest for more effective modes of community and a growing sense of the inadequacy of individualist values.[25] George Fredrickson, focusing especially on the shift away from romantic humanitarianism to a more hard-boiled approach to social problems, attributed the change to the brutalizing impact of the war itself, and the military models of discipline and efficiency it called into being.[26] Taking issue with Fredrickson on the question of timing, John Higham argued that the change was already underway in the 1850's and was merely reinforced by the war experience. The cause, he thought, lay primarily in the forces of urbanization and industrialization.[27]

Frank Sanborn's social attitudes and sensibilities are not adequately represented by the "transcendentalist-abolitionist" model, but extend beyond it to reflect the shift of mood that was just getting underway as his generation reached maturity in the 1850's. He would never be able to go as far as the next generation in construing man as a creature dependent on and shaped by society. But neither could he share the blithe unconcern with, or disdain for, external constraints of environment demonstrated by his mentor, Ralph Waldo Emerson.[28] Although his thinking remained profoundly individualistic in comparison with later generations, and although his personal style was certainly antinomian to the very end, he and his Social Science Association were nonetheless in the mainstream of the movement toward institutional consolidation—and indeed, the ASSA even spearheaded that movement on certain fronts.

Sanborn's ambivalent posture toward institutionalism is apparent as early as July 1855, when he delivered his commencement oration upon

[25] Wilson, *In Quest of Community*. See esp. 10–11, 26–31.
[26] Fredrickson, *The Inner Civil War*. See esp. chs. VI and VII.
[27] John Higham, *From Boundlessness to Consolidation: The Transformation of American Culture, 1848–1860* (Ann Arbor: William L. Clements Library, 1969), 15. See also John L. Thomas, "Romantic Reform in America, 1815–1865," *American Quarterly*, 17 (Winter 1965), 656–681.
[28] Emerson himself may have been moving away from his earlier anti-institutionalism by the time Sanborn entered his circle: see Bledstein, "Cultivation and Custom," 104, and Wilson, *In Quest of Community*, 11. Wilson's entire chapter on "The Plight of the Transcendent Individual" is pertinent, and I have profited greatly from it. As Wilson recognizes, it would be erroneous simply to equate transcendentalism with individualism. Emerson rejected the "false involvements" of chamber and society for the sake of a higher form of association, not for dissociation in itself. For all their practical hostility, the institutionalist and the anti-institutionalist have much in common, for both sense that the existing relation of individual to society has become problematical—that the relation requires *decision* and cannot be left to tradition or common sense. Once this is acknowledged, the rather naive repudiation of institutions that characterized the first half of the century can be seen to have exactly the same roots as the general effort to "reconstruct" society and revivify its institutions that characterized the latter half of the century.

graduating from Harvard. Eager to choose a meaningful career and inspired by Emerson's confidence that "all men of power and originality nowadays make their own profession," the young Sanborn presented his classmates with a trenchant capsule analysis of their society. His words are interesting enough in their own right to justify lengthy quotation, and doubly pertinent because they define a main theme in the life of the Social Science Association.[29]

Sanborn began with a historical account of the advance of individual freedom and the anti-institutional impulse:

> Within the last hundred years have occurred the most astonishing political revolutions which authentic history records. . . . But the social revolution which has been accomplishing itself in the same period is quite as remarkable, though perhaps, less obvious; and the tendency of all these changes, whether in the church, the state, or the community, has been to develope [sic] and fortify individual freedom at the expense of established institutions.
>
> In politics this strong individualism weakens the authority of the state, making men revolutionists in Europe, and followers of the Higher Law in America; in religion it loosens the bands of the church, giving rise to all manner of protest and dissent; in philosophy it manifests itself as Transcendentalism, which is the stronghold of the individual against authority and against numbers. Let me briefly point out some of the results of this same principle in education.
>
> No one . . . can have failed to notice the increasing importance of the secular teacher. . . . Wherever a clergyman maintains his authority, it is not so much by virtue of his office as by his personal weight of character. None sees these things more clearly than the clergyman himself; and he therefore either quits his pulpit for the lecturer's or the schoolmaster's desk or assumes a double office. All the week days in the winter months he is a minister-at-large, and rushes from lecture room to lecture room with the zeal of an apostle. What ideas he has he puts in his lectures rather than in his sermons, being fully persuaded that his highest duty is towards his *audience*, not his *congregation*.
>
> Thus has lecturing become a profession. . . . In our villages the schoolmaster supplants the lawyer and the clergyman,—and is become the shepherd of the people. . . .

With admirable clarity, Sanborn saw that transcendentalism was the intellectual manifestation of a disintegrative tendency that had long been at work, prying the individual free from traditional institutions. Up to this point in his text Sanborn seems to approve of the disintegrative tendency—but then a new note creeps in:

> To what then are we tending? Evidently to a state of individual teaching and discipleship, corresponding to the individualism in politics and reli-

[29] Sanborn, *Recollections*, II, 316–317.

gion which I have noticed. Corresponding, and yet leading to widely different results,—for while the one hastens to dissolve organizations, and to compel each man by himself to his own guidance, the other—Individual Education—reunites and reconstructs society. In exalting school-keeping to the rank of a dignified profession, I see that we are preparing for that fortunate time when the cultivated, the earnest and religious men and women shall find their proper place as guides and teachers of those around them. . . .

I have said that the Teacher's office is becoming the most important one in the community. . . . What man holds so responsible, so influential, so sacred a place, as he who is to form the characters of the young. . . .[?][30]

There is no celebration here of the transcendent individual, standing above the law and outside society. Sanborn admired transcendentalism; we cannot doubt that. But he believed that the tendency toward individualization had gone far enough. Indeed, his aim, in his own words, was to "reunite" men and "reconstruct" society.

Sanborn's attitude in 1855 anticipated a general intellectual retreat from individualism that became conspicuous only after the Civil War. There developed in the postwar years a widespread concern for community, an appreciation of organization, and a fascination with the mechanics of social control that is far removed from the anarchic individualism espoused by so many antebellum intellectuals. The shift of mood is most sharply represented in the thought of George William Curtis, who became a president of the ASSA and a leader of the civil service reform movement after the Civil War. In 1843, during his Brook Farm days, Curtis wrote to his friend Isaac Hecker that

. . . the wise man lends himself to no organization. . . .

This great negro slavery—this reeling drunkenness that staggers and totters through the land, is not the sin that affects me. I am mainly concerned with that want of faith which cannot pierce the slavery and the intoxication to the centre and so reconcile them with all.

So wiser man lends himself to no organization. He is his own society and does his own reforms . . . the individual life is the only life he knows, and . . . to him, men, women, the world are experiences which affect him more or less deeply.[31]

By 1881 Curtis had changed his mind. "Organization," he then said, "is the lens that draws the fiery rays of conviction and enthusiasm to a focus and enables them to bury a way through all obstacles."[32]

[30.] *Exhibition and Commencement Performances* (1845–1855), XV, No. 44, Harvard Archives, quoted in Hickok, "Political and Literary Careers of F. B. Sanborn," 93–95. Italics in the original.

[31.] Quoted by Gordon Milne, *George William Curtis and the Genteel Tradition* (Bloomington: Indiana University Press, 1956), 16–17.

[32.] Quoted by John G. Sproat, *The "Best Men": Liberal Reformers in the Gilded Age* (New York: Oxford University Press, 1968), 60.

Sanborn's ambivalent program of "individual education" stands in a transitional position between the genteel anarchism of the 1840's and the organizing fever of the 1880's. In his commencement address Sanborn did not go so far as to reject the typical transcendentalist stress upon a full-blown individualism, nor was he ever to do so. But his early concern for the cohesive element in society should not be ignored. His conception of "individual education" is an awkward and unstable compromise between contradictory impulses. On the one hand he projects from current tendencies a future in which all institutional supports for intellectual and moral authority will dissolve and all that will be left are spontaneous bonds between disciples and teachers. Each thinker's authority will derive from, or at least be recognized and warranted by, a spontaneous *"audience,"* not an institutionalized *"congregation."*

Yet this antinomian dissolution of institutional supports is to have a conservative—even institutionalist—outcome. It will "reunite" and "reconstruct" society because it does not leave each man to his own guidance, but places him in discipleship to a teacher. And who will be the teachers? Sanborn was serenely confident that they would be "cultivated," "religious," and "earnest." Presumably they would acquire these traits naturally, without any institutional preparation or selection. Indeed, Sanborn seems to have believed that it was *only* by this natural means that those whose "proper place" it was to "guide" and "teach" the masses could gain the authority to do so, and thus help knit society together, reversing the tendency toward atomization.

No doubt Sanborn's model for "individual education" was his own discipleship to the sages of Concord. The contrast between Concord and Harvard's structured formalities must have reinforced his disdain for institutional supports. But as teacher of Concord's elementary school he had already undertaken a far more highly institutionalized educational task, and when he speaks of "exalting school-keeping to the ranks of a dignified profession," we see the seeds of a much stronger emphasis on the necessity of institutional structure. Similar aspirations for professional development would echo time and again at the meetings of the ASSA.

Throughout his life Sanborn maintained an uneasy balance between admiration for spontaneous discipleship and a concern that regular channels be established by which society would grant authority to those most qualified to hold it. The latter concern allied him with a great many people after the Civil War who saw in the cultivation of expertise, within a framework of professional institutions, a way to improve the quality of American life. The disjunction of "audience" and "congregation" that Sanborn had so perceptively noted before the war proceeded with ever greater intensity afterward, not just in religion, but in every aspect of life. It required extravagant optimism to assume that each emerging "audience" would choose as its teacher the "proper" person. Even Sanborn (not to mention his more elitist friends) would lose his serene confidence and take measures after the war to guarantee a favorable outcome. Professional institutions were a way to *insure* that each audience would find its proper guide; that moral and intellectual authority would be possessed only by those who deserved it.

CHAPTER IV

The Antebellum Origins of the Movement to Establish Authority

The American Social Science Association has long been recognized as a major center of humanitarian reform sentiment in the Gilded Age.[1] This reputation is not erroneous, but it is incomplete, and therefore misleading. Closely interwoven with the Association's humanitarian impulse is another reform aspiration that might at first glance seem unrelated or even antithetical to humanitarianism—namely, a Tocquevillean impulse to defend authority, to erect institutional barriers against the corrosive consequences of unlimited competition in ideas and moral values in an interdependent mass society. The effort to institutionalize sound opinion, to insure (in Sanborn's terms) that each emerging "audience" would find its "proper" guide, was at least as central to the identity of the ASSA as humanitarian reform. By sponsoring measures to institutionalize sound opinion, the ASSA gave voice to the central preoccupation of the college-educated gentry of late nineteenth century America, and made itself the instrument of a profoundly important but little-understood movement of conservative reform.[2]

The movement in which the ASSA played such an important role has been aptly named by John Higham a movement to "establish authority." Higham spoke specifically of the founding of the American Historical Association, which was in fact accomplished under the auspices

---

[1] L. L. Bernard and Jessie Bernard, *Origins of American Sociology: The Social Science Movement in the United States* (New York: Russell and Russell, 1943).

[2] My understanding of the gentry class has been shaped by Stow Persons, *The Decline of American Gentility* (New York: Columbia University Press, 1973), esp. his chapter on "Gentry Politics." However, his definition of the gentry as a functional elite composed of "anyone who cared to pattern himself upon . . . the norms of gentility" seems to me unnecessarily ephemeral, since it appears to rule out any criterion of wealth, occupation, or education (vi). In practice he often treats a college degree as a badge of gentility (133–136), and I follow suit. Scattered evidence suggests that even in the early and mid-nineteenth century a substantial majority of the graduates of the better colleges entered professional life: if this proves generally true, the gentry may turn out to be

of the ASSA. But as he recognized, "the rise of a professional outlook in the field of history was an integral part of a broad movement for the establishment of authority in American intellectual life. In almost all fields of cultural endeavor, associations that defined standards and goals appeared in the late nineteenth century." This drive for cultural consolidation, "comparable to the trust movement in American business," was spearheaded at every stage by men of the gentry and professional class.[3]

The closest thing the movement had to a "headquarters" was the American Social Science Association. However, the movement was too pervasive and diffuse a phenomenon to be contained within or coordinated by any one organization. Although the modern social science disciplines and their professional practitioners were not, strictly speaking, the intended goal of the movement to establish authority, they are one centrally important outcome of it. The most conspicuous political manifestation of the movement was the introduction of the merit system in the civil service, but the Pendleton Civil Service Act of 1883 was only a passing victory in a larger campaign whose major triumph was the creation of the modern American university and a system of professional and quasi-professional functional elites built upon it. The common aim of the men and women who participated in the movement was to construct safe havens for sound opinion.

The history of the movement to establish authority is yet to be written, and the present essay is no substitute for that larger work. Only when the movement has been described as a whole will the place of the ASSA within it be clear. But a start can be made.

The present chapter has two purposes, both openly didactic and unavoidably schematic in execution. First, we will examine three

---

composed mainly of professional men. See Gerard W. Gawalt, "Sources of Anti-Lawyer Sentiment in Massachusetts, 1740–1840," *American Journal of Legal History*, 14 (Oct. 1970), 300, where he shows that 62% of all Yale graduates from 1806 to 1815 became doctors, lawyers, or ministers. Other useful studies of genteel culture include Richard Hofstadter, *Anti-Intellectualism in American Life* (New York: Knopf, 1963); Sproat, *"The Best Men"*; John Tomsich, *A Genteel Endeavor: American Culture and Politics in the Gilded Age* (Stanford: Stanford University Press, 1971); and Geoffrey Blodgett, *The Gentle Reformers: Massachusetts Democrats in the Cleveland Era* (Cambridge: Harvard University Press, 1966).

[3.] John Higham with Leonard Krieger and Felix Gilbert, *History: The Development of Historical Studies in the United States* (Englewood Cliffs, N.J.: Prentice-Hall, 1965), 8–9. See also Higham's *From Boundlessness to Consolidation*, and "The Reorientation of American Culture in the 1890's," in his *Writing American History*, 73–102. One *locus classicus* of the Victorian impulse to defend authority is Matthew Arnold, *Culture and Anarchy*, ed. and intro. by J. Dover Wilson (Cambridge: Cambridge University Presss, 1971), 82, 85, 95–96, 110. See also Burton J. Bledstein's splendid *Culture of Professionalism: The Middle Class and the Development of Higher Education in America* (New York: Norton, 1976), which appeared too late for me to use.

groups of men who articulated profound anxieties about authority in antebellum America–men of science, university reformers, and the traditional tripartite professional class composed of ministers, lawyers, and physicians. Each group contributed leadership to the ASSA; men of science also contributed a widely imitated strategy of institutional reform. The second task of the chapter is to suggest some of the ways in which the anxieties of these men stemmed from their experience in an increasingly interdependent society, and to explain why their efforts to defend authority led them to foster the development of social science.

The Community of the Competent

The twentieth century has adopted a cynical attitude toward professionalization, refusing to recognize that in the nineteenth century it was part of a broad movement to establish or reestablish authority in the face of profoundly disruptive changes in habits of causal attribution, in the criteria of plausibility, in the relation of the man of knowledge to his clientele—finally, changes in the very notion of truth itself.

These changes and the dangers they seemed to pose inspired the almost heroic enthusiasm with which late nineteenth century intellectuals set about constructing and reinforcing professional institutions. Professionalization in the nineteenth century was not merely a pragmatic and narrowly self-seeking tactic for enhancing occupational status, as it often is today; instead it then seemed a major cultural *reform,* a means of establishing authority so securely that the truth and its proponents might win the deference even of a mass public, one that threatened to withhold deference from all men, all traditions, and even the highest values.

The movement to establish authority expressed itself in a bewildering variety of forms, in law, medicine, politics, education, scholarship, and virtually every other field of cultural endeavor. The specific measures taken to consolidate authority in the various fields are so diverse that contemporaries seldom shared any clear sense of participation in a common effort. Even with the benefit of hindsight historians have only recently begun to see a pattern in these far-flung activities.

In the field of natural science one sees most clearly the common pattern of professionalizing action that underlies all the diverse manifestations of the movement. In this field there occurred the most successful of all nineteenth-century professionalizing efforts—efforts to build an institutional framework that would identify individual competence, cultivate it, and confer authority upon the individuals who possessed it. Men of science were so successful in this institution-building enterprise that by the end of the century the word "scientific" seemed to epitomize the very essence of well-founded authority. Through either imitation or independent innovation, all who participated in the movement to establish authority worked toward essentially the same institutional apparatus that scientific men began developing in the 1840's.

The institutional apparatus pioneered by men of science may be designated by the term "community of the competent."[4] The concept was as old as the professions themselves, but it took on a new relevance in the nineteenth century. In science, rather than in the classic professions, it was applied with most spectacular success.

The need for the community of the competent and the explanation of some of its essential functions were articulated by the leading scientific figure in antebellum America, Joseph Henry, physicist and head of the Smithsonian Institution. "We are over-run in this country with charlatanism," declared Henry in 1846. "Our newspapers are filled with puffs of Quackery and every man who can burn phosphorous in oxygen and exhibit a few experiments to a class of young ladies is called a man of science."[5] Anxious for the future of scientific authority, Henry struggled to conceive of an institutional device that would function to discredit charlatans and deny incompetent practitioners an authoritative place in the public eye. Once freed of the dead weight of the marginal practitioner, the general level of science would rise. The authentic man of science would reach his audience without distraction and be able to speak with the authority that he deserved and the truth required.

Henry's institutional device had to serve two essential functions. The authentic man of science needed not only a secure and uncluttered avenue of influence over the public, but also a degree of protective insulation against public opinion. "No one can be learned in all branches of thought," observed Henry, "and the reputation of an individual therefore ought to rest on the appreciation of his character by the few, comparatively, who have cultivated the same field with himself."[6]

What Henry described — implicitly, by its essential functions — was a community of the competent. Like men in many other occupations in

[4.] The term is Francis E. Abbot's, who was an associate of Frank Sanborn and a member of the "Metaphysical Club," where Charles Peirce and William James worked out the basic ideas of pragmatism in the 1870's. See ch. XI below, and Stow Persons, *Free Religion: An American Faith* (New Haven: Yale University Press, 1947), 31, 125–129; Philip P. Wiener, *Evolution and the Founders of Pragmatism* (New York: Harper and Row, 1965), 41–48 and *passim*; and Wilson, *In Quest of Community*, ch. II.

[5.] Quoted by Howard S. Miller, "Science and Private Agencies," in *Science and Society in the United States*, ed. D. Van Tassel and M. G. Hall (Homewood, Ill.: Dorsey, 1966), 195.

[6.] Quoted by Howard S. Miller, *Dollars for Research: Science and Its Patrons in Nineteenth Century America* (Seattle: University of Washington Press, 1970), 23.

the coming decades, he was appalled by what he perceived to be a rising tide of false and pretentious claims to authority. Recoiling against this haphazard competition, he proposed that the competent partially secede from society, remaining physically within it but insulating themselves from it. The ultimate tendency of his strategy for the establishment of authority was to insulate the practitioner of science from those persons least competent to judge him, and simultaneously to bring him into intimate contact with—and competitive exposure to—those most competent to judge him. Independence from the general public was to be purchased at the nominal expense of intense and competitive dependence upon one's certified peers. The aim was not to escape competition altogether, but to regularize it, permitting only the best competitors to compete.

To bring a community of the competent into existence, or to strengthen and consolidate imperfect communities which already existed in law, medicine, and divinity, was a practical matter that required different instruments in different fields of action. But whatever concrete measures were chosen to create and consolidate such communities, certain effects had to be achieved if Henry's goal of defending authority was to be realized. If the man of science (or any other aspirant to authority) wished to enhance his intellectual authority, he would have to alter his social identity and pattern of social affiliations. In the thinking, deliberative side of life that presumably mattered most to him, the man of science would have to exchange general citizenship in society for membership in the community of the competent. Within his field of expertise, the worth of his opinions henceforth would be judged not by open competition with all who cared to challenge him, but by the close evaluation of his professional colleagues. In this critical aspect of life he would rise above the identity conferred upon him by an island community and take his place in a trans-local community based on occupational competence.

By thus changing his affiliations, the individual practitioner would enhance his own authority in the eyes of the general public, assuming that he survived the rigorous judgment of his professional peers. The community would now certify and guarantee the soundness of his views. Even more important, by orienting himself to the trans-local community of the competent, the individual practitioner would contribute to the defense of merit, the preservation of the very principle of authority. Indeed, he would help insure the triumph of truth itself at a time of growing intellectual crisis.

Near the end of the century, the philosopher Charles S. Peirce formulated an elegant epistemological rationale for the whole movement to establish authority. Peirce recognized the social origin of doubt and stressed also the social basis for the struggle to escape doubt and achieve certainty. In so doing, he showed how and why the creation of communities of competent inquirers countered the spread of uncertainty in an interdependent mass society.[7] But long before Peirce supplied the movement with a philosophical justification, Joseph Henry acted as if he understood how great the stakes were. So did other leading men of science, such as Benjamin Peirce, father of the philosopher, who would one day become president of the ASSA. Their tacit understanding is reflected in their creation of the American Association for the Advancement of Science and the National Academy of Science, two prototypical efforts to realize the ideal of the community of the competent.

Establishing Authority in Natural Science

The word "scientist" was not coined until 1840, only six years before Joseph Henry spoke of the need for institutionalizing authority in American science. The word was not commonly used until late in the century, but even in 1840 the practitioners of science began constructing the institutional apparatus that ultimately would transform the mid-nineteenth century "man of science" into a "scientist."[8] The former pursued a gentlemanly vocation; the latter possessed a recognized professional role. This initial step toward the institutionalization of scientific authority in America was the formation in 1840 of the country's first professional scientific society, the Association of American Geologists.[9]

If one recognizes the significance and the catholic scope of the movement to establish authority, it will appear as no coincidence that the first president of the American Social Science Association, William Barton Rogers, was also among the three organizers of this first professional scientific society twenty-five years earlier. Scientific men of all kinds flocked to the geologists' meetings, and by the time of its third meeting the group had expanded its scope to become the Association of American Geologists and Naturalists. The snowball effect continued in 1847, when the expanded organization, now under the presidency of Rogers, enlarged itself once more to become the American Association for the Advancement of Science. From that date to the present, of

---

[7.] See below, ch. XI, and Murray G. Murphey, *The Development of Peirce's Philosophy* (Cambridge: Harvard University Press, 1961), chs. VI and VII; C. Wright Mills, *Sociology and Pragmatism: The Higher Learning in America*, ed. I. L. Horowitz (New York: Payne-Whitman, 1964), 159–160; and Wilson, *In Quest of Community*, ch. II.

[8.] Miller, "Science and Private Agencies," 192.

[9.] A competing group of no lasting consequence, the National Institute for the Promotion of Science, was formed the same year. See Sally Kohlstedt, "A Step Toward Scientific Self-Identity in the United States: The Failure of the National Institute," *Isis*, 62 (Sept. 1971), 339–362.

course, the AAAS—modelling itself originally on its British namesake—has been the umbrella organization covering all the disparate fields of science in America.[10]

Rogers does not provide the only direct link between the Social Science Association and the antebellum movement to establish authority in natural science. Equally direct and even more important are the roles of Louis Agassiz and Benjamin Peirce, who were instrumental in the creation of the American Association for the Advancement of Science and even collaborated in the drafting of the Association's 1847 constitution.[11] Both men later became singularly important leaders of the ASSA, saving it, in fact, from dissolution in the 1870's. Furthermore, Peirce, as acting president of the ASSA in 1878, initiated a proposal to merge the Association with the new Johns Hopkins University. Though it failed, Peirce's merger proposal was the ASSA's most fateful effort to gain a foothold in the fertile soil of the American university system.

Rogers, Peirce, and Agassiz shared a desire to establish authority which first was manifested in their work for the Scientific Association and which they later applied to broader cultural spheres in the Social Science Association. In spite of the common aims of the three men, Peirce and Agassiz often found themselves in opposition to Rogers. Perhaps because Rogers's views had been shaped by his association with Madison and Jefferson at the University of Virginia early in the century, he was more sensitive to the charge of elitism. The difference was one of degree only, however. Rogers's egalitarian scruples did not prevent him from sponsoring a proposal to designate two classes of membership in the AAAS: one for those who could show evidence of actual scientific work, and another for "associates," who would be ineligible to vote or hold office.[12]

Peirce and Agassiz were among the leaders of a boldly elitist faction of scientific men known as the "scientific Lazzaroni." Joseph Henry, whose observations on the need for institutionalized authority have already been noted, was another prominent member. Though Henry at first doubted that a popular organization like the AAAS would serve

---

[10.] For the early history of the AAAS, see Sally Gregory Kohlstedt, *The Formation of the American Scientific Community* (Urbana: University of Illinois Press, 1976), as well as her "Step Toward Scientific Self-Identity"; also Dirk Jan Struik, *The Origins of American Science (New England)*, 2nd ed. (New York: Cameron Association, 1957), 199; H. L. Fairchild, "The History of the American Association for the Advancement of Science," *Science*, n.s. 59 (1924), 365–368, 385–390, 410–415; Edward Lurie, *Louis Agassiz* (Chicago: University of Chicago Press, 1960), 132; A. Hunter Dupree, *Science in the Federal Government: A History of Policies and Activities to 1940* (New York: Harper and Row, 1957), 115–119.

[11.] Dupree, *Science in the Federal Government*, 115. For the work of Agassiz and Peirce in the ASSA, see below, ch. VI.

[12.] Robert V. Bruce, "Democracy and American Scientific Organizations," typescript of paper delivered at 1971 American Historical Association convention, 13.

his purpose, he viewed it as a first, faltering step at least toward the creation of distinctive communities for the scientifically competent.[13] He and his Lazzaroni colleagues were willing to advance more rapidly toward that goal, both inside and outside the AAAS, than Rogers's democratic scruples would permit.

The Lazzaroni were in the vanguard of the movement to establish authority. They identified themselves as neapolitan beggars to reflect their status as researchers, perpetually dependent upon the munificence of patrons. The loose and informal group's stated purpose was simply to "eat an outrageously good dinner together,"[14] but their significant purpose was to organize science as a profession, with themselves at the top. The Lazzaroni regarded themselves as the best, perhaps the only *real*, men of science in America. Their claim to superiority was inflated, in view of their exclusion of several major scientific figures such as Rogers and Asa Gray, but they were probably correct in believing that there was no more eminent circle of scientists in antebellum America.[15] Alexander Dallas Bache, head of the U.S. Coast Survey, was "chief" of the Lazzaroni. Other regular members, besides Henry, Agassiz, and Peirce, were the chemists Oliver Wolcott Gibbs and John F. Frazer, the astronomer Benjamin A. Gould, and geologist James Dwight Dana. Others who were closely associated with the Lazzaroni include Benjamin Silliman, Jr., Yale chemist and coeditor of the *American Journal of Science*; political theorist Francis Lieber; and Samuel B. Ruggles, a layman and benefactor of both physical and social science.[16]

Joseph Henry's double-edged professionalizing strategy–insulation from the public, coupled with intense competitive interac-

---

[13] As recently as 1838, Henry had opposed any imitation of the British Association in this country. He was convinced that "a promiscuous assembly of those who call themselves men of science in this country would only end in our disgrace:" Henry to A. D. Bache, 9 Aug. 1838, in Nathan Reingold, ed., *Science in Nineteenth-Century America: A Documentary History* (New York: Hill and Wang, 1964), 88.

[14] Dupree, *Science in the Federal Government*, 118.

[15] For a different view of the Lazzaroni, see Mark Beach, "Was There a Scientific Lazzaroni?" in *Nineteenth Century Men of Science: A Reappraisal*, ed. George H. Daniels (Evanston: Northwestern University Press, 1972), 115–132. Despite his title, Beach seems not to doubt that there was indeed a scientific Lazzaroni, but he takes pains to show that the group was not a monolithic entity with a party line. Its members were strong-willed individuals who went their own way and gathered for the conscious purpose of simply enjoying each other's company. One need not envision them as a band of blood brothers to recognize that they acted more nearly in concert than any other circle of scientific contemporaries; that they shared a conviction (not unique to them, certainly) that the scientific community in America was disorganized and afflicted with too many hangers-on; and finally that as individuals they held positions of power and were in fact instrumental in building an institutional framework for American science.

[16] This list is compiled from several sources that do not fully agree. See Richard J. Storr, *The Beginnings of Graduate Education in America* (Chicago: University of Chicago

tion within a community of peers—was first implemented by him and his Lazzaroni colleagues in the AAAS. Having led in the drafting of the constitution, the Lazzaroni found it easy to dominate the AAAS in its early years. The presidency was at first a Lazzaroni prerogative: Henry, Bache, Agassiz, Peirce, and Dana succeeded each other in the office in the early 1850's. The chief means of influencing the level of discussion and competition inside the AAAS was the Standing Committee, which arranged the program. The Lazzaroni usually dominated the committee in the 1850's, but when they used it to reject what they deemed to be unworthy papers, they found themselves embattled by the forces of "democracy," led by William Barton Rogers. Control seesawed back and forth in the mid-fifties, but by the outbreak of the Civil War the Lazzaroni had clearly lost control of the AAAS.[17]

As a platform upon which to elevate the authentic man of science above the clamoring masses, the AAAS was faulty in its very foundations because its membership policy was comparatively indiscriminate. Like the British Association, the AAAS admitted any member of a scientific society which published transactions. Unlike the British Association, it also admitted any college professor of any of the "theoretical and applied sciences," a fairly generous standard in Jacksonian America.[18]

The AAAS was useful to the friends of authority as a means of shaping the general character of science in America, but it was too large and inclusive to define the boundary between legitimate science and "charlatanism" with any strictness. Moreover, *within* the still ill-defined boundaries of legitimate science, the AAAS could do very little to sharpen the more delicate line between excellence and mediocrity. Without some institutional means of defining these boundaries, Henry's vision remained unfulfilled. Legitimate science was discredited, and scientific excellence was deprived of its rightful force. The competent still had to compete with the incompetent; the best still went unheard in the confusion of voices.

To achieve a closer approximation to their ideal of a community of the competent, the Lazzaroni turned to the idea of a National Academy of Science. As early as 1851 Bache, already familiar with the potential

Press, 1953), 67–84; Dupree, *Science in the Federal Government*, 118–119, 135 and *passim*; Lurie, *Louis Agassiz*, 182–184, 323–331; Miller, *Dollars for Research*, 8, 66–67 and *passim*; Robert V. Bruce, "Universities and the Rise of the Professions: Nineteenth Century American Scientists," typescript of paper delivered at 1970 American Historical Association convention. 3; Bruce, "Democracy and American Scientific Organizations"; and Beach, "Was There a Scientific Lazzaroni?"

[17.] Bruce, "Democracy and American Scientific Organizations," 10; George H. Daniels, "The Process of Professionalization in American Science: The Emergent Period, 1820–1860," *Isis*, 58 (Summer 1967), 151–166.

[18.] Bruce, "Democracy and American Scientific Organizations," 10.

uses of the federal treasury through his supervision of the Coast Survey, proposed the creation of "an institution of science, supplementary to existing ones . . . to guide public action in scientific matters." He assured his audience of AAAS members that it was but a common mistake to confuse such an academy with "monarchical institutions."[19] But by its very existence an academy of government advisers would serve to set standards and shape scientific reputations. It must have been clear to both Bache and his listeners that the federal government could only take advice from (and give money to) a tiny elite of the scientific men of the country. To warrant the government's trust, the organization would need to exclude anyone who might voice unsound opinions.[20]

Bache's idea of a national scientific academy bore fruit during the Civil War, when the government's need for scientific advice became self-evident even to laymen and politicians. Joseph Henry doubted that Congress would create such an academy, for "it could be opposed as something at variance with our democratic institutions."[21] But Bache, Agassiz, Peirce, and Gould pushed ahead with the plan in spite of Henry's skepticism.

With the help of Senator Henry Wilson, a radical Republican and fellow member of the Bird Club with Frank Sanborn and Samuel Gridley Howe, the Lazzaroni wrote up a bill and a list of fifty distinguished scientists. (Most of them were not even consulted about the project in advance.) Wilson pushed the bill through in the closing hours of the thirty-seventh Congress, relying on the pressure of adjournment to prevent debate. Once formed, the Academy recapitulated the factional battles that had been waged in the AAAS in the 1850's, with William Barton Rogers once more heading the opposition to the domineering ways of the Lazzaroni. Again, the Lazzaroni lost. By 1867, when Bache died, both the Academy and the AAAS were independent organizations beyond the control of any clique, and the Lazzaroni had ceased to be a prime influence in American science.[22] Although both the Academy and the AAAS were initially created by men seeking institutional support for their own particular viewpoint, neither organization long remained in the hands of any narrow party.

The AAAS and even the Academy were only crude approximations of the community of the competent as Henry envisioned it in 1846, and as Charles Peirce elaborated it in the 1870's. But reality always falls

[19] Dupree, *Science in the Federal Goverment*, 117.
[20] For the failure of an earlier scientific academy that did not win the support of the nation's scientists, see Kohlstedt, "Step Toward Scientific Self-Identity."
[21] Dupree, *Science in the Federal Government*, 136.
[22] *Ibid.*, 138–148; Miller, *Dollars for Research*, 66; Bruce, "Democracy and American Scientific Organizations."

short of the ideal. Simply by providing an occasion for meeting and talking together, the two organizations helped crystallize a pecking order among the nation's scientific men. And with that pecking order there necessarily came a preferential hierarchy of theories, methods, and procedures, as the work of the "best" was held up for emulation. Furthermore, the existence of these two overarching professional structures facilitated the formation of collegial networks of men working on closely related, specialized problems; on this smaller scale, the ideal of the community could be realized in quite pure form. By 1900 such networks had gradually proliferated and dovetailed to form a truly professional institutional structure that enabled science to tower over the rest of the intellectual world. The word "scientific" then seemed to epitomize the very essence of the professional idea—expert authority, institutionally cultivated and certified.

To be self-sustaining the community would have to do more than gather the competent and exclude the incompetent. It would also have to actively recruit, seeking out young candidates for admission and equipping them with the intellectual tools of competence. The natural arena of recruitment and training was the college or university, which also, of course, offered an occupational base for mature practitioners. The Lazzaroni and others eager to establish authority in science recognized higher education's importance to their goals and worked hard to expand the place of science in the traditional college curriculum. Their success is legendary. The rising prestige of science in the late nineteenth century put the defenders of the fixed classical curriculum on the defensive, opening the doors of the university to every subject from analytical chemistry to sociology to home economics. Some of the major reformers of higher education in the late nineteenth century were men of science. Rogers himself became founder and first president of the Massachusetts Institute of Technology, and a chemist, Charles William Eliot, became Harvard's most notable president of the century. Both men held office in the Social Science Association.

The Lazzaroni too were active university reformers. Although they failed in their bid to found a research-oriented national university at Albany in the 1850's, they succeeded in outmaneuvering the defenders of the classical curriculum in the nation's most prestigious schools, Harvard and Yale.[23] At Harvard, Benjamin Peirce drew up the plans that eventually led to the creation of Lawrence Scientific School in 1847. At Yale, Benjamin Silliman, Jr., a close associate if not a member

---

[23.] Miller, *Dollars for Research*, 40, 43; Storr, *Beginnings of Graduate Education*, ch. VII; Robert Silverman and Mark Beach, "A National University for Upstate New York," *American Quarterly*, 22 (Fall 1970), 701–713. See also file marked "National University" in Benjamin Peirce Papers, Houghton Library, Harvard University.

of the Lazzaroni, was the chief promoter of the Sheffield Scientific School which was formed in the late 1840's and early 1850's.[24] Both of these pioneering enclaves of science contributed leadership to the Social Science Association. David A. Wells, a student of Peirce and Agassiz and one of the first four graduates of the Lawrence School, became absorbed with economic questions during the Civil War and developed into one of the nation's leading economic writers.[25] He served as president of the ASSA in the mid-1870's. Another ASSA president, Daniel Coit Gilman, founding president of the nation's first real university, the Johns Hopkins, was closely associated with the Sheffield School during its early years. A recent graduate of Yale, he was hired in 1855 to raise funds and act as publicity agent for the new school of science. For nine years he served in Sheffield's administration and taught courses in geography, history, and political economy. As the leading figure in the professionalization of academic life in late nineteenth century America, Gilman kept alive the spirit of the Lazzaroni (though never a member) and made his university a fulfillment of their long-frustrated dreams. He occupied the presidency of the Social Science Association at a particularly critical time in its history and gave serious consideration to Benjamin Peirce's proposal that the ASSA be merged with Johns Hopkins.[26]

Joseph Henry's observations about the need for establishing authority in science would be relevant simply because he articulated a basic strategy of institutional reform that the ASSA and many other friends of institutionalized authority later acted upon. His comments turn out to be doubly pertinent because many of the same individuals who labored with him to turn the "man of science" into a "scientist" later worked in the ASSA to transform the "social thinker" into a "social scientist." It is testimony to the continuity of the movement and suggestive of the ASSA's place in it that so many leaders of the Association can be traced back to antebellum science.

Broader Implications in Educational Reform

Of course, many pioneers of the movement to establish authority were not men of science at all. The idea of reinforcing or extending the role of trained intellect in American life was attractive to any number of men who felt alarmed by the leveling tendencies at work around them

---

[24] Miller, *Dollars for Research*, 74–82, 87–97.
[25] For Wells's reputation, see Dorfman, *Economic Mind in American Civilization*, III, 135–136 and *passim*.
[26] Fabian Franklin, *The Life of Daniel Coit Gilman* (New York: Dodd, Mead, 1910), 73, 40–43 and *passim*. For the merger proposal, see ch. VII, below.

in the Jacksonian era. At this more general level, the movement could have asked for no better textbook than Alexis de Tocqueville's *Democracy in America,* the two volumes of which were published in 1835 and 1840, just as the movement got underway. Anyone who knew of Tocqueville's work—and most college-educated Americans did—had good reason to fear for the future of independent thinking and truth itself in an egalitarian society.[27]

Tocqueville taught Americans that theirs was a society peculiarly prone to conformity and intellectual sterility. Severed from the organic ties that bound him to his fellow man in traditional societies, the individual was naked prey to the arbitrary tyranny of public opinion in a democracy. "I know of no country in which there is so little true independence of mind and freedom of discussion as in America," declared Tocqueville.[28] The problem arose from that most basic condition of American life, equality. Upon what ground, he asked, could the merely equal individual justify his dissent from the consensus of a mass of equal others?[29]

Tocqueville's concern for authority in democratic societies was grist for the mill of several generations of university reformers. It is no coincidence that the greatest academic professionalizer of the nineteenth century, Daniel Coit Gilman, brought out a new edition of Tocqueville's classic in 1908. Another university reformer of an earlier era, George Ticknor, was a personal friend of Tocqueville—and also, incidentally, of Agassiz and Peirce. Ticknor exemplified what Oliver Wendell Holmes, Sr., called the "Brahmin caste of New England." Men of Ticknor's temperament provided the advice on which Tocqueville relied in writing his analysis of America, and such men continued to find his analysis relevant late in the century.

Ticknor was no scientist, but he was as eager as his Lazzaroni friends to enlarge the orbit of the disciplined mind in America. As a specially privileged professor of modern languages at Harvard in the 1820's, he tried to cultivate intellectual competence by proposing a series of

[27.] Anna Haddow, *Political Science in American Colleges and Universities, 1636–1900,* ed. W. Anderson (New York: Appleton-Century, 1939), 248.

[28.] Alexis de Tocqueville, *Democracy in America,* trans. H. Reeve (New York: Schocken Books, 1961), I, 310.

[29.] On the tyranny of the majority, *ibid.,* 309–312; on the lack of theoretical development in American science, II, 47–55. Tocqueville saw in "voluntary association" a partial escape from the tyranny of the majority. In a manner that Tocqueville had not anticipated, it was precisely this institutional device to which Joseph Henry and other defenders of authority resorted. The community of the competent is, after all, a special kind of voluntary association, one which offers its members protection against the tyranny of public opinion, even as it compels their submission to professional opinion. When the individual joins such a community, he is elevated above the mass and made independent of it; but at the same time he is deliberately made more dependent on his peers and rendered less able to resist the consensus of the competent.

institutional reforms directed against dilettantism among faculty and students alike. He would have permitted professors to pursue narrower specialities and organize themselves in departments; students would have been given certain elective courses, and the whole student body would have been ranked and graduated on the basis of demonstrated achievement alone, rather than on the automatic four-year class system.[30]

Francis Wayland, who is generally conceded to be the leading reformer of higher education in antebellum America, supplied the Social Science Association with no fewer than three of its presidents—his sons, Francis Wayland, Jr., and Heman Lincoln Wayland, and his prize pupil, James Burrill Angell, president of the University of Michigan. The elder Wayland, as president of Brown University and the most influential of all the nation's moral philosophers, proposed in 1850 a general renovation of American higher education along professionalizing lines. His *Report to the Corporation of Brown University* called for an elective system, a curriculum tailored to the needs of a practical-minded nation, a student body drawn from all sectors of society, and other reforms that became commonplace only after the war.[31]

Wayland was far more at ease with egalitarian manners than Ticknor, but his educational reforms were similarly motivated by a wish to transform the university so that it could confer authority upon an elite class of leaders for American society. Like many other members of the gentry class, Wayland envied England's aristocracy and wondered if a suitably republican version of that organized, responsible elite could be transplanted to American soil. The elite of England, wrote Wayland in 1842, constitutes "one great family. London, 'that mighty heart,' sends out its pulsations to every extremity of the empire, and is in turn receiving from every extremity the life blood which it vitalizes and sends back again. Every man of distinction is expected to report there during some part of the 'season' and he must do it in order as Walter Scott says 'to keep himself abreast of society.' "[32] Professional

---

[30.] David B. Tyack, *George Ticknor and the Boston Brahmins* (Cambridge: Harvard University Press, 1967), 194, 212, 85–128.

[31.] See Richard Hofstadter and Wilson Smith, eds., *American Higher Education: A Documentary History* (Chicago: University of Chicago Press, 1961), II, 478–487; Storr, *Beginnings of Graduate Education in America*, 44–45, 60, 170–171; Wilson Smith, *Professors and Public Ethics: Studies of Northern Moral Philosophers Before the Civil War* (Ithaca: Cornell University Press, 1956), 38; Persons, *American Minds*, 196. For Wayland's primacy among moral philosophers, see Meyer, "American Moralists," 7, 58–65, and *The Instructed Conscience: The Shaping of the American National Ethic* (Philadelphia: University of Pennsylvania Press, 1972).

[32.] Francis Wayland, *Thoughts on the Present Collegiate System in the United States* (reprint ed., New York: Arno Press, 1969), 38–39.

conventions and journals would one day serve a purpose similar to the London "season," enabling professional men to "keep abreast" of their special communities.

Before such an elite could exist in America, a change would have to take place in the nation's system of higher education. In the United States, complained Wayland, "the College or University forms no integral and necessary part of the social system. It plods on its weary way solitary and in darkness." But in England, Oxford and Cambridge form the very keystone of the social system. "There the youthful aristocracy meet and become acquainted with each other. Thither do the bar, the pulpit, and the senate look for the young men who have there made it known that nature has marked them for distinction. . . . To obtain rank there, is to place oneself immediately in a position in society. . . ."[33] To make the American university an integral and necessary part of the social system was the goal of Wayland, Ticknor, and the triumphant generations of university reformers that followed them. No phase of the movement to establish authority was more important than the modernization of the university, nor was any more successful.

The Crisis of the Professional Class

The preceding brief survey of the closely related fields of natural science and university reform, touching as it does on the early careers of Rogers, Peirce, Wells, Gilman, Angell, and the two Wayland brothers, accounts for nearly half of the men who served as presidents of the ASSA in the nineteenth century. The other leaders of the Association were committed to the same general goals, as will be shown in later chapters. In view of these direct links between the postwar Association and the earliest antebellum stirrings of a movement on behalf of authority, one can hardly doubt the continuity of the movement or the ASSA's central place in it. Having observed these palpable links, it will be useful now to view the movement from a larger perspective, taking into account the sources of the whole movement rather than just those beginnings that produced direct connections with the ASSA.

Men like Henry, Ticknor, and Wayland were, in effect, articulate theoreticians of the community of the competent; however, the driving force of the movement to establish authority lay in the three classic professions—divinity, medicine, and law—rather than in science or university reform. As the principle of deference came under general attack in Jacksonian America, professional men (especially doctors and lawyers) feared for their authority and searched for ways of defending

[33.] *Ibid.*

it. Their participation made the movement on behalf of authority a force to be reckoned with. The founders of the ASSA recognized that the "grassroots" of their elite movement lay in the professions, and accordingly they assigned each secular profession a special place in the organizational structure of the Association. The participation of professional men was essential to the purposes of the Association. Although in the twentieth century social science has become one tiny area of specialization in a professional class that has lost any sense of unity, "social science" in the nineteenth century began as a measure to preserve professional unity and reestablish the authority of *all* professional men on an unshakable foundation—a *science* of human affairs.

To discuss the history of the professions is to enter largely uncharted waters. Even the rather conspicuous crisis of the legal and medical professions in the era of Andrew Jackson has received little attention from historians; Arthur Schlesinger, Jr., in his *Age of Jackson* said hardly a word about it, and nothing is added by the most recent survey of the period.[34] Historians have grown wary of the contemporary stereotype, articulated by the professional class itself, that cast antiprofessional sentiment in the role of a barbarian invasion, but still there exists no general assessment of the meaning of the crisis or even of its dimensions.[35] Among a number of recent, quite penetrating studies, only Daniel Calhoun's splendid *Professional Lives in America* goes beyond the history of a single profession to treat the professional class as a whole.[36]

As Calhoun shows, there was one central problem to which each profession had to respond in its own particular way; it involved the increasing pressure of competitive market forces on communal life. Over the course of the century "communal values and action yielded before a growing market orientation."[37] The problems of adaptation

[34.] Arthur Schlesinger, Jr., *The Age of Jackson* (Boston: Little, Brown, 1945); Edward Pessen, *Jacksonian America: Society, Personality, and Politics* (Homewood, Ill.: Dorsey, 1969).

[35.] Maxwell Bloomfield, "Lawyers and Public Criticism: Challenge and Response in Nineteenth Century America," *American Journal of Legal History*, 15 (Oct. 1971), 269.

[36.] Daniel H. Calhoun, *Professional Lives in America: Structure and Aspiration, 1750–1850* (Cambridge: Harvard University Press, 1965). For the tight-knit quality of the early nineteenth century professional class in one important state, see Gawalt, "Sources of Anti-Lawyer Sentiment in Massachusetts." Two particularly useful studies are Joseph F. Kett, *The Formation of the American Medical Profession: The Role of Institutions, 1780–1860* (New Haven: Yale University Press, 1968), and William G. Rothstein, *American Physicians in the Nineteenth Century: From Sects to Science* (Baltimore: Johns Hopkins University Press, 1972). See also Anton-Hermann Chroust, *The Rise of the Legal Profession in America* (Norman: University of Oklahoma Press, 1965); Rosemary Stevens, *American Medicine and the Public Interest* (New Haven: Yale University Press, 1971); and Perry Miller, *The Life of the Mind in America* (New York: Harcourt, Brace & World, 1965).

[37.] Calhoun, *Professional Lives*, 13.

were especially traumatic for the fee-taking professions of law and medicine, but even the clergy, which ever since the Reformation had shunned open dependence on fees, felt the strain of changing market conditions. In New England by the 1820's and 30's the core of communal life had been severely eroded by the growing awareness of townspeople that they could rid themselves of an unwanted minister, and by the minister's awareness that he could pursue a career elsewhere. The advance of a more fluid and interconnected society opened up a wider range of options for both parties and led to a decline of ministerial permanency, the very symbol of organic community.[38]

Fee regulation and the ethics of the client-practitioner relationship were only two among a horde of problems that plagued doctors and lawyers. Both professions experienced a painful breakdown of the institutional mechanisms for conferring authority upon new recruits. Under the old apprenticeship system, candidates for admission to the professional community received authorization to practice from a local bar association or medical society after a period of tutelage under an established practitioner. But the rise of proprietary and university-based professional schools in the early nineteenth century introduced a competing emblem of competence, the medical or law degree.[39] The result was a degrading free-for-all between proponents of the two systems. Self-serving greed and sincere altruism became impossible to separate. The recurring campaigns to "raise the standards of the profession" that became a hallmark of nineteenth-century professional life were susceptible of many interpretations, then and now. Depending on the particulars of the case, "raising standards" might mean:

(a) a sincere effort to improve the quality of professional service (which was often miserably low, especially in medicine);

(b) an effort to close the doors of the profession in order to minimize competition and maximize profits;

(c) an effort by proprietary schools to discredit the apprenticeship system, thus boosting their own enrollments;

(d) an effort by established proprietary schools to drive marginal ones out of existence;

(e) an effort by proponents of the apprenticeship system to set standards unrealistically high so as to close down proprietary schools and return to themselves the income from student fees and labor.[40]

[38.] *Ibid.*, ch. IV.

[39] Rothstein, *American Physicians*, 63–80, ch. V; Kett, *Formation of the American Medical Profession*, ch. II; Gerard W. Gawalt, "Massachusetts Legal Education in Transition, 1766–1840," *American Journal of Legal History*, 17 (Jan. 1973), 27–50.

[40] Rothstein, *American Physicians*, ch. VI; Kett, *Formation of the American Medical Profession*, ch. VI.

Burdened with such a fruitful source of discord, the community of the competent was not only divided against itself, but each individual member might well be divided in his own mind as to the true nature of his own motives. Jealousy and distrust were rampant as the professions broke ranks and reformed in new patterns time after time over the course of the century. The problems were the perennial ones faced by anyone trying to sustain a community of the competent: How to identify competence? How to cultivate excellence? How to suppress idiosyncrasy without encouraging mediocrity? How to find a scale of organization that could plausibly represent the whole profession, yet be intimate enough to avoid factional fragmentation?[41] In a context of explosive growth and intensifying market orientation, these perennial problems seemed almost overwhelming. Even substantive progress could be demoralizing. When regular physicians began to abandon the misguided "heroic therapy" at midcentury, how could a doctor admit even to himself that the despised irregular practitioners of the 1830's—the Thomsonians, for example, with their harmless steam and herbs—had probably posed a smaller threat to the life of the patient than the regular practitioner?[42]

With morale slack and internal cohesion at a low ebb, the professions were confronted in the 1830's with what seemed to be a howling mob bent upon the destruction of every stronghold of aristocratic privilege. In state after state popular legislatures pulled down the barriers to professional practice, disestablished the professions, and let it be known in no uncertain terms that the boundary between authentic wisdom and charlatanism had evaporated. The results, to be sure, were by no means a disaster for the professions. The collapse of old licensing mechanisms often paved the way for new and more effective ones, but professional men could not ignore the popular contempt for expertise that accompanied this creative destruction.[43] It was the drawn-out strain of adaptation to a new social order, plus the shock of this popular attack, that provoked a whole generation of genteel professional men to embark upon a general defense of authority. They felt, as did George Shattuck of Boston in 1866, that they were living "in an age and a community where authority is misunderstood and set at nought, where the individual feels himself called upon to treat lightly the conclusions and experience of the past, and to investigate and decide anew on most important questions and interests."[44]

[41.] Calhoun, *Professional Lives*, ch. II.
[42.] Rothstein, *American Physicians*, 62.
[43.] Kett, *Formation of the American Medical Profession*, 165; Gawalt, "Massachusetts Legal Education in Transition," 49.
[44.] Kett, *Formation of the American Medical Profession*, 166.

But the reasons customarily given for the crisis of professional authority are incomplete. The anxiety that professional men like Shattuck felt at midcentury had roots deeper than the internal squabbles over licensing or the external hostility of a populistic public. The disarray of the professions was caused not only by the breakdown of ethical codes and institutions in an age of expansion, but also by the changing conditions of explanation in a society well on its way toward interdependence. The intellectual foundations of professional life were changing no less profoundly than the social and economic foundations. When men like Shattuck complained of a lack of public deference, they referred not only to lost prestige and bad public relations, but also to a change in the intellectual basis for the professional role itself.

The growing interdependence of society and the consequent recession of causation and devitalization of milieu discussed in the second chapter had profound implications for the professional, because his traditional social function was to ascertain causation and to mediate between the island community and the outside world. Esoteric questions of causal attribution had always been the métier of the professional. Even in a pre-industrial society of isolated communities, there had been a need for someone to deal with supra-local phenomena, or phenomena that would not yield to common sense and the customary ways of the community. The professional filled this need.[45]

Cosmopolitanism was inherent in the professional role. The lawyer had always mediated between the private, local interests of his clients and the demands of universal justice as manifested in laws promulgated by the sovereign central power of the state. The medical doctor possessed expert knowledge based, however imperfectly, on the universal traits of the human organism. The minister, the third of the three classic professional specialties and father of them all, was the original cosmopolite, mediating between his flock and the whole cosmos. Questions of causal attribution were central to the life of each of the professions: guilt and liability, diagnosis and prescription, the efficacy of works and sacraments, free will, and predestination all involved critical problems of judging the location of genuine causation.

[45.] Although the professional man was recognized as having the strongest claim to knowledge of supra-local affairs, there were never enough doctors or lawyers in colonial America. In consequence, ministers and domestic practitioners often supplied basic medical services, and in New England, at least, theology often shaped legal proceedings more than the science of the law. One recent study finds only fifteen trained lawyers in all of Massachusetts in 1740—one lawyer for every 10,000 persons. By 1820 the ratio had leveled off at about one to 1,100, and there were about 600 lawyers in the state. See Gawalt, "Sources of Anti-Lawyer Sentiment in Massachusetts," 285; Kett, *Formation of the American Medical Profession*, 1–9.

Since all professional men dealt in causation, all were potentially threatened by its recession. Their traditional role was to mediate between the island community and the world beyond it, so the absorption of the community into an ever denser network of social connections held profound implications for their work. Because their task was to advise laymen in matters too problematical for common sense and custom, the explosive multiplication of problematical relationships in an interdependent society compelled them to defend and reconstruct the basis of their authority. Of course, many aspects of professional life did not change in the least as the island community was penetrated by lines of influence originating outside its borders: a case of murder was still a case of murder, pneumonia still pneumonia. But all professional men were affected by changing habits of causal attribution and changing standards of plausibility among the educated lay public.

The impact on professional life of changes in habitual modes of explanation and causal attribution can be illustrated by three examples, one from each of the classical professional specialties of medicine, divinity, and law. The words of Dr. Walter Channing, an alienist active in the ASSA's Health Department, supply an especially straightforward example of the recession of causation. Dr. Channing complained in 1883 that the "causes" of insanity commonly listed in insane asylum reports did not "touch bottom" or get at the "true source" of the problem:

> Therefore, when we take up an insane hospital report and see such and such a number of cases ascribed to ill-health, intemperance, business cares, family affliction, domestic worry, and a hundred other circumstances of minor importance (a proportion of them even being ludicrous in their nature), we see at once that these things do not represent the sum total of all the elements going to make up what we might call the associated cause; but they separately represent only a single link in a chain of causes, or perhaps the last cause leading to the outbreak of the attack.[46]

If doctors were beginning to feel that the customary explanations of insanity were hollow and inadequate, it was only because an entire image of human nature based on superficial causal attribution was on the verge of collapse. No profession was more threatened by that collapse than the ministry. The nation's best-known clergyman, the Reverend Henry Ward Beecher, abjectly confessed that the minister's need for knowledge of human nature was so pressing, and his actual grasp of it so paltry, that only drastic reform of the theology cur-

---

[46.] Walter Channing, "A Consideration of the Causes of Insanity," *Journal of Social Science*, 18 (May 1884), 68–69.

riculum could save the profession from approaching disaster. In an address to the theological students of Yale College in 1872, Beecher said:

> I think that our profession is in danger, and in great danger, of going under, and of working effectively only among the relatively less informed and intelligent of the community ... We are in danger of having the intelligent part of society go past us. The study of human nature is not going to be left in the hands of the church or the ministry. It is going to be part of every liberal education, and will be pursued on a scientific basis.[47]

Beecher recognized that laymen had once consulted ministers when they wanted to know about the nature of man, but now he saw that the subject was being taken over by outsiders. He did not dare hope that the church could reclaim the subject, but he did hope that at least ministers would keep up with laymen in their practical grasp of it. Indeed, as Beecher grudgingly admitted, by adding the works of Bain, Spencer, and other scientific analysts to their training curriculum, ministers might not only keep up with their flocks, but even learn to shepherd them more expertly. In Beecher's own words, knowledge of human nature was part of a "science of management" that ministers needed as a tool of the trade: "You must know what men are in order to reach them, and that is part of the science of preaching. If there is any profession in the world that can afford to be without this practical knowledge of human nature, it certainly is not the profession of a preacher."[48]

Underlying Beecher's anxiety about knowledge of human nature was an issue of causal attribution: how is human behavior to be explained? Where do its effective causes lie? In the conscious, intending mind, or in some more remote location? Since he was aware of the threat that positivistic conceptions posed to a traditional Christian view of man, Beecher recommended Bain and Spencer cautiously, with the reservation that their work left much to be desired. But there were no reservations in the mind of E. L. Youmans, an enthusiast for Spencerian positivism who gave Beecher's confession wide circulation in his journal, *Popular Science Monthly*. "Human beings," said Youmans, "should be studied exactly as minerals and plants are studied, with the simple purpose of tracing out the laws and relations of the phenomena they present. Men should be analyzed to their last constituents,

---

[47] Henry Ward Beecher, "The Study of Human Nature," *Popular Science Monthly*, 1 (July 1872), 327–328, 330.

[48] *Ibid.*, 332, 333, 335. Beecher singled out phrenology as an eminently useful kind of knowledge about human nature, because it enabled him to predict the character of his parishioners simply by observing the shapes of their heads.

physiological and mental."⁴⁹ What Beecher fearfully suspected, Youmans glibly proclaimed: causation in human affairs lay outside the conscious willing mind of the individual.

Lawyers too felt obliged to reach beyond customary objects of causal attribution to a deeper and more complex level of causation. Nicholas St. John Green won himself a position on the Harvard law faculty in 1870 with an article attacking the conventional jurisprudential notion of causation. Green, who helped William James and Charles Peirce hammer out the basic ideas of pragmatism in the discussions of the "Metaphysical Club," urged lawyers to abandon the futile distinction between "proximate" and "remote" causation in liability proceedings. Instead of undertaking that formalistic legal ritual, said Green, the courts ought to ascertain liability according to the degree of certainty or uncertainty with which a given effect might be expected to follow from a given cause—a question to be decided on the basis of past human experience and probability of recurrence, rather than the "proximity" of cause to effect. "The term proximate and the term remote have no clear, distinct, and definable signification," wrote Green. Contrary to the simplistic assumptions of contemporary legal practice, an effect has potentially as many causes as it has connections to an interdependent environment:

> In as many different ways as we view an effect, so many different causes, as the word is generally used, can we find for it. The true, the entire, cause is none of these separate causes taken singly, but all of them taken together. These separate causes are not causes which stand to each other in the relation of proximate and remote, in any intelligible sense in which these words can be used.... They are rather mutually interwoven with themselves and the effect, as the meshes of a net are interwoven. As the existence of each adjoining mesh of the net is necessary for the existence of any particular mesh, so the presence of each and every surrounding circumstance, which, taken by itself, we call a cause, is necessary for the production of the effect.⁵⁰

These examples suggest the kinds of concrete alterations of professional practice that were engendered by life in an interdependent society. Beecher saw that the ministry was threatened by its adherence to a superficial conception of human nature, one that failed to take into account all the determinants of behavior. Channing complained that doctors too often attributed insanity to "causes" that were really last links in long, complex chains of causation. Green went all the way to the heart of the matter by showing that centuries of jurisprudence had erred in failing to recognize that all events in the universe of human

---

⁴⁹· *Ibid.*, 366.
⁵⁰· Nicholas St. John Green, "Proximate and Remote Cause," *Boston Law Review*, 4 (Jan. 1870), 214, 212. See also Wiener, *Evolution and the Founders of Pragmatism*, 232.

affairs were enmeshed in a dynamic flux of interdependent relations. In each case existing professional practices and explanations were rendered implausible by a new sensitivity to complex causation, a new reluctance to attribute autonomous causal potency to any particular thing, event, or person.

The anxiety about authority felt by professional men, who constituted the bone and sinew of the college-educated gentry (in New England, at least), gave a characteristic temper and cast of mind to a whole generation of genteel intellectuals. That temper was embodied in the American Social Science Association. The definitive task of Victorian culture in the United States, as John Higham observes, was to elaborate "a framework of order within and—so to speak—around American individualism."[51] By midcentury the spirit of ebullient boundlessness that characterized the 1830's was being visibly superseded by a somber, cautious mood of consolidation. Emerson's bold ethic of "Self-Reliance" seemed glib, a frail romantic gesture in a rising torrent of institutional proliferation. As we have seen, even Frank Sanborn, young, radical, and vibrant with the antinomian impulses of Concord, conceived of the future in institutional terms and searched for ways to confer authority on those who most deserved it.[52]

Professional men seeking a firmer base for their authority contributed greatly to this shift of cultural mood away from boundlessness toward consolidation. They had every reason to reject the frothy optimism of the Jacksonian era. Their distinctive claim to esoteric knowledge was rooted not in intuition, but in training. Far from being antinomian, they were advocates of the law; far from being anti-institutionalists, they were expressly dependent on institutions for their authority and professional identity. How could radical individualism appeal to a community of the competent? Cohesion, not autonomy, was the path to authenticity. What place was there for the Romantic temperament in a disciplined community committed to consensus as the test of validity? The professional man's whole historical and occupational situation inclined him toward consolidation. In an increasingly interdependent society, his values emerged triumphant.

ASSA: "Headquarters" for the Movement

In the years after Appomattox the ASSA provided the only central meeting place for those who were establishing authority in all the diverse areas made problematical by the growth of an urban-industrial society. Sitting side by side at the meetings of the Social Science Associa-

---

[51] Higham, *From Boundlessness to Consolidation*, 26.
[52] See ch. III, above.

tion one might find all three of the men who contributed most to the construction of the modern American university—Gilman of Hopkins, Eliot of Harvard, and Andrew Dickson White of Cornell. Down the aisle one might find E. L. Godkin, editor of the *Nation*, or Judge Simeon E. Baldwin, the man who organized the American Bar Association. Also in attendance one might expect to see John Eaton or William Torrey Harris, the nation's leading reformers of secondary education before John Dewey.

Forerunners of professional social work, eager to create a more "scientific" philanthropy, thought it natural to organize the National Conference of Charities and Correction in close connection with the ASSA. When doctors wanted to expand their profession's competence to deal with the health problems of massive urban populations, they organized the American Public Health Association in conjunction with an ASSA meeting. Civil service reform, the most ambitious attempt to establish authority and impose the merit system on American life, was launched as a movement by the ASSA. When a younger generation of social thinkers took up careers in the new university system that Gilman, Eliot, and White were building, they found the ASSA inadequate to their professional needs. However, many of them still used the ASSA as a temporary base of operations or organized their own specialized associations under its protective wing.

The final and saddest phase of the movement to establish authority was played out after the turn of the century in one of the ASSA's many offspring, the National Institute of Arts and Letters. The Institute originally hoped to organize all the "best men" of the society, in whatever field. In conjunction with the American Academy of Arts and Letters which evolved from it, the Institute was headquarters for the "custodians of culture" who defended the genteel tradition in its last hours.[53]

The American Social Science Association was the institutional focal point of the movement to establish authority, not only in the obvious sense that it provided participants with a place and the occasion for meeting together, but also because the diverse threads of the movement naturally converged in the Association's conception of "social science." To twentieth-century eyes the conglomeration of interests and activities that the ASSA tried to include under the rubric "social science" is a discordant hodgepodge. But contemporaries were not wrong to think that there was a common denominator. Each of those diverse interests and activities represented an effort to establish authority in some area of human affairs rendered newly problematical by the consolidating tendencies of an interdependent society.

[53] May, *End of American Innocence*, 78–79; Higham *et al.*, *History*, 9.

"Social science" was understood by ASSA members to refer to the whole realm of problematical relationships in human affairs, those relationships that were unfamiliar to common sense and alien to customary knowledge. One became a "social scientist" by contributing to this store of esoteric knowledge and practical expertise. New ventilation or drainage techniques for the city dweller; new legal forms for the industrial corporation; a new theory of rent or prices; a new way to care for the insane or to administer charity — all of these were equally valuable contributions to "social science." Only by accumulating and systematizing such knowledge could the alarming new social tendencies of the age be confronted and made humanly tolerable.

The movement to establish authority required that there be a science of society. Precisely because society proceeded according to discoverable scientific law, decisions could not be left in the hands of the incompetent. Precisely because there were truths that no honest investigator could deny, the power to make decisions had to be placed in the hands of experts whose authority rested on special knowledge rather than raw self-assertiveness, or party patronage, or a majority vote of the incompetent.

By the same token, "social science" could be viewed prospectively as the long-range outcome of the movement to establish authority. It was the anticipated summation that someday would integrate all the seemingly separate activities and interests represented at an ASSA meeting. It was to be a systematic synthesis of all knowledge of human affairs that transcended common sense and custom. "Social science" was, in other words, a synthesis of all the esoteric matters that had customarily been the province of the professional man.

Our analysis has now proceeded far enough to show that the ASSA is best understood as part of a movement to defend the authority of a gentry class, whose sturdiest foundations lay in the professions. The fundamental animus of the Association is most clearly revealed in its departmental structure, which conformed roughly to the tripartite division of labor of the classical professional class. The Association's organization appears peculiar and cumbersome from the vantage point of what is today called social science, but it appeared self-evidently natural to men whose conception of "social science" evolved in the context of the movement to establish authority.

It was natural because what Frank Sanborn and his ASSA colleagues wanted to do was simply expand upon what professional men had always done: supply the public with authoritative guidance in those areas of life that are so problematical that laymen cannot or should not bear the risk of decision alone. Over the course of the nineteenth century, as society became increasingly interdependent, the realm of the problematical expanded at a geometric rate, with custom and common sense failing accordingly. In their confusion laymen naturally turned first to the professional man, the advisor who was supposed to take up where common sense and custom left off. Consequently, in midcentury a traditionally organized professional class of lawyers, doctors, and ministers was compelled to address itself to a growing range of esoteric problems that really no longer fit within the traditional spheres of law, medicine, and divinity.

When laymen continued to cast the professional class in the traditional role of all-competent advisor even after that role had become impossible to fill, the inevitable result was frustration for professional men and disappointment for the lay public. The dilemma is exemplified in the 1851 complaint of a Louisiana physician that his neighbors, being planters and farmers, expected him to be knowledgeable about such non-medical matters as soil mechanics and crop rotation:

> and if the physician to whom they naturally look, and to whom they apply for correct information is unable to give it, the reflection occurs to the planter, that, after all, the doctor does not know much more than other men, notwithstanding his collegiate education, his diploma, and his high pretensions. His position as a man of scientific knowledge is necessarily forfeited, and the confidence once implicitly placed in him is greatly impaired if not entirely withdrawn.[54]

In absolute terms doctors, lawyers, and ministers were at least as competent in 1900 as in 1800, and probably more so. But they had lost the old image of omnicompetence and had been relegated to the subordinate status of specialists, capable of dealing with but a small fragment of the esoteric. Their initial response to this growing disjunction of role and performance was not resignation, but a campaign to establish professional authority on a firmer base and to extend professional performance into new areas. That stronger base for professional authority is what some of the more thoughtful professional men called "social science," and the ASSA was a major vehicle of their campaign.

Whose Authority?

What was sound opinion? The men who set out to establish authority thought they knew the answer to that question. They took it for granted that the communities of inquiry they constructed or reinforced would defend their own version of the truth and support their own authority. Many of the gentry who came to fashionable Saratoga Springs for social science meetings came primarily to enjoy each other's

---

[54.] Quoted by Kett, *Formation of the American Medical Profession*, 167–168.

distinguished company and to wax indignant over the frenetic society that spurned their offer of leadership. For them the movement to establish authority was essentially a parochial defense of class, an attempt to find a new basis for deference in a society no longer deferential.

But the nineteenth-century search for the foundations of authority was motivated by far more than selfishness, and it served purposes that transcended the interest of any particular class. With ironic regularity, the movement to establish authority rose above even its most selfish intentions to defend not class interest, but *authority*, as a general principle. The most striking feature of the institutional devices created by the movement is their impersonal neutrality. The creators of the new communities of the competent often hoped by that means to establish their own personal authority, or at least to safeguard their own values and their own version of the truth. But the institutional structures that these men built to cultivate and confer authority proved remarkably immune to permanent capture by any conspicuously parochial faith.

This ironic outcome was determined in part by the changing conditions which provoked the movement in the first place, and also, in part, by the internal competitive dynamics of the community of the competent. To be credible and persuasive in an interdependent universe of human affairs, the individual claimant to esoteric knowledge had to shed his idiosyncrasies and adopt a disciplined place within the whole community of practitioners. Likewise, the community itself, in order to be credible, had to avoid any appearance of collective idiosyncrasy or "partisanship." It was imperative that *all* the competent be embraced by the community, and that all be permitted to compete on a formally equal basis. No person or point of view could be privileged; all had to submit to the potential tyranny of the majority of the competent. Whatever pecking order emerged among the members of the community had to be understood as a product of pure merit, spontaneously recognized by equal competitors.

In other words, the community of the competent had to identify competence, cultivate it, and confer authority on those who possessed it in accordance with universalistic criteria—or, more realistically, criteria that were not in any obvious way personal, partisan, or particular. The criteria of judgment had to seem truly a product of consensus among the competent, beyond the power of any individual, clique, or party to control, and hence impersonal, objective, value-free—not mere opinion but "truth." Otherwise the communalization of inquiry would not serve its original purpose of safeguarding sound opinion and enhancing individual authority. It goes without saying that no community of the competent can be totally impersonal, neutral, or devoid of value commitments. But the strain toward impersonal neutrality must be powerful if the community's essential function is to be served. As the Lazzaroni discovered, such communities cannot be easily controlled, even by their founders. Those of the genteel professional class who built the ASSA found no lasting refuge in it, or in the other institutions they constructed to defend authority.

CHAPTER V

The Founding and Formative Years of the ASSA, 1865-69

As the movement to establish authority gained momentum in the 1860's, it naturally turned to politics and expressed itself in two principal forms: the investigatory commission and civil service reform. The ASSA was intimately involved with both developments. As the closest thing to a headquarters that the movement to establish authority ever had, it is only appropriate that the Association should have been created in the first place as an adjunct to the nation's first successful state commission, the Massachusetts Board of Charities, and that it should itself have launched the civil service reform movement.

Both the commission that gave birth to the ASSA and the reform movement that the ASSA itself sponsored held strong appeal for the gentry defenders of authority. Both measures set definite limits to the rule of majorities and party machines, yet did not raise a direct challenge to democratic principles. The members of the commissions were presumed to be right-minded, disinterested men, with special competence to oversee whatever area was entrusted to their commission. They were not elected, but selected for their knowledge and experience. They were, in short, experts, though there seldom existed any professional bodies to certify them as such. The commission was thus a means of introducing an element of expertise—authority based on disinterested knowledge rather than a majority vote—into what remained a basically democratic framework of government. Civil service reform applied an essentially similar idea to a broader spectrum of government offices, substituting institutionally certified competence for party loyalty as the proper criterion of government employment. Both measures tended to regularize competition and put authority in the hands of those who, by an impersonal standard, most deserved it.

The founding of the ASSA is the main subject of this chapter, but it properly begins with the Massachusetts Board of Charities and ends with the civil service reform movement. These two developments shed light on a central question: What were the motives and fundamental assumptions of the founders of the American Social Science Association?

The Massachusetts State Board of Charities

Frank Sanborn and Samuel Gridley Howe evidenced their hostility to established institutions during their association with John Brown, but their aversion to institutional power disappeared overnight when the triumph of the Republican party made them "ins" instead of "outs." Their opportunity came when John Albion Andrew, a fellow member of the "Bird Club" and legal counsel for the "secret six" after Harpers Ferry, became governor of Massachusetts. In 1862 Andrew invited Howe to formulate plans for the reorganization of the state's charities. The outcome was the Massachusetts State Board of Charities, a "major landmark," as Gerald N. Grob says, in the centralization and rationalization of welfare functions.[1] Indeed, the Massachusetts Board was a landmark of the bureaucratic phenomenon in general, for it not only inspired the creation of similar boards of charity in other states, but also spread the commission idea to other spheres of action such as health, railroad regulation, and the collection of labor statistics.[2] In these areas, too, Massachusetts became a model for the rest of the nation.

The Massachusetts Board of Charities came into existence as a direct result of the breakdown of the island community. The Massachusetts welfare system was built on the ancient principle of "settlement" whereby each village was held responsible for the support of its own dependents, be they poor, crippled, or insane.[3] The "wandering poor," who had no real homes, had always posed a problem for this decentralized relief system, but it worked well enough to survive until the 1840's. In that decade and the next it finally collapsed under the weight of massive immigration, mainly Irish, through the port of Boston. New York faced a similar crisis provoked by immigration. Massachusetts created a Board of Alien Commissioners in 1851 to deal with the problem, but the board proved ineffectual. As welfare expenses mounted and ethnic tensions heightened, political pressure developed behind a move to rationalize and centralize the system. A legislative committee in 1858 was the first to propose a central board to oversee all the state's charitable institutions, and Howe renewed the committee's recommendation in December 1862. What is needed, he said in his

---

[1.] Gerald N. Grob, *Mental Institutions in America: Social Policy to 1875* (New York: Free Press, 1973), 276.
[2.] *Ibid.*, 281.
[3.] Robert W. Kelso, *The History of Public Poor Relief in Massachusetts, 1620–1920* (1922; reprint ed., Montclair, N.J.: Patterson-Smith, 1969), 27, chs. VI and VII.

response to Governor Andrew, is "a board or central commission whose duty it shall be to collect and diffuse knowledge, to prevent abuses, to protect the rights of paupers, and to establish as far as may be a uniform and wise system of treatment of pauperism over the Commonwealth."[4]

A board to supervise charitable agencies no doubt seemed particularly attractive at the time because of the success of the U.S. Sanitary Commission, of which Howe was then a member and to which he was then devoting much of his energy. The Sanitary Commission concerned itself with all aspects of troop health and happiness, performing a combination of functions now borne by the regular Army Medical Corps and the Red Cross. Though a temporary wartime measure, it was the most ambitious charitable undertaking the nation had ever seen and a notable experiment in large-scale charity organization.

The vice-president of the Sanitary Commission was none other than Alexander Dallas Bache, "chief" of the Lazzaroni. Wolcott Gibbs, another member of the Lazzaroni, was also on the commission, and under its authority a third member of the group, Benjamin A. Gould, conducted important anthropometric research.[5] But the major task of the commission was to compile reports on the health and mortality of the troops while its branch offices and ladies' auxiliaries, eventually more than 7,000 in number, raised vast sums of money and shipped food and clothing to the front. So much did they ship, in fact, that Howe soon begged them to stop for fear the troops would be spoiled. Typically, he insisted on the principle of self-reliance. After the war the Sanitary Commission was often cited as proof that public relief could best be achieved by private rather than governmental agencies, but Howe would have preferred that the government supervise sanitary conditions directly, without resorting to voluntary private aid.[6]

[4.] *Ibid.*, 135-142, and Grob, *Mental Institutions*, 272-276. For the consequences of the breakdown of the old welfare system in one town near Boston, see Stephan Thernstrom, *Poverty and Progress: Social Mobility in a Nineteenth Century City* (New York: Atheneum, 1970). Quotation from *Letters and Journals of Samuel Gridley Howe*, ed. Laura E. Howe Richards (Boston: Dana Estes, 1909), II, 511.

[5.] John S. Haller, "Civil War Anthropometry: The Makings of a Racial Ideology," *Civil War History*, 16 (Dec. 1970), 309-324.

[6.] Harold Schwartz, *Samuel Gridley Howe: Social Reformer, 1801-1876* (Cambridge: Harvard University Press, 1956), 252-256; Richards, *Letters and Journals*, II, 479-498. On the significance of the Commission as a model for later social work, see Fredrickson, *Inner Civil War*, ch. VII. Fredrickson may exaggerate the contrast between Howe and his supposedly more hard-hearted colleagues on the Commission. As an old trooper who had subsisted on boiled sorrel and raw snails in Greece, Howe was appalled at the luxuries Massachusetts men carried to the front. Like the other members of the Commission, he had little respect for generous sentiments *per se;* he shared with them an admiration for sentiment carefully harnessed to reality by appropriate instruments. See Richards, *Letters and Journals*, II, 484-485, 498.

In his letter to the governor in which he proposed a permanent state charity commission, Howe had cautiously suggested that "such a board might have more power of interference in the local administration of poor laws than the spirit of municipal independence would probably permit."[7] The Board that actually went into operation on 1 October 1863 had only advisory powers in matters of administration, but it was given effective control over the transfer and distribution of dependent persons and was thus in a position to shape public opinion decisively. The Board's functions were defined as follows:

> They shall investigate and supervise the whole system of the public charitable and correctional institutions of the Commonwealth, and shall recommend such changes and additional provisions as they may deem necessary for their economical and efficient administration. They shall have full power to transfer pauper inmates, but shall have no power to make purchases for the various institutions. They shall receive no compensation for their services except their actual travelling expenses, which shall be allowed and paid.[8]

Other states in the Northeast and Middle West soon followed the example of Massachusetts. By 1874 eight other states had established such boards: Connecticut, New York, Rhode Island, Pennsylvania, Michigan, Wisconsin, Kansas, and Illinois. The explicit aim of the boards was to improve the efficiency and economy of public welfare. In concrete terms, that often meant that they were also civil service reform measures, designed specifically to divorce the administration of the growing welfare complex from partisan politics.[9]

The various state charity boards differed in structure and responsibility. Most of the boards initially covered the four fields of health, penology, mental disease, and poverty, but their responsibilities changed over the years. In Massachusetts, health was assigned to a separate board in 1869; in 1875 care of the deaf and blind was transferred to the Board of Education; in 1879 responsibility for delinquent children was transferred to a new Board of Commissioners of Prisons, the Board of Health was abolished and the Board of Charities was renamed the State Board of Health, Lunacy and Charity; in 1886 the name changed again as a separate Board of Health was reestablished; and in 1898 it was renamed yet again when insanity was shifted to a separate board. The Massachusetts Board was of the supervisory type,

---

[7] Richards, *Letters and Journals*, 511.

[8] Massachusetts, *First Annual Report of the Board of State Charities* (Boston: Wright and Potter, 1865), iv.

[9] Frank Bruno, *Trends in Social Work, 1874–1956: A History Based on the Proceedings of the National Conference of Social Work* (New York: Columbia University Press, 1957), 3, 32, 39.

with each institution under its supervision retaining its own administrative staff and board of trustees and a considerable degree of autonomy. Of the early boards, only that of tiny Rhode Island was placed in direct administrative control of the state's charities.[10]

As an instrument of reform, the state charity board was an idea congenial to men whose basic assumptions about man and society conformed to the individualistic tenets of classical liberalism. Since precisely the same assumptions dominated the thinking of the ASSA's members, they require our close attention. To be sure, in the minds of Howe, Sanborn, and other reform leaders, the liberal framework was strained to the bursting point by a vigorous humanitarian impulse on the one hand, and a conservative paternalism on the other. Both of these impulses were hostile to the doctrine of laissez-faire. But both their humanitarianism and their paternalism were controlled by the assumption that individual men were, in all but exceptional cases, masters of their own fate and therefore responsible for their own situation in society—be it high or low. Society, in the eyes of the reformers, was not a monolithic influence that shaped individuals to its needs and broke them at a whim, but an ephemeral aggregate of individual wills. To reform society, therefore, one had to reform individual men, not vice versa.

In retrospect, the charity board seems a paltry measure, hardly capable of getting at the roots of poverty and degradation in an industrializing society. At its most ambitious the Massachusetts Board was never expected to do more than publicize abuses and coordinate a decentralized system of almshouses, hospitals, and ad hoc relief.

But to contemporaries the Board of Charity seemed a bold instrument of reform, one capable of reaching all the way down to the root causes of dependency—causes which did not appear to lie very deep. In fact, contemporaries perceived causation as lying so close to the surface that the legislative committee of 1858 hoped a central commission might even solve the social problem once and for all, permitting the state to close down its welfare system and return any residual problems to the local community![11]

Samuel Gridley Howe, who became the guiding spirit of the Massachusetts Board and author of its most notable *Reports*, was not so naive as to suppose that the Board could uproot the problems of dependency at one stroke. But he too attributed causal potency to the recipients of welfare in a degree that jars twentieth-century sensibilities. We have no better index of the changing locus of perceived causation in human affairs than the attitudes of men like Howe toward

[10] *Ibid.*, 33–42; Kelso, *Public Poor Relief,* 153–154; Grob, *Mental Institutions,* 279–280.
[11] Grob, *Mental Institutions,* 273, 280.

social problems. Though radical humanitarians in their own time, they seem "callous" and "insensitive" to twentieth-century eyes—not because they were inferior in moral insight, but because they worked within a different moral *system*, based on different habits of causal attribution. Likewise, their reform instruments seem blunt and superficial today, not because they were hypocritical or stupid, but because they perceived the causes of social problems to lie close to the surface of events and therefore to require no very elaborate instruments of reform.

Howe's letter to Governor Andrew proposing the creation of the State Charity Board exemplifies the attitudes of the men and women who founded the Social Science Association. The curious blend of humane sympathy and haughty condescension that one finds in Howe reflects the persistence of the assumption that the individual is the primary, even if not the entire, cause of his own place in life.

Howe recognized the continuing need for a state-supervised welfare system, but he preferred to organize that system around small almshouses in each town, rather than around large state institutions:

> The place for the poor in a Christian community is in the home of those who are not poor. As we cannot have that arrangement, we should give them a house and home in the *midst of our own and as near like our own as may be. We should not sever* social ties, be they ever so feeble, nor break local attachments, but let the poor live on where the lines have fallen to them. This can be done without violating the plain maxim that we must not make the status of the pauper a desirable one.[12]

Throughout his career Howe's reform strategy remained essentially constant: the handicapped or dependent or deviant person was to cultivate every resource of self-reliance and thus be brought into a full range of associations with normal people. Howe's greatest achievement as a young reformer had been to show that the handicapped person often suffered more from ostracism than from any physical disability. His most famous triumph was the case of Laura Bridgman, a deaf and blind child whose seeming idiocy gave way to lively intelligence when Howe taught her to communicate. Teaching the techniques of self-mastery was the key to his reforms.[13]

[12.] Richards, *Letters and Journals*, II, 512. Italics in the original.

[13.] Schwartz, *Samuel Gridley Howe*, ch. VI. Howe allied himself with Dorothea Dix in her battles for humanitarian care of the insane, and he also supported various reforms of the penal system. In the 1840's, with Charles Sumner and Horace Mann, he advocated the seemingly illiberal "separate" or solitary prison system, as opposed to the "congregate" plan. The latter system required convicts to maintain silence, but at least allowed them to work together. Howe's system prevented any conversation or association between prisoners. It was the one exception to his usual reform strategy of breaking down barriers to association. His preference for the solitary system stemmed from his own

Howe's letter to the governor stressed the importance of improving the collection of factual data, so as to reveal the different conditions which encouraged and discouraged pauperism. He noted that most of the state's paupers were foreigners. "We have this foreign element among us; we cannot get rid of it if we would; and we should strive to fuse it into our common nationality as fast as possible. We strengthen our state by homogeneity; we weaken it by the contrary course."

Howe of course believed that there were political and moral rights of which no man should be deprived; more than that, he also argued that there was a *"right* to work . . . which is apt to be disregarded." He urged the "necessity of having all questions and all systems looked at a little from the *pauper stand-point."* But he meant this only in a limited sense that underlines the outer boundaries of mid-nineteenth-century humanitarianism. "The pauper is to be legislated for and about, and he is to be disposed of and treated as seems best for the class above pauperism; and this should be so mainly, but not entirely."[14]

Howe was a generous man, but he was not essentially different from the typical charity board member who believed, as Gerald Grob says, that "dependency was largely self-inflicted."[15] To men only beginning to adapt their thought to the condition of interdependence, it did not seem contradictory to see the individual as the nearly independent cause of his own fate—even of his own lack of independence.

Founding the ASSA

In August 1865, six months after Appomattox, Frank Sanborn on behalf of the Massachusetts Board of Charities drew up an invitation to selected persons all over the nation to participate in the founding of an organization devoted to social science inquiry. The organizational meeting was scheduled for 4 October 1865 at the State House in Boston. Sanborn described the objects of the proposed organization as follows:

> Dear Sirs,—Our attention has lately been called to the importance of some organization in the United States, both local and national, whose

incarceration in an Austrian prison, which he said would have driven him insane had he been forced to live with the other prisoners. For juvenile offenders, Howe in his later years recommended a very different system, in which offenders would not even be imprisoned but would be entrusted to the care of selected families in the community. Somewhat similar was his advocacy of family care for the insane, modeled on the town of Gheel, Belgium, where patients and townspeople mingled at the same work. See *ibid.*, 147-149, 273-274, 286-287, and Richards, *Letters and Journals,* II, 532-535. On prison reform see Rothman, *Discovery of the Asylum.*

[14.] Richards, *Letters and Journals,* II, 511, 512. Italics in the original.

[15.] Grob, *Mental Institutions,* 287. See also Grob's discussion of Howe's famous *Second Annual Report of the Massachusetts Board of State Charities,* 284-285.

object shall be the discussion of those questions relating to the Sanitary Condition of the People, the Relief, Employment, and Education of the Poor, the Prevention of Crime, the Amelioration of the Criminal Law, the Discipline of Prisons, the Remedial Treatment of the Insane, and those numerous matters of statistical and philanthropic interest which are included under the general head of "Social Science."[16]

Though the October organizational meeting was sponsored by the Board of Charities, it was convened by an independent group of Boston citizens who had taken steps several months earlier to ally themselves with the British National Association for the Promotion of Social Science. The British Association, basically Benthamite in its social views, served as a model and inspiration for the ASSA; however, no firm links were ever formed between the two organizations. Established in 1857 under the symbolic leadership of octogenarian reformer Lord Brougham, the British Association was larger and politically more powerful than its American counterpart would ever become. It was an umbrella organization, encompassing many specialized and local reform groups, which presented itself to the public as a hybrid, combining the roles of lobbyist and expert advisor to Parliament. The British Association, like the American, set out to air important issues before the public, but, unlike the American Association, it often was able to reduce the issue to a draft bill or Order in Council and see it acted upon. In 1880, when the British Association was past its peak of strength, its Council included thirty-one Peers of the Realm, forty-eight MP's, and many other eminent political figures. The American Association would never stand so close to the throne.[17]

Among the nearly one hundred people who attended the inaugural meeting of the ASSA in the Boston State House were many notables in reform and social inquiry. Dr. Edward Jarvis, pioneering statistician and student of insanity, called the meeting to order. Jarvis had been a close associate of Lemuel Shattuck and others in the founding of the American Statistical Society in 1839, and he served as president of that organization from 1852 to 1882.[18] Governor John A. Andrew was

---

[16.] ASSA, *Constitution, Address, and List of Members of the American Association for the Promotion of Social Science* (Boston: Wright and Potter, 1866), 10. For Sanborn's authorship, see his essay "The Work of Twenty-five Years," *Journal of Social Science*, 27 (Oct. 1890), xiii. The abbreviation *JSS* will henceforth be used for the *Journal of Social Science*.

[17.] Abrams, *Origins of British Sociology*, 45. In ch. IV, "Ameliorism," Abrams ranks the NAPSS quite high in contemporary influence and historical significance. For a less enthusiastic evaluation, see Lawrence Ritt, "The Victorian Conscience in Action: The National Association for the Promotion of Social Science, 1857–1886" (Ph.D. dissertation, Columbia University, 1959).

[18.] In 1840 the Society changed its name to the American Statistical Association. The Royal Statistical Society of London was founded in 1834. See Struik, *Origins of American Science (New England)*, 228–233; and "Address of Carroll D. Wright, President of the American Statistical Association, at its annual meeting in Boston, January 17, 1908," mss. in C. D. Wright Papers, Clark University.

promptly chosen by acclamation as chairman of the meeting. After Frank Sanborn and Dr. James C. White were chosen as secretaries, Sanborn was called upon to read a draft constitution for the proposed association. Among those who debated the provisions of the constitution were abolitionists Wendell Phillips and Thomas Wentworth Higginson; Amasa Walker, a noted writer on political economy (and father of Francis Amasa Walker, first president of the American Economic Association); and Caroline Healey Dall, an ardent supporter of equal rights for women. Charles Sumner was invited but could not attend.[19]

Many of the participants were men of local reputation, but there were also notable representatives from outside Boston and Massachusetts. Professor Daniel Coit Gilman of Yale's Sheffield Scientific School was appointed to the committee to nominate the first officers of the Association, as was Dr. A. B. Palmer of the University of Michigan. Other important early members of the Association included William Lloyd Garrison, Samuel Gridley Howe, Francis Lieber, Congressman George S. Boutwell, publicist James Redpath, economists Edward Atkinson and David A. Wells, President Thomas Hill of Harvard, and President Theodore Dwight Woolsey of Yale.[20]

It is appropriate in several ways that this first meeting of "social engineers" should have chosen as its president William Barton Rogers. Rogers was selected in advance of the October meeting by the committee on arrangements, of which Edward Jarvis was the spokesman and Frank Sanborn a member. Why the committee members chose Rogers is not recorded, but their choice had the practical effect of linking the Association both with natural science and with the forerunners of the movement to establish authority. Rogers's pioneering activities in the professional organization of science have been pointed out in a previous chapter. In 1865 he was well known as the founder and president of the Massachusetts Institute of Technology, which had struggled into existence only four years earlier. In the debate then raging over Darwin's theory of evolution, Rogers was a prominent defender of Darwin and an opponent (as always) of Louis Agassiz.[21]

The draft constitution prepared by Sanborn and other members of the arrangements committee was accepted with no major revisions. There was debate over several issues, however. A New York represen-

[19.] F. B. Sanborn to C. H. Dall, Hampton Falls, N.H., 1 Oct. 1865, Massachusetts Historical Society. Sumner probably was offered an office in the Association; if so, he turned it down. See F. B. Sanborn to Charles Sumner, Boston, 11 Nov. 1865, Sumner Papers, Houghton Library, Harvard University.

[20.] These last nine names appear in the first list of members published in 1866, but it is not certain that they attended the organizational meeting. See ASSA, *Constitution, Address*, 25–30.

[21.] See Rogers's obituary, written by Sanborn, *JSS*, 16 (Dec. 1882), 8–10. Rogers's part in the famous Boston evolution debate of 1860 is excerpted in George Daniels, ed., *Darwinism Comes to America* (Waltham, Mass.: Blaisdell, 1968).

tative, Dr. Oliver S. Strong, asked at the outset whether it was to be a local or a national organization. Dr. Jarvis replied that the committee of arrangements had been unable to decide this question and left it to the meeting to determine whether it was to be a Massachusetts, a New England, or a national organization. After a "spirited discussion" the meeting declared its national ambitions by naming itself the "American Association for the Promotion of Social Science."[22] Another "warm discussion" was triggered by a constitutional provision that designated Boston as the place of the regular annual meeting. Persons from Connecticut and Michigan objected without success.[23]

The American Association adopted the same four departments into which the British Association was divided: Education, Public Health, Jurisprudence, and Economy, Trade and Finance. George B. Emerson of Boston, Professor A. B. Palmer of Michigan, and others urged that a special department be added for the prevention of crime and the reformation of criminals. Caroline H. Dall, Amasa Walker, and Edward Jarvis opposed the motion as needless, claiming that the subjects proposed by Emerson could be covered by the departments of Jurisprudence and Education. The motion was finally withdrawn.[24]

The most heated debate took place over membership qualifications and fees. The draft constitution established no barrier to membership save a five-dollar annual fee, or a fifty-dollar fee for life membership. Thomas Wentworth Higginson proposed that the annual fee be reduced to three dollars and life membership to thirty dollars. His motion was amended to make the annual fee only one dollar. Then another participant, William Emerson Baker of Boston, moved that a majority vote of the executive committee be required to elect new members. Baker's surprising motion, which was immediately carried, made the Association an exclusive organization. But there followed a period of discussion and parliamentary maneuvering, in which Wendell Phillips and Caroline H. Dall took part, which resulted in the adoption of Higginson's proposal: a three-dollar fee and no other barrier to membership.[25]

Fundamental Assumptions

The constitution made it clear that the prime object of the Association was not only to understand society, but to improve it as well. No distinction was made between inquiry and reform.

[22.] ASSA, *Constitution, Address,* 26. The original long name survived as late as 1878 in the vocabulary of some members, but the name "American Social Science Association" was common from the beginning.
[23.] *Ibid.,* 28.
[24.] *Ibid.,* 27. The British Association at first had a special department for crime and reformation, but by 1865 it had only the four departments adopted by ASSA.
[25.] *Ibid.,* 28–29.

Its objects are, to aid the development of Social Science, and to guide the public mind to the best practical means of promoting the Amendment of Laws, the Advancement of Education, the Prevention and Repression of Crime, the Reformation of Criminals, and the Progress of Public Morality, the adoption of Sanitary Regulations, and the diffusion of sound principles on questions of Economy, Trade, and Finance. It will give attention to Pauperism, and the topics related thereto; including the responsibility of the well-endowed and successful, the wise and educated, the honest and respectable, for the failures of others. It will aim to bring together the various societies and individuals now interested in these objects, for the purpose of obtaining by discussion the real elements of Truth; by which doubts are removed, conflicting opinions harmonized, and a common ground afforded for treating wisely the great social problems of the day.[26]

The initial statement of purpose suggests a good deal about the character of the Social Science Association. Condescension blends with sincere humanitarianism in the underlying assumption that a man's place in society is not entirely his own responsibility—the "well-endowed," "wise," and "honest" share in the failures of others. But, on the other hand, a man's position is intimately related to his inner quality, his God-given character. The statement is typical also in its bold assumption that all of society's problems can be dealt with at once and ought to be attacked in the most straightforward way. There is no caution here, no expectation that some problems might prove unmanageable. It is noteworthy also that subjects which seem widely separated today—such as "Public Morality" and "Sanitary Regulations"—are juxtaposed with no sense of incongruity.

Most typical of all is the implicit assumption that the social order is fundamentally rational and will reveal itself to the patient inquirer. Beneath the confused surface of events there is a harmonious order, a realm of "Truth," in which the interests of all members of society blend without friction. If truth has not yet revealed itself, it is because thinkers and inquirers have not adequately communicated their thoughts and observations to each other. The most pressing and perhaps most fundamental obstacle to social improvement seemed to be simple ignorance and inadequate communication. Society's ills persist not because of any intrinsic flaw in human nature or because of any irreconcilable conflict between the component parts of society, but because of a lack of coordination and interaction among those working for social improvement.

A more elaborate statement of purpose in the same vein was published by the committee on arrangements prior to the October meeting.

[26] *Ibid.*, 3.

This Association proposes to afford, to all persons interested in human improvement, an opportunity to consider social economics as a whole.

The persons composing it are expected to meet together to read papers and pursue discussions, and to seek the assistance of those who have a practical acquaintance with reform, as well as that of purely abstract reasoners.

They are to collect all facts, diffuse all knowledge, and stimulate all inquiry, which have a bearing on social welfare. It has long since been shown that the man of science who confines himself to a specialty; who does not, at the very least, conquer the underlying principles of other branches of scientific inquiry,—is necessarily misled, and cannot avoid frequent mistakes. To have any perception of the perspective of his subject, he must see it in its relation to other subjects. Something like this is true of those who investigate the necessities of society. If they associate themselves together, they have the advantage of each other's knowledge; they do not misunderstand their own relative positions; and they insure an economy of time, labor and money.

We would offer the widest hospitality to individual convictions, and to untried theories, provided only that such convictions and theories are the fruit of a serious purpose and an industrious life. To entertain the vagaries of the indolent would be at once undignified and unprofitable.[27]

Implicit in these brave words there are assumptions which, from the vantage point of the present, can only be described as simplistic. The founders of the ASSA shared a fundamental lack of realism about the nature of society and the tasks of understanding and improving it. Their viewpoint need not be defended, but it should be understood. It was not a product of foolishness or frivolity, but a sincere effort to come to terms with the most profound changes in the material conditions of life that men have ever experienced.

One distinctive set of assumptions centered on the problem of gathering facts. Merely to gather the most elementary statistical data about society was so difficult in the 1860's that it was easy to believe that adequate information would, in itself, lead almost automatically to vast social improvement. The student of mid-nineteenth-century society was compelled to think in a vacuum, to speculate without the discipline of hard, empirical data. Vital statistics in this country were kept haphazardly, with each state, county, and municipality employing its own system (if any). Only the decennial federal census and the scattered efforts of insurance underwriters and a few commercial organizations brought a semblance of order to the tasks of data collection. If these resources were inadequate, the inquirer had to compile the data

[27.] *Ibid.,* 15.

himself or do without. Often the only way to obtain information was to correspond with strategically located acquaintances, asking them to report on local conditions.[28]

Many of the men and women who gathered in the Boston State House to create the ASSA in 1865 thought the ills of society might vanish if only enough of the right information could be assembled in one place and brought to the attention of rational minds. Given the facts, solutions would be obvious. The naive and superficial views of society typical of ASSA members in the 1860's and 1870's were largely an inevitable result of the inaccessibility of factual data and—most important—the optimistic presumption that once factual data were brought into public view, few other problems would remain.[29]

Both the Social Science Association and the Board of Charities were institutional devices designed to invigorate and systematize the collection, dissemination, and evaluation of factual information. Sanborn's chief task as secretary of the Board was to travel about the state to familiarize himself with all its charitable institutions and their modes of operation. He set out, in other words, to become a living encyclopedia of charity work. In his first year he made 165 visits to institutions in Massachusetts and another 20 visits to out-of-state institutions, traveling 7,000 miles in the process.[30] The whole idea of the investigatory commission as a popular instrument of reform hinged on the widespread assumption that once the facts were known, everyone would agree about what needed to be done.

[28.] This is in fact how Henry Villard put together ASSA's *Immigrant's Handbook* in 1869 and 1870. See below, this chapter. For a typical inquiry for data, see Villard to D. C. Gilman, Boston, [22?] and 26 June 1870, Gilman Papers, Johns Hopkins University. Sanborn in 1891 recalled that "it was this scarcity of material for the investigation of social questions, indeed, which suggested to the founders of this Association the importance of bringing together in this way the persons interested in the development of civilization here. . . ." See Sanborn, "Aids in the Study of Social Science," *JSS*, 29 (Aug. 1892), 49. My understanding of fact-gathering and its implications in this period has been heightened by conversations with Harold M. Hyman; see his *A More Perfect Union* (New York: Knopf, 1973), chs. XVIII and XIX.

[29.] The simplistic assumptions about society that characterized the founding members of the ASSA were of course not peculiar to them or to America; see Abrams, *Origins of British Sociology*, 8–52. As late as 1903, Charles Booth, referring to his massive study of the population of London, spoke in much the same terms: "The root idea with which I began . . . was that every fact I needed to know was known to someone and that the information had simply to be collected and put together." Quoted by Harold W. Pfautz, ed., *Charles Booth: On the City: Physical Pattern and Social Structure* (Chicago: University of Chicago Press, 1967), 19. Talcott Parsons points out that within a utilitarian framework (such as the founders of the ASSA shared) any departure from rational norms *must* be explained as the result of a shortage of accurate factual knowledge: *Structure of Social Action*, 65–66.

[30.] *First Annual Report of the Board of State Charities*, 5.

Precisely because the founders of the ASSA expected that all decent, honest, and intelligent men would spontaneously agree upon a course of action once the facts were known, they were permitted the luxury—denied all later generations—of supposing that inquiry and reform were a natural and inseparable unity. This union of theory and practice, inquiry and reform, could exist only as long as the facts remained relatively inaccessible. In the 1880's and 1890's and beyond, after a host of fact-gathering institutions had gone into operation and begun to bury inquirers under an avalanche of factual information, it became obvious that the social problem was not fundamentally one of access to facts but of their interpretation. Only then, when decent, honest, and intelligent men still failed to reach agreement despite a wealth of facts, did inquiry and reform begin to split apart, becoming two distinct activities—related, to be sure, but only in a problematical manner. To the founders of the ASSA in 1865, information collection still seemed the fundamental obstacle to social improvement, and therefore it seemed only natural that those who inquired into the nature of society (those who were most familiar with the facts) should proceed immediately to practical reform on the basis of their inquiries. What fair-minded man could question a course of action selected after careful scrutiny of the facts?

The exceedingly optimistic and fundamentally superficial understanding of society and its problems that characterized the ASSA's early statements of purpose was deeply implanted in the organizational structure of the Association. Indeed, the division of labor established by the Association's departmental structure is its single most revealing feature.

The ASSA was initially organized in four departments, each led by a vice-president of the Association. The original nominee to take charge of the Education Department was the Reverend Thomas Hill, president of Harvard. Samuel Gridley Howe, who had been trained as a doctor at Harvard in the 1820's, was assigned the Department of Public Health. The president of Yale, Theodore Dwight Woolsey, was asked to take charge of the Department of Economy, Trade and Finance. Professor Francis Lieber of Columbia was asked to serve as vice-president in charge of Jurisprudence and the Amendment of Laws.[31] The subject area of each department was described in detail in the

---

[31.] By 1866 Woolsey had been replaced by George S. Boutwell, and Howe's place in the department of Public Health was vacant. See ASSA, *Constitution, Address,* 5, 31. In 1867–68 the leaders of the four departments, in the order given above, were Andrew D. White, president of Cornell; Dr. Edward Jarvis; David A. Wells, Commissioner of Revenue and a noted economic thinker; and Emory Washburn, ex-governor of Massachusetts and Harvard professor of law.

report of the committee of arrangements before the first meeting. This document merits close reading.

1. Under the Department of Education will come everything relating to the interests of Public Schools, Universities and Colleges; to Reformatory, Adult, and Evening Schools; to Instruction in the Useful Arts; to Systems of Apprenticeship; to Lyceums, Pulpits, and the formation of Societies for the purposes of Public Instruction. In this department will be debated also all questions relating to Classical, Linguistic, and Scientific Studies in their proportion to what is called an English Education; and the bearing of the publication of National and Patriotic Memorials upon Popular Culture.

2. Upon the Department relating to Public Health, a very large proportion of the popular interest will naturally be fixed. All Sanitary and Hygienic matters will come before it; and what the Sanitary Commission has learned in the last four years will be made available, through its action, to the people at large. The subjects of Epidemics, of the origin and spread of Cholera, Yellow Fever, and Eruptive Diseases, will be legitimately discussed here. It will consider all questions of Increase of Population, Vaccination, Ventilation of Public and Private Buildings, Drainage, Houses for the Poor, the Management of Cemeteries, Public Baths, Parks and Public Gardens, Places of Recreation, the Management of Hospitals and Insane Asylums, the Adulteration of Food and Drugs, all questions relating to the Duration of Human Life, Sanitary Regulations for the Army and Navy, and all matters of popular interest connected with medical science. We shall look to our ablest physicians and surgeons for contributions to this department.

3. Under the head of Social Economy, we shall consider Pauperism *actual* rather than legal, and the relation and the responsibilities of the gifted and educated classes towards the weak, the witless, and the ignorant. We shall endeavor to make useful inquiries into the causes of Human Failure, and the Duties devolving upon Human Success. We shall consider the Hours of Labor; the relation of Employers and Employed; the Employment of Women, by itself considered; the relation of Idleness to Female Crime; Prostitution and Intemperance; Workhouses; Public Libraries and Museums; Savings Banks and Dispensaries. Here, too, will be discussed National Debt; the subjects of Tariff and Taxation; the Habits of Trade; the Quality of our Manufacturers; the Control of Markets; and Monopolies in the Sale of Food, or the Production of articles of common use; the Value of Gold; and all questions connected with the Currency.

4. In the Department of Jurisprudence, we aim to consider, first, the absolute Science of Right; and, second, the Amendment of Laws. This department should be the final resort of the other three; for when the laws of Education, of Public Health, and of Social Economy, are fully ascertained, the law of the land should recognize and define them all. Under this head will be considered all questions of the justice, the expediency, and the results, of existing statutes; including their administration

and interpretation, and especially their bearing on Suffrage, Property, Privilege, Debt, Crime, and Pauperism. Here, then, will come up the vexed questions of Prison Discipline and Capital Punishment.[32]

Again, one is struck by the quixotic boldness of the scheme, the glib assumption that society's problems are, in the most fundamental sense, *simple*. Imagine the conception of social reality implicit in this statement: "When the laws of Education, of Public Health, and of Social Economy are fully ascertained, the law of the land should recognize and define them all." The tone is reminiscent of Francis Bacon, who expected that the new science of the seventeenth century might answer every question worth asking within a decade or two.[33]

But most interesting are the assumptions implicit in the institutional structure of the Association. The founders copied the four-department structure of the British Association, evidently believing it just as suitable for the United States as for Great Britain. They did not doubt that all necessary questions could find a place within one of the four departments; nor did they doubt that the four departments adequately classed like with like and distinguished between things not alike.[34]

The ASSA's four departments of Education, Health, Jurisprudence, and Finance established a division of labor among social inquirers, allocating to each department a certain segment of the total universe of human affairs. To divide the realm of inquiry among specific groups of inquirers is to make a fundamental decision that has wide ramifications. Today, the same universe of human affairs is divided into radically different segments corresponding to the disciplines of History, Economics, Political Science, and Sociology (leaving aside Psychology, Anthropology, and other disciplines less relevant to the ASSA). What are we to make of these two very different divisions of labor?

Of course it would be anachronistic to expect the ASSA to have hit upon the modern division of labor. Though taken for granted today, the present division was not intellectually or logically foreordained and was partly a product of happenstance. Certainly it had not yet crystal-

---

[32] *Ibid.*, 15–16. Italics in the original. It is not clear when the third department was renamed "Economy, Trade and Finance," but that title, rather than "Social Economy," appears in the original constitution. There probably were political considerations involved in the selection of a title for this department. When Sanborn became the regular secretary of the Association in 1873, he created a fifth department, named "Social Economy," in which he always worked. By his understanding the term was synonymous with the cause of the poor. See below, ch. VI.

[33] Herbert Butterfield, *The Origins of Modern Science, 1300–1800*, rev. ed. (New York: Free Press, 1965), 116.

[34] The minority effort at the organizational meeting to create a fifth department for crime and reformation does not constitute a significant exception to this claim.

lized, in the university or anywhere else, in 1865. In the moral philosophy classes of the mid-nineteenth century all the topics of what is today called social science floated in happy, undifferentiated confusion.[35] Even at the most formal, rigorous levels of scholarship there were a number of traditional foci of research that overlapped and contended with each other for disciplinary status, without conforming closely to any of the present disciplines. Among these research traditions, History and Political Economy were probably the best established intellectually, but these two rubrics were far looser and more inclusive in the 1860's than they are today. To see how flexible the boundary was even between these comparatively well established disciplines, one need only look at the German historical school of economics which crossed and re-crossed the boundary between them in the 1870's and 1880's.[36] The modern division of labor in social science did not finally crystallize until the three decades following 1880; it was then given concrete form by the new professional associations created by historians (1884), economists (1885), political scientists (1903), and sociologists (1905).

Why did the founders of the ASSA divide the universe of human affairs between departments of Education, Health, Jurisprudence, and Finance? Is there some circuitous path of evolution from the ASSA departments to the modern disciplines, or are we dealing with two irreconcilable modes of dividing up the task of social inquiry? If we are, what significance should be attached to the shift from one to the other? What does the ASSA's division of labor tell us about its members' basic assumptions about "social science?" And about our own?

The key to understanding the ASSA's division of labor is to see that its departments conformed, as much as possible, to the existing pattern of specialization within the professional class. Ministers were left out, but each of the major secular specializations at the professional level was given a place in the Association. Doctors naturally worked in the Department of Health. Lawyers took charge of the Department of Jurisprudence. College teachers lacked the organization and secure status of the three classic specialties, but this quasi-profession (upon which the other professions depended for their training) naturally took charge of the Department of Education. The only Department not built on an existing professional specialty was the one variously

---

[35.] On moral philosophy, see Meyer, *Instructed Conscience;* Smith, *Professors and Public Ethics;* Bryson, *Man and Society,* and "The Emergence of the Social Sciences from Moral Philosophy," *International Journal of Ethics,* 42 (Apr. 1932), 304–323, and "The Comparable Interests of the Old Moral Philosophy and the Modern Social Sciences," *Social Forces,* 11 (Oct. 1932), 19–27, and "Sociology Considered as Moral Philosophy," *Sociological Review,* 24 (Jan. 1932), 26–36.

[36.] Herbst, *German Historical School in American Scholarship;* Small, *Origins of Sociology.*

known as "Social Economy" or "Economy, Trade and Finance." Later this department was split into a liberal Department of Social Economy and a more conservative Finance Department. The very confusion of titles and politics suggests the lack of an established profession to staff the department. Only in connection with this field were the founders of the ASSA compelled to look beyond the existing division of professional labor.

The division of labor made good sense in the context of the times. The founders of the ASSA thought it was possible to establish social inquiry on a more authoritative basis simply by improving the skills of men who already occupied positions of authority in society. They did not dream of fabricating out of whole cloth an entirely new corps of full-time inquirers. Erroneously but understandably, they took it for granted that social inquiry would continue to be the responsibility of men and women like themselves, college-educated members of the gentry class, preferably equipped with a rigorous postgraduate training in one of the professional specialities. Professional life was in fact the occupational mainstay of the college-educated gentry class. A study by Charles Warren in 1872 showed that of 4,218 graduates of four New England colleges between 1833 and 1860, over 80 percent became professionals—33 percent lawyers, 20 percent ministers, 14 percent teachers, and 15 percent physicians.[37] The categories of Warren's study were the same as the ASSA's departments, except for the obvious necessity of including the ministry and excluding the as yet unprofessionalized field of economics. Warren's categories correspond closely with the ASSA's departments because both he and the Association wished simply to reflect the existing division of labor in the gentry class.

What could have been more natural? The role of the professional had always been to provide answers to esoteric questions that defied common sense and custom. As the explosive growth of interdependent relationships in nineteenth-century society increasingly thwarted common sense and magnified the realm of the problematical in human affairs, it was only natural for the public to turn for aid initially to the person who had always dealt with esoteric problems of remote causal attribution—the professional. Who was better equipped by both experience and training to relocate causation in an increasingly problematical social universe? Accordingly, the founders of the ASSA, like the founders of the British National Association for the Promotion of

---

[37.] Charles Warren, "An Inquiry Concerning the Vital Statistics of College Graduates," Bureau of Education, *Circular of Information* (Mar. 1872), 17, cited by Walter P. Rogers, *Andrew D. White and the Modern University* (Ithaca: Cornell University Press, 1942), 109.

Social Science before them, adopted a departmental structure and a division of labor geared to the existing professional class.[38]

It was not accidental therefore that the ASSA served as a central meeting place for men seeking to extend the competence of their professions. The modernization of the existing professional specialties was the very essence of the Association's mission, its prime *raison d'être* The departmental division of labor suited the ASSA ideally to the ambitions of a classic, tripartite professional class, that still hoped to regain intellectual mastery over an increasingly problematical society, without making major concessions to novelty. By organizing its "social scientists" in departments conforming to the division of labor in the professions, the ASSA effectively defined "social science" as a synthesis of the existing fields of professional knowledge in their most progressive, advanced form. From the vantage point of 1865, "social science" was the convergence (as yet incomplete) of the hitherto separate fields of knowledge upon which professional men based their expertise. To fully develop and articulate a science of society would be to establish the authority of the professional class on the firmest of foundations.

Of course, the departmental division of labor was only a point of departure for the ASSA. The founders did not set out with the conscious purpose of imitating the professional division of labor; they merely allocated the tasks of inquiry in what seemed to them the "natural" way. What seemed "natural" was controlled by the existing division of labor in the professional class. Certainly there was no thought of *restricting* membership to professional men. The leaders of the Health, Education, and Jurisprudence Departments were nearly always trained professionals, but individuals attached to no profession —like Frank Sanborn—were also prominent in the affairs of the Association.

In addition to professional lawyers, doctors, and educators, the meetings of the Association attracted two other main groups: humanitarian reformers and businessmen. The presence of reformers is not surprising; the presence of businessmen should not be, either. Merchants were the first to live under conditions of interdependence. The people at the very heart of the expanding network of dependencies in the nineteenth century were businessmen. The elaboration and extension of techniques first developed by business was for many years

---

[38.] It is noteworthy that the British National Association for the Promotion of Social Science was founded by George Woodyatt Hastings, son of the founder of the British Medical Association: see Ritt, "Victorian Conscience in Action." The NAPSS was basically Benthamite. For an interpretation of utilitarianism as a means of equipping men for the novel intellectual environment of urban-industrial society, see D. J. Manning, *The Mind of Jeremy Bentham* (New York: Barnes and Noble, 1968).

the main task of social science. As late as the 1890's, the practical experience of thoughtful businessmen was valued highly both in the ASSA and among academic social scientists.[39] It is also noteworthy that the role of professional educators in the Association was greater than that of any other profession and by no means limited to the Education Department. Among the first nominees to head the four departments, only Samuel Gridley Howe was not a professional educator. The departments of Jurisprudence and Health were frequently led by men who taught law or medicine in professional schools, rather than by practitioners. The ASSA was oriented toward the established professions, but it included non-professionals and naturally drew upon the educational sector of each profession for its leadership.

The position of leadership occupied by educators in the ASSA was a sign of the future. In 1865 college professors did not try to monopolize the whole subject of man and society. Such a claim would have seemed implausible even to them in 1865, for it conflicted too sharply with the claims of the traditional professions, which, after all, had been offering the public advice about human affairs for centuries. Only later, when the social universe seemed to have developed a new density and the classic professions proved unable to penetrate beneath the surface of it, did professors lay claim to the whole subject of social science solely on their merits as researchers and teachers. Intellectually, there is no real line of evolution out of the ASSA departments to the modern disciplines, but institutionally the decline of the ASSA and the rise of the professional academic disciplines can be seen as the result of the breakdown of the traditional division of professional labor and the fragmentation of the customary leadership class in an interdependent society.

An Eclectic Style of Inquiry

Having adopted a constitution in the morning, those persons attending the organizational meeting reassembled in the afternoon of the same day to begin the ASSA's routine business of inquiry and discussion. The tone of the first meeting and the ones immediately following is broadly representative of the climate of opinion, the style of thought, and the areas of concern that characterized the Association for the

[39] A. D. White in 1891 urged the ASSA to solicit papers from "men in active contact with the greater lines of business in the country, and especially with the management of banks, railways, express companies, forwarding companies, insurance companies, and the like"—White, "Instruction in Social Science," *JSS*, 28 (Oct. 1891), 6. White's list is interesting for its stress on enterprises oriented to transportation and other inherently supra-local phenomena. Lee Benson notes that "at a time when knowledge had not yet become the exclusive property of specialists, merchants in the seaboard towns were in a

remainder of its lifetime. The irreducible eclecticism of the Association proscribes any concise characterization of its style; one can only turn to a chronological description of topics and speakers. If the following discussion of the ASSA's routine business seems disjointed and unfocused, that is an accurate reflection of the reality described.

The first paper read at an ASSA meeting, by Professor William P. Atkinson of the Massachusetts Institute of Technology, was a plea for the development of an inductive science of education. Noting that the history of education was strewn with the "wreck of theories," Atkinson called for an empirical examination of "the effects of different modes of education on the after-life of the pupil." Even in advance of such empirical studies, he felt sure that students ought to spend more time studying "God's great book of nature" and the science of society. "Had political science been assigned its rightful place," asked Atkinson, "would we have needed a bloody war to open the eyes of the nation to its plainest duties?" Atkinson attributed the immaturity of educational theory to its dependence on another immature field, a "science of the mind" that was "in great measure yet to be created."[40]

The only other paper heard at the first meeting was delivered by Dr. Henry G. Clark of Boston, on the subject of cholera. A wide-ranging discussion followed, in which Dr. Jarvis, Professor A. B. Palmer of Michigan, Professor Atkinson, and several others spoke.[41]

The second meeting of the Association was held on 27 December 1865 in the hall of the Lowell Institute in Boston. The first paper, again on the subject of education, was delivered by President Thomas Hill of Harvard. His views, drawn from the regnant school of Scottish commonsense philosophy, typified those of educated Americans at midcentury.

> The ultimate ends of common sense, of philosophy, and of science are the same. They may be summed up in one,—it is the reading of God's thought. The order of the universe is rational, intelligible. The conviction of this primal truth springs up upon occasion for its use in every human mind. No mind capable of scientific labor ever doubts that all phenomena are subject to law, that is, that all phenomena succeed each other in an order which can be understood and expressed in the formulae of human

strategic position to analyze the nature of the rapid changes in society . . . much of the basic thinking [about society] in the decades after the Civil War can be attributed to cosmopolitan merchants in close touch commercially and intellectually with Europe"—Benson, *Merchants, Farmers and Railroads*, 266n. Men associated with the insurance industry were especially prominent in the early years of the Economic Science section of the AAAS and the American Statistical Association. E. C. Kirkland estimates that half the members of the ASSA in 1877 had business connections: *Dream and Thought in the Business Community, 1860–1900* (Ithaca: Cornell University Press, 1956), 15–16.

[40] ASSA, *Constitution, Address*, 31–33. From an abstract of Atkinson's paper.
[41] *Ibid.*, 33–34.

speech. To discover this order, to comprehend it, to express it in words and teach others to see it, this is the labor and the play, the work and the wages of the human intellect.[42]

Upon the conclusion of Hill's paper, the vice-president of the Department of Health read a paper submitted by Professor A. B. Palmer on the importance of "Sanitary Science" as a branch of college instruction. Palmer reported that his students at the University of Michigan were extremely interested in the subject and profited from its practical character and the element of variety it introduced into the curriculum.

The star attraction of the second meeting was Henry C. Carey, Esq., of Philadelphia, one of the nation's most famous economic writers and a long-time advocate of protective tariffs. Predictably, Carey's address was a tribute to the Morrill Tariff of 1861. His audience included representatives of Boston's predominantly free-trade business community, some of them not regular members of the Association, who contributed to a lively debate at the end of the address.[43]

The meeting continued on the following day with a paper by Frank Sanborn on "Prison Discipline in Europe and America." Sanborn advocated the "Irish System," based on a sentence of labor rather than time, of which the chief spokesman in England was Miss Mary Carpenter.[44] Dr. Isaac Ray, probably the nation's most influential psychiatrist, followed with a paper on the "Isolation of the Insane," which raised the question of possible abuses of the power of commitment of the insane to asylums. Ray cautiously suggested that in exceptional cases physicians may have issued a certificate of insanity unwisely, or through professional error, or even through personal bias. To safeguard the rights of the person committed, Ray proposed a general law providing for review "on the representation of respectable parties."[45]

Professor W. P. Atkinson, who had discussed education at the first meeting, now delivered a paper on "Competitive Examinations for the Civil Service," based on the reports of the English civil service commissions. Unfortunately, this first discussion of civil service reform in the ASSA was not recorded. The only point noted was Atkinson's conclusion that the method of civil service examinations in England had not been wholly successful.

The famous urban reformer Charles Loring Brace delivered a paper on "The Sanitary Legislation of England and the Effect of Sanitary Science." Brace stressed the great progress made in England since

[42] *Ibid.*, 36. From an abstract.
[43] *Ibid.*, 40, 43.
[44] *Ibid.*, 44; Blake McKelvey, *American Prisons: A Study in American Social History Prior to 1915* (Chicago: University of Chicago Press, 1936), 25 and *passim.*
[45] *Ibid.*, 46; see also Grob, *Mental Institutions*, 146 and *passim.*

1848 by the improvement of drainage, the prohibition of cesspools in close proximity to wells, restrictions against letting cellars as lodging rooms, and the practice of removing inhabitants from infected districts. Brace declared that the resulting improvements extended "to the morals as well as the health of communities, and [also had the effect of] greatly reducing poor rates."

Dr. Edward Jarvis read a paper on "Vital Statistics," which delineated "the causes which shorten human life and lessen the powers of its abbreviated existence, and the methods of securing sounder health and greater longevity." A statement from an Australian correspondent was read, showing the virtues of the eight-hour day in his country.[46]

In June of the following year the Social Science Association sponsored a three-day conference of managers of reform schools and "other persons interested in the training of orphan and vagrant children and juvenile delinquents." Representatives came from Maine, Rhode Island, Connecticut, New York, New Jersey, and Illinois, as well as Massachusetts. They discussed such topics as the history of state care for neglected children, the causes of vagrancy, the comparative advantages of the congregate and the family systems, the reformatories of Europe, and the advisability of transporting delinquents from major cities to the west. The representatives also toured several Massachusetts reformatories. Frank Sanborn and Samuel Gridley Howe were active in running the conference.[47]

The third general meeting was held in New Haven on 9 and 10 October 1866. The program was an imposing one, consisting of fourteen addresses on a wide variety of topics. The list of speakers read like a who's who of respectable reformers and social thinkers. Edward M. Gallaudet, Sanborn, and several others spoke on the education of deaf mutes and the blind. The labor question was discussed in round-table fashion by Amasa Walker, David A. Wells, Emory Washburn, George Walker, Frank Sanborn, Caroline Healey Dall, Dr. Edward Jarvis, and Professor Arthur L. Perry, among others. Prison reformer Enoch C. Wines spoke on the contract system of prison labor. The employment of women was discussed by Robert M. Hartley of New York. David M. Wilder of Boston delivered a paper on tenement houses. Other topics ranged from "Early Training as a Means of Preventing Pauperism and Crime" to "Prison Holidays" and "The Application of Design to American Manufactures."[48]

[46.] *Ibid.*, 47, 48.
[47.] *Ibid.*, 51–64.
[48.] Most of these papers were not published by ASSA, but the program for the third, fourth and fifth general meetings (1866, 1867, and 1868) is printed in American Social Science Association, *Address before the American Social Science Association at the Fifth General Meeting, New York, November 19, 1867, by Samuel Eliot, Corresponding Secretary* . . . (Boston: Wright and Potter, 1867), 81–84. This publication is paged continuously with *Constitu-*

The fourth general meeting was combined with the third "annual," or business meeting, held in Boston on 9 October 1867. It was a very modest affair at which only five papers were read. But the fifth general meeting, held in New York City on 19, 20, and 21 November 1867, surpassed all previous gatherings of the Association and set a model for the future. Twenty-seven papers were read. For the first time, the department organization seems to have been fully operative. The Department of Jurisprudence, heretofore not well represented in the program, provided three papers: "The Sphere of Law in Social Reform" by Professor John Bascom of Williams College; "The Equation between Crime against Property and Its Punishment" by Richard L. Dugdale, future author of the famous Jukes study; and a letter on "Prison Discipline" by Enoch C. Wines. Professor Francis Lieber of Columbia, Josiah Quincy of Boston, and Alexander Delmar of Washington, D.C., were on the agenda of the Department of Trade and Finance. Samuel Gridley Howe provided the keynote for the Department of Health with a paper on the unique town of Gheel, Belgium, where insane patients were left free to work and live among the townspeople. There were several other papers on insanity and one on the highly successful Metropolitan Board of Health of New York City by Edward B. Dalton, M.D. The Department of Education provided eleven papers, including an address on the Federal Bureau of Education by its commissioner, Henry Barnard. Andrew Dickson White, newly appointed president of Cornell, spoke on university education.

There was another round-table discussion of the labor question in which David A. Wells, Edward Atkinson, Simon Stern, George Walker, Amasa Walker, Henry Wilson, Charles Moran, and others took part. A separate paper on "Co-operative Labor" was given by E. L. Godkin, editor of the *Nation* and intellectual doyen of the Mugwump class for decades to come.

Without anyone giving much careful thought to it, the Social Science Association in its first few years adopted a settled pattern of operation, a routine of inquiry and discussion, and, most important, a range of questions deemed worth asking. In these basic matters the Association would change very little in the coming years. Even the names of the participants in ASSA meetings changed less than one might imagine. Gone would be Edward Jarvis, Amasa Walker, Samuel Gridley Howe, and other men of their prewar generation who lent fame and dignity but not energy or ambition to the founding of the ASSA. But Sanborn,

---

tion, Address, and the two documents are sometimes bound together under the title *Documents Published by the Association* (Boston: Wright and Potter, 1866–67).

David A. Wells, and Andrew Dickson White—men of the generation that reached its youthful prime in the Civil War years—dominated the Association until they were too old to attend its meetings.

Civil Service Reform

Public events like the ASSA's general meetings required hard preparatory work behind the scenes. In the first few years most of this work was done by Frank Sanborn, recording secretary, and Samuel Eliot, corresponding secretary. As president, William Barton Rogers rapidly became a mere figurehead, thus setting a precedent from which few of his successors departed.[49] Most of the work in the early years probably fell on Frank Sanborn's shoulders. When he completed his first term with the Board of State Charities in October 1868 and left Boston to become an editor of the Springfield *Republican,* the Social Science Association was compelled to find a permanent administrator.

When Sanborn left Boston, the executive committee of the Association invited William Lloyd Garrison's son-in-law, Henry Villard, to become permanent secretary. Villard was offered a regular salary and placed in practical charge of all the affairs of the Association. At the same time, the ASSA opened an office in Pemberton Square, an area close to the Boston State House which was frequented by legislators and lobbyists.[50] Aside from his famous father-in-law, Villard was well known for his journalistic coverage of the Lincoln-Douglas debates. The recent immigrant from Germany would ultimately be remembered for his role in the completion of the Northern Pacific Railroad. His *métier* proved to be high finance, rather than social science.

With the leadership of a regular administrator, the Association was able to undertake projects that went beyond the routine operations of holding meetings and publishing occasional papers. One such project was a handbook for immigrants, designed to ease the newcomer's introduction to American society. Villard began assembling material for the handbook in early 1869, but it was completed by others after

---

[49.] Rogers, who obviously had his hands full as president of MIT, apparently wanted to resign in 1866 but was persuaded to stay to save the Association from embarrassment. See Rogers to Sanborn, Lunenburg, Mass., 8 Oct. 1866, Sanborn Papers, Concord Free Public Library; Sanborn to Rogers, Sing Sing, New York, 15 Oct. 1866, Clarkson Collection (original in Boyd B. Stutler collection, Charleston, W.Va.); and Rogers to Sanborn, Boston, 2 Nov. 1867, Sanborn Papers.

[50.] The committee considered the possibility of locating the ASSA's office in New York: see Sanborn to A. D. White, Boston, 1 Sept. 1868, Clarkson Collection (original in A. D. White Papers, Cornell University). On Villard's appointment, see Samuel Eliot to H. Villard, Boston, 31 Oct. 1868, and Caroline H. Dall to H. Villard, Boston, 31 Oct. 1868, Villard Papers, Houghton Library, Harvard University. It is not entirely clear, but the last letter seems to indicate that Villard's application was sponsored by Caroline H. Dall upon the recommendation of William Lloyd Garrison.

declining health forced him to leave the Association in 1870. The ASSA *Handbook for Immigrants to the United States* (1871) was addressed to the educated immigrant, supplying maps and detailed information on everything from clearing customs to buying land and understanding the American character. Villard also put together the first issue of the *Journal of Social Science* in 1869. Scattered pamphlets and occasional papers such as the ones the Association published during its first four years would henceforth be augmented by a regular annual volume in a more or less standardized format.[51]

Villard's primary task in his two short years with the Association was to try to transform the ASSA into a national movement. This was probably one reason why the Association hired a permanent secretary in the first place. In the same letter which officially informed Villard of his election as secretary, President Eliot reminded him of the importance of being in Chicago on 10 November—only ten days hence—in order to establish an alliance between the ASSA and a like-minded group forming in that city.[52] The Chicago trip failed. Despite the efforts of Villard and John W. Hoyt, an agricultural editor and later governor of Wyoming, the Western Social Science Association which formed in Chicago in 1868 refused to ally itself with the ASSA. Villard made another attempt to affiliate in June 1870 which also fell through. The short-lived Chicago Association refused to become a branch of the ASSA because of the "too narrow or sectional policy" of the Boston-based Association.[53]

As the Social Science Association tried to expand its influence and membership in other cities, it also gave the civil service reform movement its first organized base of operations. A few months after becoming secretary of the Association, Villard invited Congressman Thomas A. Jenckes, the recognized leader of civil service reform, to come to Boston to speak under the auspices of the ASSA. Villard told Jenckes

---

[51.] ASSA, *Handbook for Immigrants to the United States: Prepared by the American Social Science Association* (New York: Hurd and Houghton, 1871). Frank Sanborn continued to oversee the operation of the Association even after leaving for Springfield. He sent Villard to H. B. Wheelwright of the State Board of Charities for information on Boston immigration and told him where to find data on New York immigration. He also recommended speakers for a series of Lowell Institute lectures allocated to ASSA. Sanborn advised Villard on the preparation of the *Journal* and sent a considerable amount of material to be published in it without his name in evidence. When the first volume was printed, Sanborn cut up his copy for excerpts to be printed in the Springfield *Republican*. See Sanborn to [Villard], Springfield, 6 June 1869, and Sanborn to Villard, Springfield, 23 Apr. and 9 Dec. 1869, Clarkson Collection (originals in Concord Free Public Library).

[52.] Eliot to Villard, Boston, 31 Oct. 1868, Villard Papers, Houghton Library, Harvard University.

[53.] The "sectional policy" was probably free trade. By 1874 the Chicago Association was dead. See *JSS*, 7 (Sept. 1874), 377, and 3 (1871), 200.

that his audience would be composed of "professional and business men."⁵⁴ To Frank Sanborn, Villard wrote that "the invitation to Mr. Jenckes was given with the idea of giving a little éclat to our enterprise. A number of businessmen outside the Association have taken hold of the matter and we shall have a good meeting."⁵⁵

Within a week of Jenckes's very successful Boston visit, Villard traveled to New York to organize a similar meeting there. E. L. Godkin provided him with letters of introduction. A select list of 1,200 persons was invited to hear Jenckes speak on 16 January 1869. Villard reported that he "found everywhere the strongest sympathy for our movement."⁵⁶

The "movement" to which Villard referred was as much civil service reform as social science. In a letter to Jenckes in February, Villard revealed the virtual equivalence, in his mind, of social science and civil service reform: "Our Association is now rapidly expanding its organization throughout the country and we propose, with the aid of our branches, to carry on a regular campaign for reform of the civil service between now and the next winter in the Eastern as well as the Western States."⁵⁷ In the words of the historian of the movement, this adoption of civil service reform by the Social Science Association meant that "civil service reform was no longer an undirected interest." By becoming a "pressure group" on behalf of civil service reform, the Association made the reform a serious political force.⁵⁸

At Jenckes's request, the ASSA addressed reform memorials, signed by the "best men" of Boston and New York, to the collectors of the ports of the two cities. In June Villard went to New York to make arrangements for the Association's fall meeting in that city, and he engaged civil service reformer George William Curtis to be the keynote speaker. Once Villard was satisfied with arrangements in New York, he traveled to Philadelphia to head off the creation of a separate reform group there. Hundreds of the "best men" of that city had signed a petition written by the notable historian Henry C. Lea, which urged Philadelphia's congressmen to support the Jenckes bill or a similar measure. Villard consulted with about fifty prominent Philadelphians,

---

⁵⁴· Quoted by Ari Hoogenboom, *Outlawing the Spoils: A History of the Civil Service Reform Movement, 1865-1883* (Urbana: University of Illinois Press, 1961), 55.

⁵⁵· H. Villard to F. B. Sanborn, Boston, 23 Dec. 1868 [written on a printed invitation to the Jenckes address], Sanborn Papers, Harvard University.

⁵⁶· Hoogenboom, *Outlawing the Spoils*, 55. Hoogenboom notes that "'movement' here probably refers to the Social Science Association and its general objectives," but the important point is that civil service reform and the "social science" objectives of the Association were virtually indistinguishable to Villard.

⁵⁷· *Ibid.*, 56.
⁵⁸· *Ibid.*

organized the framework of a branch association, and made arrangements for a fall meeting. Leaders of the Philadelphia group were Lea, Joseph Rosengarten, and James Miller McKim, abolitionist and founder of the *Nation*.[59]

The Philadelphia Social Science Association was officially founded as a local branch of the ASSA on 17 November 1869. Of all the branches of the ASSA, the Philadelphia branch was the only one of any real substance. The New York and Chicago groups evaporated within a few years of their formation; other so-called branches in New Haven, St. Louis, Detroit, and various other cities either dissolved shortly after their founding, or never were anything more than discussion clubs and salons. The Philadelphia branch held public meetings and printed its papers in *The Penn Monthly Magazine,* copies of which were for a time distributed to all Social Science Association members. In turn, members of the Philadelphia branch received ASSA publications. In the early 1870's the membership of the Philadelphia branch stood at about 160, half of the ASSA's membership. However, the Philadelphia Association's status as a "branch" was not much more than a formality, even though it did contribute $450 to the treasury of the parent organization in 1870. It acted quite independently of the ASSA and drew even farther away from it in later years. In 1890 the Philadelphia Association was absorbed into the American Academy of Political and Social Science, which still flourishes today.[60]

Like the parent organization, the Philadelphia Association more or less equated social science with civil service reform. Papers presented to one of the early meetings in 1870 included "The Theory of Civil Service Reform" by Joseph G. Rosengarten; Lorin Blodgett on "The Waste of Existing Social Systems"; Joseph Wharton, later founder of the Wharton School of Finance and Commerce at the University of Pennsylvania, discussing "International Industrial Competition"; and Supreme Court Justice William Strong, the first president of the Philadelphia branch, speaking on "Social Science." Other papers were

---

[59] *Ibid.* McKim's daughter was married to Wendell Phillips Garrison, Villard's brother-in-law. That the Philadelphia group was forming spontaneously before Villard arrived is evident from a letter written to Villard by Samuel Eliot, president of the ASSA: "It was a happy thought to enlist Mr. McKim, and the names on the newspaper slip which you enclose are of the highest promise. I need not say that I am glad the danger of an independent organization in Phila. was successfully averted. We must be national." S. Eliot to H. Villard, Torquay [?], 8 Dec. 1869, Villard Papers, Houghton Library, Harvard University.

[60] A brief history of the Philadelphia Association to 1873 is given in *JSS,* 5 (1873), 202–204. See also 8 (May 1876), 36, and 7 (Sept. 1874), 340. For a more complete history, see Joseph G. Rosengarten, "Work of the Philadelphia Social Science Association," *Annals of the American Academy of Political and Social Science,* 1 (1890–91), 708–719.

presented on "The Evidence of Experts," "Method of Study in Social Science," and "Minority Representation."[61]

The American Social Science Association made civil service reform its main business in 1869 and 1870. After 1870 the Association's involvement with the movement diminished slightly because it lost the services of Henry Villard, and because of political developments within the reform movement over which the ASSA had no control. But civil service reform remained a major interest of the Association throughout the 1870's, 1880's, and even later. George William Curtis, the leader of the movement in the 1870's, served as president of the Association in 1873-74. In 1877 Frank Sanborn observed that the ASSA had worked for civil service reform ever since its founding in 1865:

> It is nearly twelve years since the Social Science Association began to labor for a reconstruction of the civil service, and we have not yet got very far in that direction. . . .
>
> It is with this more than with any other measure that our Association has been identified, and it was by our members that the unwelcome topic was forced upon the attention of Congress and the executive. . . .[62]

By aligning itself with a movement quite popular in its day, at least among the prosperous classes, the Association gained many new members, new financial resources, and a place in the public eye that it might not otherwise have enjoyed. Strategically the alliance was valuable. But was it intellectually honest? What, if anything, did civil service reform have to do with social science?

The question occurs naturally to any twentieth-century mind and therefore must be asked, but the preceding pages are meant to show what a misleading and anachronistic question it is. In the context of the times it was only appropriate for Villard virtually to equate civil service reform with social science, for they were in fact two aspects of a single phenomenon, the movement to establish authority. To the early leaders of the ASSA, civil service reform was not merely one peripheral category in the whole universe of subjects concerning man and society, something that a modern social scientist might file under "public administration" or "bureaucratization." Instead, it was the vital center of social science itself. To reform the civil service was to carry the movement to establish authority to the front lines of politics. And for

---

[61] Rosengarten, "Work of the Philadelphia Social Science Association," 708-709.

[62] *JSS*, 9 (Jan. 1878), 5. After the Pendleton Act became law in January 1883, Sanborn recalled that "it was by this Association that the question was first publicly presented in the United States and by our members that the agitation in its favor was for some years effectively maintained." *JSS*, 18 (May 1884), 194.

either civil service reform or the larger movement on behalf of authority to succeed, there had to exist a science of society.

Civil service reformers sought to do at all levels of the government bureaucracy what the investigatory commission had already done at the top: introduce an element of expert authority into what was basically a democratic, majoritarian system. Their aim was to exclude certain decision-making powers from the haphazard competition of politics and to shift those powers into the hands of men whose competence was institutionally certified. The authority of these men would not depend upon majority vote or party patronage but upon merit, as measured by certain standard criteria. They would not, of course, be professional men, contributing members of a community of the competent. But in their modest way these men—bureaucrats, as they would be called in the twentieth century—would carry out the highest ideals of the movement to establish authority. That movement had no more cherished goal than to confer authority upon those who most merited it.

Historians have been all too quick to dismiss civil service reform as a panacea for gentlemen of declining force and fortune. The reformers are typically portrayed as timid men, unable to take the rough-and-tumble of democratic politics, who pinned their frail hopes for political purity on a single, naive reform measure. Either that, or they are characterized as cynical politicians ready to adopt any cause that will return them to office. Both portraits dismiss them as inconsequential.[63] Such a dismissal is too hasty, if for no other reason than because the introduction of the civil service system was by no means a trivial reform if viewed sociologically as a phase of bureaucratization.

But, more important, the conventional view is misleading because civil service reform was not a self-contained movement operating in a vacuum. Instead, like the exposed tip of the proverbial iceberg, it was only the most conspicuous manifestation of a broad movement of cultural reform, the full scope of which historians are only now beginning to appreciate. Civil service reform was significant in itself, but it was part of a larger movement for the establishment of authority; the movement's major triumph was not the Pendleton Act, but the con-

---

[63.] The power of the conventional stereotype of civil service reform is so great that it was adopted even by Robert Wiebe, who had every reason to reject it: "Although many topics of late nineteenth century reform reappeared after 1900, most of the old issues had changed beyond recognition. Civil Service, for example, had once been a negative, absolute goal, self-contained and self-fulfilling. Now the panacea of the patrician had given way to the administrative tool of the expert, with efficiency rather than moral purity its objective"—*Search for Order,* 170–171.

struction of the modern American university and a system of professional and semi-professional elites built upon it.[64]

C. Wright Mills was not entirely wrong when he described the ASSA as an "attempt 'to apply science' to social problems without resort to explicit political tactics. Its members, in brief, sought to turn the troubles of lower-class people into issues for middle-class publics."[65] But Mills's characterization is finally misleading. Was the ASSA primarily important for its efforts to politicize social questions — to create and publicize issues where none had been before? Or was it important for its efforts to *de*politicize social questions—to raise them up out of the political marketplace altogether and assign them to the care of experts? It is true that the ASSA deliberately and formally set out to publicize what its members regarded as the correct solutions to social problems, and that it aimed to influence public opinion and bring about reform. But in working out "correct solutions," and in seeking to buttress their claim to correctness so as to seem credible and persuasive in competition with other such claims, the members of the Association were repeatedly and often reluctantly driven toward a conception of social reality which denied that the untutored politician, much less the average voter, had access to significant truth. They finally were driven toward the discomforting conclusions that even they themselves, as amateur social thinkers, were not fully competent to understand human affairs, and that only full-time professional specialists could speak authoritatively about such matters in a modern society.

[64.] Bernard Crick has observed the intimate connection between civil service reform and the rise of the university: "By the 1870's the harsh growing pains of industrialism and mass immigration had so changed, or, as many thought, corrupted, the style and tempo of early Republican politics, that the old 'political science' of precept and example in the practice of statesmanship was now felt to be insufficient. Political science must now be directly taught, not merely as 'civics' to the city offspring of the 'melting-pot', but also, though in a broader manner, to the native Americans in the colleges, men who, it was feared, had lost their earlier spontaneous political skill, responsibility and wisdom. Reform of the universities and reform of the Civil Service were to go hand in glove"— *American Science of Politics*, 32. Laurence R. Veysey has noted the same connection: "Harvard University, rather than Boston or the United States, became the body politic of the Mugwump's dream. With Harvard as their greatest achievement, it could not be said that the genteel reformers of post-civil war New England labored in vain"—*Emergence of the American University*, 98.

[65.] Mills, *Sociological Imagination*, 84.

CHAPTER VI

The 1870's: Near Collapse

The 1870's were turbulent years for the American Social Science Association. The Republican reform consensus of the 1860's evaporated, leaving the Association without a clearly defined public and short of financial resources. Deprived of Henry Villard's secretarial services and unable to rely on Frank Sanborn, who was still in Springfield working for the *Republican,* the Association came perilously close to total collapse in 1871 and 1872. New life was poured into the enterprise in 1873 by the return of Sanborn and by the vigorous efforts of two elderly Lazzaroni, Louis Agassiz and Benjamin Peirce, who saw the Association as a means of extending their campaign on behalf of authority beyond natural science to the whole of American culture. Buffeted by criticism from rival interpreters of human affairs on the one hand, and men who distrusted all theoretical speculation on the other, the Association struggled throughout the decade to expand its influence and cultivate new and more authoritative knowledge about man and society. It achieved only modest success.

The difficulties of the seventies were manifestations of deep-seated weaknesses that made the ultimate failure of the Association almost inevitable. But, as the next chapter will show, the Association came closest to success near the end of the decade, when Peirce tried to arrange a merger with the new Johns Hopkins University. Had the Association succeeded in hitching itself to the rising star of the university, it might have survived as an institutional entity, the course of professional development in the social sciences might have been substantially altered, and the role of the Lazzaroni in the rise of professional social science might have been direct, rather than only indirect. As it was, Peirce's proposal set the stage for the transition to modern social science. A younger man cast in the same mold as the Lazzaroni, Hopkins's president Daniel C. Gilman, finally rejected the ASSA and inspired the organization of new, specialized communities of social inquiry securely based in the university.

To a casual observer it must seem odd that the nation's leading zoologist (Agassiz) and mathematician (Peirce) played a pivotal leadership role in the ASSA, an organization dedicated not to science, but to social science. Within the context of the movement to establish authority, however, the role played by these two men appears perfectly natural.

The Association was delighted to accept the leadership of Agassiz and Peirce because they had led the nation's most successful project to institutionalize authority. The American Association for the Advancement of Science and the National Academy of Science were testimony to their labors, and the rising prestige of science added lustre to their personal accomplishments. For their part, the two Lazzaroni thought it only natural to adopt the ASSA as a base of operations because it was virtually the only institutional focal point for the growing movement to establish authority. Encouraged by the spirit of discipline and consolidation generated by the Civil War, Agassiz, Peirce, and many others greatly broadened their efforts on behalf of sound opinion after 1860. They hoped that the nationalist energies engendered by war would carry the United States forward to a new level of cultural development, comparable to the mature nations of Europe. Not just science, but the whole of American culture was now the target of reform.[1]

The triumph of the Republican party meant that Agassiz and Peirce, like Samuel Gridley Howe and Frank Sanborn, suddenly had a friend in power. Within a week of receiving Howe's recommendation for a Board of Charities, Governor John A. Andrew also found in his mail a letter from Agassiz, marked "Private," that recommended an even bolder undertaking. Though never carried out, Agassiz's plan illustrates the scope of his broad campaign on behalf of institutionalized authority. Agassiz wanted the state of Massachusetts to take control of all the postgraduate facilities of Harvard and transform the school into the nation's first true university. What Daniel Coit Gilman did at Johns Hopkins in 1876, Agassiz wanted to do at Harvard in 1862.

Agassiz proposed that the state make available the unheard-of sum of two million dollars to secure for Massachusetts "intellectual supremacy over the whole country." He was contemptuous of Harvard's current level of attainment. The undergraduate college, he wrote, "has more the character of a high school than of a university & the special Schools have in no instance yet reached the true character of University faculties." The schools of law, divinity, medicine, and even the scien-

---

[1.] Lurie, *Louis Agassiz,* 350; and *Nature and the American Mind: Louis Agassiz and the Culture of Science* (New York: Science History Publications, 1974).

tific school and the observatory, said Agassiz, are generally considered to be mere "accessories," or even "excrescences," of the college. His ideal university would dispense entirely with the petty discipline so typical of the American college. It would be free so that even "the poorest in the community" might attend. Professors would be elected at stated intervals, and any competent individual could open courses in opposition to the regularly appointed professor. The university would present to its students "the most advanced progress of the human mind in every direction."

There were to be five distinct faculties in Agassiz's university: Divinity, Law, Medicine, Letters, and Science. The Divinity School would admit professors from every church. The School of Science would include engineers and agriculturalists. Agassiz did not go into great detail about the course offerings he envisioned in each school, but it is worth noting that what we think of as the antecedents of social science occupied no special place in his scheme. Moral philosophy is not mentioned, though he presumably had it in mind when he wrote "philosophy and all its branches" under the faculty of Letters. To the same school he assigned "history in all its ramifications." Under the Law School he listed "Roman law, comparative jurisprudence and diplomacy," and then went back to insert "political economy" above the line.[2]

Social science bulked larger in his mind a short time later. Upon the invitation of Charles Sumner in May 1864, Agassiz submitted a plan to organize two new honorary institutions sponsored by the federal government: an Academy of Letters and an Academy of Moral and Social Sciences. More than half of the names he proposed for the Academy of Letters were fellow members of the Saturday Club. His Cambridge and Boston friends who could not be included in the literary group found a place under social science. Ralph Waldo Emerson, Oliver Wendell Holmes, George Ticknor, Charles Francis Adams, Jared Sparks, and Samuel Gridley Howe were among those named to the two academies. Agassiz wished that "all the eminent men of the country" could be grouped together in similar societies. His dream was carried out at the end of the century, when the National Institute of Arts and Letters was organized under ASSA auspices.[3]

[2.] Agassiz to Andrew, Cambridge, 16 Dec. 1862, Massachusetts Historical Society. See also Lurie, *Louis Agassiz,* 327–329.

[3.] Sumner to Agassiz, Washington, D.C., 11 May 1864, Agassiz Papers, Houghton Library, Harvard University. Quotation from Lurie, *Louis Agassiz,* 335. The founding of the NIAL is discussed below, ch. X. For the parallel activities of Ralph Waldo Emerson, see Daniel Aaron, *The Unwritten War: American Writers and the Civil War* (New York: Oxford University Press, 1973), 34–36.

Neither Agassiz's academies nor his project for upgrading Harvard ever came to pass. The power of the Lazzaroni waned rapidly after the Civil War, even in Cambridge, where the succession of Charles William Eliot to the Harvard presidency in 1869 deprived them of much of their previous influence. But even men whom the Lazzaroni regarded as opponents, like Eliot, were by this time working for many of the same professional goals. During Eliot's long presidency, Harvard would go far toward building advanced professional schools at the graduate level, and of course it is with his name that the elective system is most often associated.[4]

Even before Eliot began the era of reform at Harvard, Andrew Dickson White began moving in the same direction at Cornell. Agassiz spent several months in Ithaca to assist at the founding of the new school, and he was overjoyed with its plan of organization. He thought it half a century ahead of Harvard in every respect but the prestige of its faculty. "I was amazed," wrote Agassiz to Benjamin Peirce, "to find here that in the college course every branch of the physical & mathematical & natural historical sciences are absolutely on a level with the litterary [sic] studies.... Many features of the whole concern are either borrowed from the plan you once laid out, or if not borrowed are almost identical."[5] Agassiz placed several of his own ex-students as professors at Cornell and advised White on the selection of other staff members. Years later White recalled a day in 1868 when he and Agassiz discussed candidates for chairs in science. "As we discussed one after another of the candidates he suddenly said: 'Who is to be your Professor of Moral Philosophy? That is a far more important question than all the others.' "[6]

Everyone who knew Agassiz testified as White did to his "deep ethical and religious feeling," a personal commitment that contributed to his rejection of Darwinian evolution and prompted his concern for moral philosophy and social science. His dear friend "Benny" Peirce shared with him the conviction that nature was God's puzzle, and therefore that scientific research was akin to reading the divine scripture. Students remembered Peirce in a characteristic moment, briskly stepping back from a blackboard crammed with algebraic proofs to declare: "Gentlemen, there must be a God!"[7] Peirce's mathematician

[4.] Regarding the antagonism between Eliot and the Lazzaroni, see Lurie, *Louis Agassiz*, 326–331, 361–363.

[5.] Agassiz to Peirce, Ithaca, N.Y., 26 Oct. 1868, B. Peirce Papers, Houghton Library, Harvard University. See also Mark B. Beach, "Andrew Dickson White as Ex-President: The Plight of a Retired Reformer," *American Quarterly*, 17 (Summer 1965), 239–247.

[6.] Andrew D. White, *A History of the Warfare of Science with Theology in Christendom* (New York: Appleton, 1896), I, 70n; Lurie, *Louis Agassiz*, 360.

[7.] Quoted in Wilson, *In Quest of Community*, 37.

son, James Mills Peirce, published some of his father's last lectures under the revealing title, *Ideality in the Physical Sciences* (1881). Like most of the leading members of the ASSA, both Peirce and Agassiz were closer to the idealism of Emerson than the positivism of Spencer. Between Agassiz and Emerson there was, in fact, a special fondness and cordiality.[8] Peirce exemplified the idealist core of both men's scientific viewpoint in saying that the "object of geometry" was "to penetrate the veil of material forms, and disclose the thoughts which lie beneath them."[9]

Agassiz and Peirce were not among the founders of the ASSA in 1865, perhaps because of lingering ill feeling between the Lazzaroni and ASSA president William Barton Rogers. But they always maintained close connections with the Association. Their loyal friend, the Reverend Thomas Hill, president of Harvard, was a founding vice-president of the Association and the original chairman of the Department of Education. By 1867 another friend, Andrew White, had become chairman of the Education Department, but he hardly had time to supervise it in view of the imminent opening of Cornell. In January 1869 the more or less dormant Education Department was taken over by Peirce and Agassiz, Peirce becoming chairman. The next few years were remembered as busy ones by other Department members, as the two elderly Lazzaroni used the Association as a sounding board on behalf of postgraduate education, professional education for educators, and other reforms of a similar nature.[10]

The Crisis of Liberal Republicanism

For the Association as a whole, however, the years from 1871 to 1873 were a period of near collapse. In September 1870 Henry Villard resigned his post as secretary, complaining of ill health. In hopes that he might return, and also for financial reasons, the Association did not immediately select a replacement. From September 1870 to October 1873 the Association had no permanent secretary and therefore no effective leadership. It appears that Dr. David F. Lincoln, chairman of the Health Department, informally took over Villard's duties for a time. In November 1872 one of the Directors of the Association, James M. Barnard, volunteered to perform secretarial duties until a permanent secretary could be found. In February 1873 the Executive Com-

[8.] Lurie, *Louis Agassiz*, 202–203.
[9.] Quoted in Wilson, *In Quest of Community*, 37.
[10.] Hill owed his Harvard presidency in part to the influence of Peirce and Agassiz: see Lurie, *Louis Agassiz*, 320. For Agassiz's and Peirce's contributions to the Education Department, see the glowing report of Mrs. Emily Talbot, a friend of Vassar's Maria Mitchell and a moving spirit behind Boston University, in *JSS*, 10 (Dec. 1879), 34–42.

mittee appointed Barnard, Frank Sanborn, and J. S. Blatchford as a committee to fill the secretarial office, but Barnard seems to have continued to do what little work was accomplished.[11]

The Association failed to hold its yearly "General Meeting" in 1871 and again in 1872. For practical purposes, it had no public existence in these years.[12] The resources of the treasury fell so low that some of the papers presented at the general meeting of May 1873 could not be included in the *Journal*. The secretary of the Health Department, Dr. David F. Lincoln, had to report that as of the end of 1872 no departmental meeting had been held in nearly two years, though six meetings were held between January and October 1873. The inactivity of the Department of Finance was such that there was no report to present in 1873. Even Peirce's Education Department had nothing significant to report.[13]

The ASSA deteriorated to such an extent in the early 1870's that the question of disbanding it was actively debated. At an adjourned annual meeting held on 9 November 1872, it was decided to continue the work of the Association. The lease on the office at number 5 Pemberton Square was renewed, but still there was no secretary. Not until October 1873, when Frank Sanborn was made permanent secretary, did the revival of the Association begin in earnest.[14]

The immediate cause for the aimless drift and inactivity of the ASSA was lack of the care and management that only a permanent secretary could provide. But underlying that unfilled post was a failure of purpose and united will among the leading elements of the Association. The *Journal of Social Science* tells us remarkably little about the reasons for the ASSA's near collapse, and the private correspondence of the leading members is not much more informative. Though the record is sparse, it seems reasonable to assume that the ASSA's decline in these years was a function of the unsuccessful Liberal Republican movement.

When the Social Science Association was formed in 1865, all reformers marched together under the banner of the Republican party. The Association aspired to national scope and took for granted a reform consensus; it began as an organization for all interested Republicans of respectable intellect. But by 1872 the Republican party had split apart. There was a yawning gap between party regulars and reformers, and a series of further splits between various factions of reformers. The slave power having been defeated, men now divided on new issues, with

[11] *JSS*, 3 (1871), 200–201; 5 (1873), 136; 7 (Sept. 1874), 339.
[12] *Ibid.*, 6 (July 1874), 13.
[13] *Ibid.*, 7 (Sept. 1874), 340–342.
[14] *Ibid.*, 339; 8 (May 1876), 30.

the tariff, the currency question, and civil service reform being the most important.

The Liberal Republican convention of May 1872 which nominated Horace Greeley to the presidency perhaps did more to divide reformers than to unite them. Many New England reformers were repelled by Greeley's eccentricity and protectionism. Some would have much preferred to see Charles Francis Adams win the nomination; others had hopes for Charles Sumner. Ari Hoogenboom, historian of the civil service reform movement, suggests that the campaign of 1872 split reformers into three factions. First there was a group, represented by Carl Schurz and Horace White, which was disappointed in the Cincinnati convention but nonetheless supported Greeley against Grant as the lesser of two evils. The second group, to which E. L. Godkin and Charles Eliot Norton belonged, was composed of men who swallowed their reform sentiments and returned to Grant rather than support Greeley. A third group, exemplified by George William Curtis, stayed with Grant from the beginning, believing that reform would be best served by preserving the institutional integrity of the Republican party. What Hoogenboom says of reformers in general very likely applies to ASSA members in particular, with one qualification: the third group, of men who refused to break with the party, was probably underrepresented in the Association. Certainly Curtis did not look as attractive to the Association in 1871 and 1872 as he did in 1873, when he finally lost patience with Grant and resigned his post on the Civil Service Commission. In the same year he was made president of the Social Science Association.[15]

The growing dissension among Republicans between 1868 and 1872 can be traced in the correspondence of Frank Sanborn. In February 1868 he wrote to his friend Moncure Conway to say that he soon would become "first mate" of the Springfield *Republican*. Sanborn was looking forward to his new position so that he could "aid the election of Grant as President." Sanborn had no reservations: "We want a military President for the next four years to keep the country from anarchy, and Grant is as free from ambition and intrigue as any man can well be."[16] A year later Sanborn confided to Conway that "Grant is not all we bargained for."[17] By the spring of 1872 Sanborn was actively working to unseat Grant by urging Massachusetts Republicans to attend the

---

[15.] Hoogenboom, *Outlawing the Spoils*, 111–116, 122–123. The only account of the Liberal Republican movement is Earle Dudley Ross, *The Liberal Republican Movement* (New York: Henry Holt, 1919).

[16.] Sanborn to Conway, Boston, 9 Feb. 1868, Clarkson Collection (original in Columbia University).

[17.] Sanborn to Conway, Springfield, Mass., 10 Apr. 1869, Clarkson Collection (original in Dickinson College).

Cincinnati convention.[18] After the convention Sanborn apparently supported Greeley, as did his boss, Samuel Bowles, editor of the *Republican*.[19]

Reformers fell to quarreling not only about national policies, but also about the internal affairs of the Association. In 1870 E. L. Godkin of the *Nation* led an effort to raise the level of discussions in the ASSA by means of a restrictive classification of membership. Sanborn opposed this change, as he always opposed measures designed to exclude "cranks" and "eccentrics." He also vigorously opposed a movement in 1870 to remove women from the governing board of the Assocation; he had insisted from the beginning on a prominent place for women in the affairs of the ASSA. Furthermore, he was displeased with several of the speakers chosen for recent sessions. Despite his complaints, he was unable to attend meetings of the directors because of his work for the *Republican* and other tasks, such as organizing the Cincinnati Prison Congress of 1870.[20]

When the ASSA's leading members gathered on 9 November 1872 to decide whether or not the Association should continue in existence, the recent debacle of Greeley and the Liberal Republicans could not have been far from their thoughts. The Association had been founded upon a consensus of the "best men"; now that foundation had washed away. Sanborn recalled several years later that Louis Agassiz and Benjamin Peirce were prominent in the decision to keep the Association alive.

> At the time, in 1872–3, when the practical discontinuance of the Association was favored by many members, by reason of the difficulties attending its work, Professor Peirce was one of those who most earnestly urged its continuance; and it was mainly owing to his remarks and those of Professor Agassiz, at one or two public meetings in Boston, that the Association remained in activity during the years of panic and political change that followed the reelection of General Grant in 1872.[21]

In October 1873 the Association moved out of the doldrums decisively by making Frank Sanborn its permanent secretary. He was

---

[18.] See a handwritten circular addressed to "the Republicans of Massachusetts" signed by Sanborn, F. W. Bird, and many others; originally dated 11 Mar. 1872 and then corrected to read "April, 1872"—in the Clarkson Collection (original in the Massachusetts Historical Society). The Bird Club may have had hopes for Sumner: see Bird to Sanborn, East Walpole, Mass., 12 Feb. 1872, Sanborn Papers, Houghton Library, Harvard University.

[19.] See William Dean Howells to Sanborn, Cambridge, Mass., 25 Aug. 1872, Sanborn Papers, Concord Free Public Library. Howells wrote to withdraw an earlier invitation to write for *The Atlantic Monthly* in view of Sanborn's support for Greeley. Also see Hoogenboom, *Outlawing the Spoils*, 116.

[20.] See Sanborn to Samuel Eliot, Springfield, Mass., 26 Sept. and 5 Nov. 1870, Clarkson Collection (originals in Concord Free Public Library).

[21.] *JSS*, 12 (Dec. 1880), x; 8 (May 1876), 30.

offered a salary of $1,200 a year with the expectation that he would give one-third of his time to the work of the Association and continue his other two jobs as secretary of the Board of State Charities and as a columnist for the Springfield *Republican*. By 1875 Sanborn was able to report that "we seem to have entered once more upon a busy and prosperous career."[22]

Problems of Men and Money

With the appointment of Sanborn as permanent secretary, the sheer survival of the Association was no longer in doubt. But serious problems remained. There were concrete difficulties of money and membership that no one could ignore, and larger problems of credibility and authority that were harder to pin down but even more threatening in the long run.

The financial position of the Association was precarious throughout the 1870's. The Association's main source of income was the five-dollar fee it charged for membership and a subscription to the *Journal of Social Science*. That source proved to be inadequate. When Sanborn took over as permanent secretary in 1873, he hoped to publish the *Journal* on a semi-annual or even quarterly basis.[23] But by the time of the Saratoga meeting of September 1877, Sanborn's own salary was six months in arrears, membership had not markedly increased, and once again the Association lacked funds to publish all of the papers presented to it. The Executive Committee refused to guarantee a secretarial salary at all after October, and it was proposed that the Association's Boston office be closed. Sanborn threatened to resign if the Association could not arrive at some adequate method of supporting itself.[24]

Only a few years earlier the Association had possessed ample resources. In the year ending 12 October 1868 its regular receipts totaled $6,583, and in addition there was a special fund of $4,606 for the immigrant's handbook. The regular income for 1869 was reported to be $4,868, and special mention was made of the liberality of the Association's "friends in New York."[25] In contrast, in 1878 the Association had no funded property and its income amounted to less than

---

[22.] *Ibid.*, 8 (May 1876), 30. On Sanborn's arrangements with the Association, see Sanborn to Gilman, Boston, 20 Sept. 1877, Gilman Papers, Johns Hopkins University, and Sanborn to William Torrey Harris, Boston, 16 Oct. 1873, Clarkson Collection (original in Concord Free Public Library).

[23.] *JSS*, 6 (July 1874), iii, 3.

[24.] Sanborn to D. C. Gilman, Boston, 20 Sept. 1877, Gilman Papers, Johns Hopkins University; and Sanborn to C. H. Dall, Boston, 30 May 1877, Clarkson Collection (original in Massachusetts Historical Society).

[25.] *JSS*, 8 (May 1876), 30; 2 (1870), v-vii; 3 (1871), 201.

$1,800 a year. The cost of publishing three numbers of the *Journal* (600 pages) was $1,000; another $2,000 was needed for the expenses of the secretary, office rent, clerk hire, cost of the annual meeting, and miscellaneous items. Expenses exceeded normal income by more than a third. It is not recorded who made up the difference: "It has been necessary to rely upon sources which may be called extraordinary, and which may fail us at any time."[26] One of the sources appears to have been Sanborn himself, for he seems to have drawn only a fraction of the $1,200 salary originally offered to him. In 1881 he received only $428. In the fourteen-month period ending in February 1879 the Association paid out for all salaries a total of only $638.[27] "Our assured income," said Sanborn, referring to membership fees, "is . . . about $1,500 a year. We can get along with $3,500, and with $5,000 we should have no difficulties of a pecuniary character. Does it not seem to you that this sum can be raised in the whole United States?"[28]

Of course, the ASSA was not the only organization of its kind in financial difficulty in the 1870's. As Sanborn pointed out, most learned societies had been in trouble since the panic of 1873.[29] In 1879 the American Academy of Arts and Sciences, ancient and very well endowed indeed compared to the ASSA, had an income of only about $1,900 from membership assessments plus $1,200 in interest from its general fund.[30] The American Association for the Advancement of Science was sufficiently hard pressed that the circular announcing its 1877 meeting carried, in italics, an appeal for the early remittance of dues, so that publication could begin on schedule. Then followed in boldface type the statement: "THE SECRETARY PERSONALLY REQUESTS YOUR CONSIDERATION OF THE FACT THAT 'HARD TIMES' ARE MAKING IT VERY DIFFICULT FOR HIM TO MEET THE DEMANDS UPON THE ASSOCIATION."[31]

The Social Science Association never resolved its financial difficulties in an entirely satisfactory manner, although a series of special subscription funds successfully lured patrons to its aid during the 1880's and 1890's. The paying membership of the Association never greatly exceeded 300 until the turn of the century, when the character of the organization changed and it adopted high-pressure recruiting

[26.] *Ibid.*, 9 (Jan. 1878), v.
[27.] *Ibid.*, 14 (Nov. 1881), vii; 11 (May 1880), xv.
[28.] Sanborn to D. C. Gilman, Boston, 20 Sept. 1877, Gilman Papers, Johns Hopkins University.
[29.] *JSS*, 9 (Jan. 1878), 12.
[30.] See circular dated 26 Dec. 1879 in file marked "American Academy of Arts and Sciences" in B. Peirce Papers, Houghton Library, Harvard University. The Academy was far stronger than the ASSA, however, for it could boldly solicit a permanent publication fund of $50,000 on the occasion of its centennial.
[31.] Also in the Peirce Papers.

techniques. The lack of money and members in the 1870's was merely the most concrete aspect of a more general problem having to do with the authority and credibility of the Association. Even the most dedicated members were uncertain about the viability of their enterprise, and their uncertainty made it difficult to attract new members or patrons.

Problems of Credibility and Authority

The task which the Association so optimistically undertook in 1865—to improve society and inquire into its essential nature—seemed much more difficult in the 1870's. The network of human relationships in society was already denser and more problematical; authoritative social opinion was correspondingly harder to formulate and to defend. Reform, which once had seemed to be an almost automatic consequence of coupling rational minds with accurate facts, seemed a more elusive goal after the corruptions of the Grant administration were exposed to view. Frank Sanborn recalled, "with a certain regret .... the warmth and eagerness with which we then [1865] launched for the voyage, and anticipated noble results from our venture. . . . There was little we did not fancy ourselves capable of achieving. I fear we must confess now that we rather overestimated our powers. . . ."[32]

In this chastened mood, Sanborn searched for ways to make the Association more effective. One of his first acts upon becoming permanent secretary was to invite some of the leading members to propose new strategies of action for the Association. The replies were disappointing. Most of the respondents focused on the obvious problems of money, membership, and prestige. The more serious problem of their own questionable credibility eluded all of them except Louis Agassiz, and even he could offer no practical solution.

Emory Washburn, Harvard professor of law and past governor of Massachusetts, emphasized the desirability of building the Association on a firm foundation of "smaller bodies, local associations . . . where the members know each other and are content to come together and work without the stimulus and *éclat* of numbers."[33]

John W. Hoyt, a Wisconsin agricultural editor, educator, and future governor of Wyoming Territory, suggested that the Association's chief difficulty was its regional character. It was really a New England organization. "With all due respect," said Hoyt, "I feel bound to say

[32.] *JSS*, 8 (May 1876), 23.
[33.] *JSS*, 7 (Sept. 1874), 375–376.

that there is less than a just appreciation of what the West and South could do for it, if once really enlisted."[34] Hoyt complained of the "too narrow or sectional policy" of the Association, which had thwarted the efforts of Henry Villard and himself to establish a close connection between the ASSA and the "Western Social Science Association" of Chicago in June 1870. The $100 minimum expense of traveling to Boston precluded active membership for Westerners, so he urged that meetings be held nearer the center of the country.[35]

Daniel Coit Gilman, at that time president of the University of California at Berkeley but a New Englander at heart, seemingly wrote in direct opposition to Hoyt. The travel difficulties in the West were so great, thought Gilman, that there really was no way to nationalize the ASSA. "The American Association, as now constituted, is substantially what is wanted. It has done good work, has acquired a good name, is managed by sensible and judicious persons. I should be sorry to see its essential characteristics very much modified just yet."[36] Gilman thought that local societies with varying degrees of affiliation ought to be encouraged, and to that end he had taken some steps toward the formation of a San Francisco Association. It is noteworthy that Gilman in 1874 proposed nothing more drastic than a broader base of local associations. Ten years later he would encourage members of his Johns Hopkins faculty to organize the specialized academic associations that ultimately made the ASSA obsolete.

The will-o'-the-wisp of local branch associations appealed to Frank Sanborn, too. "We need to give Social Science a foothold in as many sections of the country as possible," said Sanborn, "and to do away with the fancy that it is a New England invention and monopoly,—a sort of Yankee notion. To do this, we must have local Associations, and must entrust to them eventually the control of the parent Association."[37] Not all of ASSA's leading members would so willingly sacrifice New England's monopoly. The Association was indeed a regional body: in 1878–80 about a third of its members resided in Massachusetts and more than a quarter were from New York. Out of a total membership

---

[34] *Ibid.*, 376–378. Hoyt was founder and president of the Wisconsin Academy of Sciences, Arts and Letters and helped reorganize the University of Wisconsin to include an agricultural college. After serving as governor of Wyoming Territory in 1878–82, he returned as first president of the University of Wyoming in 1887–1890. In the 1890's he was the chief promoter of a National University in Washington. See *Dictionary of American Biography*, IX, 321.

[35] *JSS*, 7 (Sept. 1874), 376–378. For Villard's efforts to bring about an affiliation with the Western Association, which formed independently of the ASSA, see *ibid.*, 3 (1871), 200. The Chicago Association had collapsed by the time Hoyt spoke.

[36] *Ibid.*, 7 (Sept. 1874), 378–379.

[37] Sanborn to T. D. Woolsey, Boston, 13 Apr. 1874, Woolsey Papers, Yale University Library.

exceeding four hundred, there were only a few scattered names from the South and West.[38]

Benjamin Peirce also addressed himself to the problem of enlarging "the area of our influence" and thought in that respect that "something might be borrowed from the Scientific Association." He proposed that the ASSA hold two annual meetings instead of one, each lasting about a week, in cities all over the country, even in the far West. "Persons in the vicinity of each meeting," thought Peirce, "will be drawn into the meeting and will suddenly find themselves gifted with powers to aid the progress of thought in directions stimulated by the Society."[39] His optimistic plan contained only two measures intended to control the quality of ASSA discussions: the Association council would screen abstracts of each paper before delivery, and it would solicit papers from certain eminent figures to set the tone of the discussions. These were the same techniques the Lazzaroni had employed in the American Association for the Advancement of Science.

Most of those who replied to Sanborn's query took it for granted that they and their colleagues in the ASSA already held sound opinions about man and society; the problem as they saw it was not how to formulate sound opinion, but how to propagate it. Only Louis Agassiz pointed to the more serious problems of credibility and intellectual rigor: "I value the success of our Association as much as anything in which I ever had a part; and yet I feel, as you do, that we are not succeeding as we should. The fault lies, I am sure, in the fact that we have no one who is truly, by life-long training, a student of Social Science, and who can direct our action. Good will is insufficient for that purpose."[40] In a quiet way, Agassiz's words were a confession that the Association was unfit to carry out its stated mission: the cultivation of authoritative opinion in the study of man and society. The Association presented itself to the world as an authoritative advisor in matters of social improvement, but why should the public accept its advice? What warrant could it show a skeptic? Unlike the secular professions on which it parasitically depended for its departmental leaders, the ASSA itself had no recruiting or apprenticeship program, no systematic plan of inquiry, no intellectual discipline to impose on its members. Really it was not a community of the competent, as doctors and lawyers were, but only a loose aggregate of individuals. Instead of conferring authority upon its individual members, it drew what institutional authority it

---

[38] Bernard, *Origins of American Sociology*, 549. The membership rolls (published in most volumes of *JSS*) upon which the Bernards based their figures are quite inaccurate. For example, in a letter to the Executive Committee in September 1877, Sanborn speaks of "about 300" paying members; the membership list for 1878 shows 441 names.

[39] *JSS*, 7 (Sept. 1874), 376.

[40] *Ibid.*, 375.

possessed from the personal achievements of its members as individuals.

Agassiz sensed that truly authoritative commentators on man and society would have to posssess some warrant for their authority more compelling than good character, generous intentions, and a genteel upbringing. But he could suggest no practical course of action, save to avoid frivolous subjects.

> Every topic concerning civilization is a proper subject for communications and discussions; but I know too little of the men best qualified to present papers, and those I would recommend are probably too busy to prepare special papers. We ought forever to discard rambling addresses and discourses on topics involving human nature in its totality. The Academy of Sciences, in Paris, assumed its commanding authority from the day they excluded discussions upon the system of the Universe. We might well follow their example,—have people speak and write of what they do know, and not of what they feel or believe.[41]

The problem of authority was intimately bound up with the more obvious problems of money and membership. The Association's difficulties could not be resolved singly; no solution that ignored the question of authority could hope to put the Association on a firm footing, but the enhancement of authority would itself require a stronger financial and institutional base. After the death of Louis Agassiz in December 1873, it was Benjamin Peirce who most clearly perceived these needs. His plan to connect the ASSA with Johns Hopkins, conceived in 1878, was the most imaginative and promising effort to strengthen the Association.

However, for several years before Peirce conceived of the merger project, Sanborn had been leading the Association in the opposite direction—away from academia, toward the buzzing confusion of practical experience. This is the significance of the close liaison he established between the ASSA and the National Conference of Charities and Correction. By bringing ASSA members together with men and women who had direct, practical experience in trying to mitigate the most painful effects of social change, Sanborn broadened the perspective of the Association and improved the quality of its inquiry. But the practical officeholding reformers that the Conference of Charities brought to ASSA meetings in the 1870's were too close to the flux of experience to have much tolerance for theoretical questions. They discouraged the development of the highly generalized knowledge that the practitioners of a profession must possess. They brought the ASSA valuable insights, but not professional authority.

[41.] *Ibid.*

The latter would require not only familiarity with fact, but also a degree of insulation from fact that only the university could provide.

The National Conference of Charities and Correction

Sanborn thought of the National Conference of Charities and Correction as an offshoot of the Social Science Association, particularly the Association's Department of Social Economy. This department, which Sanborn added to the original four when he became permanent secretary, was generally understood to be his own personal province. The name "social economy" was a matter of some significance. One of the original five departments of the British Association for the Promotion of Social Science had borne that name; some of the founders of the ASSA had expected the American Association to follow suit, but instead the original constitution established a Department of "Trade, Economy and Finance." The more humanitarian connotations of "social economy" had been established in 1830 by John Stuart Mill, who defined political economy by distinguishing it from the larger field of which it was a branch, "social economy," or "speculative politics." The larger study treated "the whole of man's nature as modified by the social state," whereas political economy considered man only in his role as wealth-getter.[42]

Sanborn's formation of a Department of Social Economy in 1873–74 was an important anticipation of later attacks on orthodox political economy. Young German-trained economists such as Richard T. Ely and Edmund J. James made themselves famous in the 1880's and 90's by denouncing orthodox political economy for its constricted view of human motivation. Man was more than a wealth-getter, they argued, and no theory of political economy could afford to ignore his real complexity. In 1880, before the new generation of economists had attracted much attention in this country, Sanborn said that his interpretation of social economy "made it synonymous with the cause of the poor."[43] After the turn of the century Sanborn adopted the rhetoric of the new generation of economists, explaining the formation of the department in words that might have been used by Ely or James: "The cold deductions of political economy as then taught (in 1873–74),

---

[42.] John Stuart Mill, "On the Definition of Political Economy; and on the Method of Philosophical Investigations in that Science," *London and Westminster Review,* 26 (Oct. 1836), 11. In the first issue of the *American Journal of Sociology* Lester Ward traced the origins of sociology to Mill's conception of social economy: "The Place of Sociology," *American Journal of Sociology,* 1 (July 1895), 22–23. See also White, *Social Thought in America,* 22.

[43.] *JSS,* 11 (May 1880), xxiv.

could not satisfy those who wished actively to aid in social reformations or to diminish human suffering."[44]

Sanborn understood social economy to be "the feminine gender of Political Economy, and so, very receptive of particulars, but little capable of general and aggregate matters.... Social welfare, therefore, and not wealth in its wide and compound sense, is what we consider...."[45] In practice this meant that the department was concerned primarily with such mundane matters as public recreation, life insurance, vocational education, and, more than any other single topic, "homes for the people": various schemes such as cooperative saving and loan associations by which the laboring man might save enough to own his home.[46] In every case, the primary concern of Sanborn and his department was not with theory and not with rigorous data collection, but with reform techniques of a very immediate and practical nature.

The same particularistic and practical kinds of concerns that led Sanborn to form the Department of Social Economy also led him to ally the ASSA with the National Conference of Charities and Correction. "Scarcely had the new Department organized in the spring of 1874," he said, "when it proceeded to evolve from itself another organization, the National Conference of Charities, which has since grown to such a vigorous life that it quite eclipses its parent, the Department of Social Economy."[47] Sanborn's parental claim is accurate enough, but it should be noted that the independent-minded president of the Wisconsin Board of Charities, Andrew E. Elmore, once declared that the origins of the National Conference went back to 1872. In that year the Wisconsin, Illinois, and Michigan boards held the first of three joint meetings without any assistance from the ASSA. These early meetings in the midwest actually drew more participants than the New York meeting of 1874, which was attended by representatives from only Wisconsin, Massachusetts, and New York. The meeting organized by Sanborn was nonetheless the first to establish the national principle.[48]

The Conference of Charities and Correction continued to meet with the ASSA for most of the decade. The intimacy of the relationship between the two organizations is illustrated by the fact that the proceedings of the Conference were printed in the *Journal of Social Science*

[44.] *Ibid.*, 39 (Nov. 1901), 158. On the origins of the Social Economy Department, see also *ibid.*, 7 (Sept. 1874), 408; 8 (May 1876), 27; and 16 (Dec. 1882), 98.
[45.] *Ibid.*, 23 (Nov. 1887), 21.
[46.] *Ibid.*, 16 (Dec. 1882), 98. Sanborn's continuing interest in "homes for the people" is discussed in a later chapter.
[47.] *Ibid.*, 11 (May 1880), 87.
[48.] Andrew E. Elmore, "President's Address," *Proceedings of the Ninth Annual Conference of Charities and Corrections, held at Madison, Wisconsin, August 7–12, 1882* ed. A. O. Wright (Madison: n.p., 1883), 10.

in 1874 and 1875. From 1876 through 1881 the *Proceedings* of the National Conference were printed separately, but the ASSA contributed a share of the printing cost and distributed copies to each of its members. Frank Sanborn was the editor of both the *Journal* and the *Proceedings*.[49]

The ASSA's relationship with the Conference of Charities proved to be a marriage of only temporary convenience. As early as the second National Conference, in 1875, the Wisconsin delegation demanded that the Conference sever its connection with the ASSA. At first a majority of the Conference delegates preferred to retain the connection, however, and the Wisconsin delegation finally had its way only after boycotting the 1878 meeting. From 1879 on, the Conference met separately from the Social Science Association.[50] Sanborn continued to devote much of his time to the Conference, and up to the end of the century no one contributed more often than he did to its published *Proceedings*.

According to those Conference members who initiated the split between the NCCC and the ASSA, the reason was a lack of truly common interests. Theory and practice just would not mix. To the members of the Wisconsin delegation, it seemed that "the practical questions of what we shall do with the poor, whom we always have with us, what we shall do with our insane and criminal classes, were of sufficient importance to absorb *all* our time at the conference; that many of the questions discussed at the meetings of social science were very interesting to listen to, some of them really valuable, but did not meet the demands of *to-day* with us." Referring to the Detroit meeting of 1875, the Wisconsin delegates said "they went there seeking information, expecting to find able teachers on the question of how to provide best for the chronic insane. Theory was abundant. They departed disappointed."[51] Untheoretical and reformist though the Social Science Association was, it was too abstract, too far removed from application, to suit the delegates of the National Conference of Charities and Correction.[52]

[49.] For financial arrangements, see Sanborn to T. D. Woolsey, Boston, 1 June 1875, Clarkson Collection (original in Yale University) and Sanborn to D. C. Gilman, Concord, 19 Sept. 1879, Gilman Papers, Johns Hopkins University. Alexander Johnson, long-time secretary of the NCCC, reported that the *Proceedings* were originally financed by the state boards, each of which purchased 100-200 copies annually for distribution in its own state. See Alexander Johnson, *Adventures in Social Welfare* (Fort Wayne, Ind.: Fort Wayne Printing, 1923), 270.

[50.] Elmore, "President's Address," 13-14.

[51.] *Ibid.*, 13, 14. Italics in the original.

[52.] The NCCC was an amorphous organization that passed through three phases before World War I. At first it was a creature of the state boards. In the 1880's it was taken over by the Charity Organization Societies, whose members stressed the voluntary,

The Positivist Critique

If charity workers found the Association too abstract, there were others like Edward L. Youmans who thought its orientation excessively practical and reformist. When Youmans began publication of *The Popular Science Monthly* in 1872, in the very first issue he drew an invidious contrast between the ASSA's approach to social science and that of his mentor, Herbert Spencer. "The term social science has indeed come into vogue," observed Youmans, "and large associations have assumed it; but, as thus applied, it fails to connote any distinctive or coherent body of principles such as are necessary to constitute a science." A few months later Youmans chastised the Association for its "vagueness" and inability even to define the concept of social science. Its proceedings, he said, consisted mainly of "philanthropic projects and reformatory schemes for public improvement—plans for repairing the defects of society—which would be better described as social art than social science."[53]

Youmans's low opinion of the Association did not improve with time or familiarity. He became an officer of the Association in 1874. Although he admired its goal and thought many of the papers at that year's meeting "interesting" and "able," still he found it wanting in "pure investigation . . . the strict and passionless study of society from a scientific point of view." Most of its members, he complained, were "hot with the impulses of philanthropy." The Association might be worthwhile as a "general reform convention" or "an organization for public action," but it failed in its supposed aim of promoting social science, thought Youmans, as long as it refused to recognize "that social science is but a branch of natural science."[54]

By 1875 Youmans's irritation had turned to gall; he was no longer an officer of the Association and he now regarded it as an enemy, parading "under false colors."

> The name of the Association is entirely misleading: it avows one object, and pursues others; it professes to do a certain work of very great public

---

private principle and looked down on government functionaries. Shortly after the turn of the century these private charity workers were in turn displaced, often with contempt, by the settlement house workers led by Jane Addams. The NCCC's Committee on Occupational Standards, appointed by Addams in her first year as president, had a hand in formulating the platform for Theodore Roosevelt's Progressive party in 1912. See Jane Addams, *The Second Twenty Years at Hull House* (New York: Macmillan, 1930), 24–29. The organization still exists under the title National Conference of Social Work.

[53] E. L. Youmans, "Editor's Table," *Popular Science Monthly*, 1 (May 1872), 117; (Sept. 1872), 626. On Youmans see William E. Leverette, Jr., "E. L. Youmans' Crusade for Scientific Autonomy and Respectability," *American Quarterly*, 17 (Spring 1965), 12–32, and Hofstadter, *Social Darwinism*, 22–23, *passim*.

[54] *Popular Science Monthly*, 5 (July 1874), 367–368.

importance, and then, by totally neglecting it and doing something else under its name, it produces a mischievous confusion in the public mind. . . .

. . . We strenuously object to any perversion or misappropriation of the term to illegitimate uses. We object to its employment as merely a dignified title for miscellaneous speculations on human affairs . . . and we protest against its use as a kind of imposing category for the schemes of philanthropists and the projects of reformers. . . .

So long as the term "social science" is employed to characterize the heterogeneous and discordant opinions of unscientific men upon the most intricate and refractory problems of civilized life, it will be discredited in its true application.[55]

Youmans's effort to wrest the name "social science" away from the ASSA is readily understandable. As Spencer's American impresario, he had a vested interest in discrediting rival conceptions. But his critique went beyond that petty motive to reflect honest differences between positivism and idealism, the two major approaches to the study of man and society in the late nineteenth century. As an advocate of Spencer's positivist system, Youmans viewed social science not as an adjunct to ameliorative reform but as a branch of natural science, not essentially different from geology or botany. "Human beings," said Youmans, "should be studied exactly as minerals and plants are studied, with the simple purpose of tracing out the laws and relations of the phenomena they present. Men should be analyzed to their last constituents, physiological and mental."[56] Like Spencer, Youmans was profoundly fearful of the state, no matter how good its intentions, and he thought humanitarian reform essentially futile, whether carried out by the state or by other agencies. Man could no more escape the economic laws of the marketplace or the natural laws of evolutionary development than defy gravity or suspend the laws of motion.[57]

Idealism was not necessarily reformist in its implications; indeed, it could be quite conservative. However, because it rejected mechanistic determinism and insisted on the reality of human freedom, it at least left open the possibility of reform. The members of the ASSA were not immune to the vogue of Spencer, but they were not eager to adopt such a bleak world view. Like probably the majority of intellectuals in the West in the late nineteenth century, they struggled to find an intellectual position that would acknowledge the indisputable explanatory power of positivism while retaining the spiritual comforts of idealism. To some, like William James, this meant an effort to preserve the

[55.] *Ibid.*, 7 (July 1875), 365–367.
[56.] *Ibid.*, 1 (July 1872), 336.
[57.] Peel, *Herbert Spencer*, 244–245.

principle of human freedom against the claims of a "block universe." To others it meant a strident effort to keep the teachings of science congruent with traditional Christian doctrine. The effort to mediate between idealism and positivism achieved its greatest triumph in the philosophy of James and the other pragmatists at the end of the century. Their triumph required that idealism surrender much ground to its opponents, but the principle of voluntarism seemed to be saved.[58]

The nation's most formidable proponent of philosophical idealism, William Torrey Harris, lent his considerable weight to the ASSA in the late 1870's. Editor of the *Journal of Speculative Philosophy* and leader of the St. Louis Hegelians, Harris became a faithful contributor to the Association's meetings and its publications. No one but Frank Sanborn published more articles in the *Journal of Social Science*. In 1889, after serving for eleven years as chairman of the ASSA's Department of Education, Harris was selected to be U.S. Commissioner of Education, a post he held for the next seventeen years. Morris R. Cohen described him as the "intellectual leader of the education profession in the United States" from 1867 to 1910—that is, before John Dewey.[59]

Harris's hostility toward positivism was well known. As early as 1866 Sanborn had supported his unsuccessful efforts to publish a lengthy critique of Spencer in the *North American Review*.[60] Harris took up the anti-positivist cudgel again in his first paper for the ASSA, on "Method of Study in Social Science," in 1878. Not surprisingly, the paper quickly drew fire from Youmans. His response to Harris is especially revealing because it illuminates not only the differences that separated positivists from idealists, but also the common ground that ultimately permitted the intellectual convergence of the 1890's.

Harris and Youmans found common ground in the idea of interdependence. Both men defined scientific method as a habit of mind that discovered interrelationships and dependencies which ordinarily were hidden from view. The error of the unscientific mind, according to both men, was to succumb to the isolative tendency of common sense, which ascribed a false autonomy and particularity to the objects of perception. To be scientific, in contrast, was to see through this illusion of autonomy and perceive all the ways in which the object was related to and dependent upon the rest of the universe. In Youmans's words, the very essence of the scientific method was to pass beyond "the

---

[58.] On the convergence of idealism and positivism, see the first and last chapters of this study and the works cited there.

[59.] Morris R. Cohen, *American Thought: A Critical Sketch* (Glencoe, Ill.: Free Press, 1954), 265. See also Kurt F. Leidecker, *Yankee Teacher: The Life of W. T. Harris* (New York: Philosophical Library, 1946).

[60.] Leidecker, *Yankee Teacher*, 325.

mere sensible properties of objects to their relations." He then quoted approvingly Harris's elaboration of the same idea:

> No object can be understood by itself, and even the weather of to-day is found to be conditioned upon antecedent weather. . . . Science sees the acorn in the entire history of the life of the oak. It sees the oak in the entire history of all its species. . . . We must learn to see each individual thing in the perspective of its history . . . as part of a process. . . . The ordinary habit of mind occupies itself with the objects of the senses, and does not seek their unity; . . . the scientific habit of mind chooses its object, and persistently follows its thread of existence through all its changes and relations.[61]

Harris and Youmans agreed about the method of natural science but disagreed about its application to the universe of human affairs. To Youmans man and nature were one, while to Harris they were radically distinct and therefore required different methods of analysis. "Social science deals with man," said Harris:

> Man has a natural being as a mere animal, as well as a spiritual being of intellect and will. . . . Man is not only an animal, having bodily wants of food, clothing, and shelter, but he is a spiritual being, existing in opposition to nature. . . . His true human nature is reason; his actual condition is irrational, for it is constrained from without, chained by brute necessity, and lashed by the scourges of appetite and passion. There is thus a paradoxical contrast between nature and human nature. . . . As man ascends out of nature in time and space into human nature, he ascends into a realm of his own creation. . . .

Harris's conclusion that "the application of scientific method to the explanation of human institutions in the ordinary form is not valid" provoked Youmans to an exasperated gesture of dismissal. "His mode of dealing with the subject," declared the Spencerian, "would seem to leave us no social science at all."[62]

From the vantage point of the twentieth century, neither Youmans nor Harris was entirely right—or entirely wrong. The view of human affairs that began to crystallize in the 1890's drew on both traditions. To arrive at that intellectual position positivists would have to provisionally give up their reductionist tendency to explain all human behavior as a reflex of physiological, biological, or even non-organic

---

[61] Youmans, "Harris on Social Science," *Popular Science Monthly*, 15 (1879), 703. William Torrey Harris, "The Method of Study in Social Science," *JSS*, 10 (Dec. 1879), 28-34.

[62] *Ibid.*, 703, 704, 702. For a close examination of another idealist variety of social science, espoused by an occasional participant in ASSA meetings, see Robert A. Jones, "John Bascom 1827–1911: Anti-Positivism and Intuitionism in American Sociology," *American Quarterly*, 24 (Oct. 1972), 501-522.

causes. They would have to acknowledge that men were not only effects, but also causes, possessed of a degree of freedom. To arrive at the position of convergence, idealists would have to admit severe limits to the extent of human freedom and abandon their radical, ultimately theological, distinction between man and nature. They would have to admit the propriety of cause-and-effect language in the discussion of human affairs, just as in natural science. But it is important to recognize that the idealists' distinction between man and nature would be sustained in the twentieth century, albeit in a pale and secular form. The concept of culture and a host of variations upon that concept would permit twentieth-century social scientists to admit that man is a creature of circumstance, but then to rule out extreme reductionism by defining the circumstances themselves as a human product, susceptible to deliberate control.

The ASSA lost nothing by refusing to hop on Youmans's positivist bandwagon. The Association finally withered and died not because its views of man and society were incompatible with a scientific worldview, but because its views were not grounded in a disciplined community of inquiry. No view of man and society could be sure of survival in the corrosive intellectual conditions of the late nineteenth century unless it could strike deep institutional roots. Louis Agassiz sensed this in the early 1870's, and his Lazzaroni colleague Benjamin Peirce took steps in 1878 to give the ASSA those roots by trying to merge it with the new Johns Hopkins University.

CHAPTER VII

The Proposed Merger of the ASSA and the Johns Hopkins University

The following sequence of events—briefly summarized here, but subjected to detailed examination in this chapter and the next—was pregnant with implications for the development of professional social science. Indeed, in retrospect it is clear that these events signaled not only the beginnings of an irreversible transit of authority out of the Social Science Association into new centers of social inquiry, but also the final breakdown of the traditional division of professional labor and the inauguration of a new phase of the movement to establish authority.

The sequence began in June 1878 when Benjamin Peirce proposed a merger of the American Social Science Association with the nation's newest and most ambitious university, the Johns Hopkins, then only two years old. President Daniel Coit Gilman of Johns Hopkins, a charter member of the ASSA and for many years one of its executive officers, gave the merger proposal serious consideration. Without either accepting it or finally rejecting it, he permitted himself to be named president of the Association in 1879–80, a position which enabled him to explore further the problems and potentialities of the proposal. However, Gilman completed his term as ASSA president without consummating the merger.

As he stepped down in 1880, Gilman recommended that the Association add to its existing departmental structure a new department devoted to historical studies. History (a field whose boundaries were not readily distinguishable from those of political science or even political economy at this time) was the only field within the general area of social science in which Johns Hopkins already possessed a reasonably strong faculty. Gilman no doubt expected that an ASSA Department of History would afford his faculty and students an avenue of influence and a platform on which to display the fruits of their research. Though favorably disposed toward Gilman's recommendation, the Association did not immediately act upon it.

Four years later, in 1884, historians convened to organize themselves on a national basis, as Gilman had hoped they would. But they did not organize as a department of the ASSA. Instead they formed the independent American Historical Association, the first of the modern academic specialist associations in the field of social science. In deference to the ASSA, the historians met with the old Association at its annual Saratoga Springs convention and held their charter meeting under its official auspices; nevertheless, they turned aside all requests to organize as a subordinate department. By organizing separately they established a pattern that was to be followed by all the other social science disciplines—a pattern that ultimately doomed the ASSA.

The separate organization of historians came as no surprise to Gilman; indeed, he actively encouraged the plan and may well have initiated it. Sometime between 1880 and 1884 Gilman evidently decided that the ASSA was not an adequate framework for social inquiry. Had he still wanted in 1884 the historical department he had proposed for the ASSA four years earlier, he probably could have had his way, for the organizer of the American Historical Association was Herbert Baxter Adams, a young professor at Johns Hopkins who was hardly likely to defy Gilman's wish.

Moreover, Adams's labors in the field of history were only part of a general campaign inspired by Gilman to organize the American academic world on a professional basis. The campaign may be said to have begun in 1883, when a Johns Hopkins professor of literature sparked the formation of the Modern Language Association. Then came the AHA in 1884. One year later Johns Hopkins professor of economics Richard T. Ely organized the American Economic Association. Hopkins faculty and students were also prominent leaders in the organization of the American Political Science Association in 1903 and the American Sociological Society in 1905. Indeed, in the pivotal quarter-century following 1883, Johns Hopkins people were instrumental in creating most of the specialist associations that demarcate the modern division of labor in the academic world.

Gilman has long been recognized as a singularly important figure in the professionalization of academic life in the United States.[1] The above sequence of events further underscores his importance and reveals him to be the key link between the amateur social science of the ASSA and the professional social science of the twentieth century.

[1.] See, for example, Hugh Hawkins, *Pioneer: A History of the Johns Hopkins University, 1874–1889* (Ithaca: Cornell University Press, 1960); Francesco Cordasco, *Daniel Coit Gilman and the Protean Ph.D.: The Shaping of American Graduate Education* (Leiden: E. J. Brill, 1960); Abraham Flexner, *Daniel Coit Gilman: Creator of the American Type of University* (New York: Harcourt Brace, 1946).

In fact, the whole crisis of authority provoked by the increasingly interdependent circumstances of an urbanizing, industrializing society came to a sharp focus in the life of this one man in the years between 1878 and 1884. To the admittedly limited extent that any process as complex as professionalization can be said to depend on the actions of one man, the path of institutional development in the social sciences did, for a brief moment, hinge upon those of Gilman.

Gilman's decision, sometime in the early 1880's, to reject the ASSA as an institutional framework for social inquiry and to undertake an alternative scheme of organization is obviously a matter of paramount significance. Accordingly, the remainder of this chapter is devoted to a close examination of the merger proposal and Gilman's reaction to it. The next chapter examines the first steps toward the modern scheme of social inquiry, the formation of the American Historical Association and the American Economic Association.

The Mathematician and the President

After the death of Agassiz in December 1873, the influence of the Lazzaroni in the ASSA continued unabated in the person of Benjamin Peirce. As director and later vice-president in the mid-1870's, the mathematician actively supported the Association's continuing efforts to formulate and propagate authoritative opinion about man and society. His influential position within the Association is reflected in the selection of David Ames Wells as president in January 1875. Wells, an economic writer appointed special Commissioner of the Revenue by Lincoln, was an ex-student of Agassiz and Peirce at Harvard's Lawrence Scientific School, one of the early Lazzaroni triumphs in educational reform.[2]

Peirce himself became acting president of the Association in 1878, when Wells fell ill. As presiding officer at the annual meeting held that year in Cincinnati, Peirce delivered an address on "The National Importance of Social Science." The address is distressingly vague, as many ASSA orations were, but it reveals Peirce to have been a conservative exponent of established authority, very much in the cultural

---

[2.] Wells succeeded Theodore Dwight Woolsey, ex-president of Yale, who very reluctantly served as acting president after the resignation of civil service reformer George William Curtis in the autumn of 1874. Woolsey at first declined the office, and may never have accepted, but the executive committee regarded him as interim president anyway. He had formed a branch association at New Haven. See Sanborn to D. C. Gilman, Boston, 23 Jan. 1875, Gilman Papers, Johns Hopkins University, and Sanborn to T. D. Woolsey, Boston, 15 Oct. and 10 Nov. 1874, Clarkson Collection (originals in Woolsey Papers, Yale University Library). See also Fred Bunyan Joyner, *David Ames Wells: Champion of Free Trade* (Cedar Rapids, Iowa: Torch Press, 1939), 139–150.

mainstream of his era. He had a profound faith in Progress; his attitude toward modern civilization was celebratory. Running through his address is an urgent quest for harmony, a horror of disorder. Peirce was confident that the inviolable laws of God were "indelibly engraved upon each man's nature" so they could not be "ground out by despotism, burned out by communism, nor voted out by the ballot box." His inclusion of the "ballot box" as a threat comparable to communism and despotism is an illuminating comment on the whole movement to establish authority. Peirce began and ended his address with an allusion to the Tower of Babel:

> ... The only solid basis for an enduring republic is the Rock of Ages. Any other foundation is unstable and insecure as the sands of the seashore. Let the tower be built in obedience to God's laws, and it will reach unto heaven; the children of men will reunite in permanent harmony; science and religion will coincide; and the one universal speech will be God's word written on the sun, moon, and stars, on the solid earth itself, and in the Gospel.[3]

Like most educated men of his era, Peirce felt sure that there was only one correct way for man to conduct himself in society. That way had not been fully revealed to man—it was not more obvious, certainly, than the laws of geometry, some of which still eluded discovery—but it would yield to patient and sincere investigation. Though the right way of organizing society was not fully known, certain policies and practices seemed to possess an air of indisputable validity. Men like Peirce, Wells, and Agassiz really could not doubt that the proper society would conform to the teachings of the Bible; nor could they seriously question the ultimate correctness of more specific and much more vulnerable policies, such as free trade or sound currency. They could not believe that research and analysis would lead far away from their own ideal visions of the good society. For them, it was reassuring to believe that there was a science of society, for they felt certain that science would sustain the values they ranked highest. "Science and religion will coincide," as Peirce put it.

It would be a mistake, however, to conclude that Peirce and men like him set out to erect a static defense of received opinion. On the contrary, they felt more keenly than most of their contemporaries the necessity for research and analysis of social questions and human relationships. One does not analyze existing relationships if one perceives them as given and eternal. For many of Peirce's contemporaries it still seemed abhorrent even to raise questions about man and society,

---

[3.] Benjamin Peirce, "The National Importance of Social Science in the United States," *JSS*, 12 (Dec. 1880), xii, xxi.

for these were given, even sacred, matters. As men of science, Peirce, Agassiz, and Wells could not help but realize that to engage in research is to risk a painful reversal of opinion. Their confidence that scientific investigation would vindicate their own opinions and values should not obscure their willingness to submit these matters to investigators over whose conclusions they would have no final control. But investigators entrusted with such a sensitive task would have to be true scholars, professionals who devoted full time to their investigations, who policed each other's opinions, and who stood ready to denounce charlatanism. It was for these reasons that the university had to assume command of social inquiry.

The man to whom Peirce addressed his merger proposal, Daniel Coit Gilman, was then presiding over the new Johns Hopkins University, the most exciting development to occur in American higher education during the nineteenth century. Gilman's educational reform plans took shape during many frustrating years at Yale, where his work with the Sheffield Scientific School won him a nationwide reputation as an academic administrator. When he lost the presidency of Yale to traditionalist Noah Porter in 1871, he went west to become president of the University of California at Berkeley, a position he had turned down two years earlier, just as he had turned down a similar offer from the University of Wisconsin in 1867.[4] When in 1874 the trustees of Johns Hopkins earnestly queried the leading educators of the day to help them decide who ought to preside over the new university they were building at Baltimore, one candidate was named by Eliot of Harvard, Porter of Yale, White of Cornell, and Angell of Michigan: Daniel Coit Gilman was the man.[5]

Gilman proposed to the Hopkins trustees what White had previously recommended to them, and what some of them seem to have favored from the beginning: namely, that they use Hopkins's generous bequest, the largest yet made in America, to construct an institution exclusively for graduate education. Gilman wanted a research institution, one concerned not with the instruction of undergraduates but devoted to the advancement of the frontiers of knowledge and the training of professional scholars. Actually, before the school opened Gilman was forced to compromise and admit undergradutes, but even so, his university was a dream come true for those who had struggled to establish authority by advancing university education in the United States.

Peirce regarded Hopkins as the fulfillment of hopes that he and Agassiz and other friends of sound opinion had cherished for decades.

[4.] Franklin, *Life of Daniel Coit Gilman*, 101–109.
[5.] Hawkins, *Pioneer*, 14–15.

PROPOSED MERGER

In the words of his son, James Mills Peirce, the elder Peirce considered Hopkins "as a great advance in the university system of this country, and as the only institution where the promotion of science is the supreme object, and the trick of pedagogy is reckoned as of no value."[6] In assembling a faculty at Baltimore, Gilman consulted with Peirce on several occasions and appealed to his authority when trying to persuade the Hopkins trustees of the importance of hiring a few scholars of world-wide eminence.[7] Gilman, like Peirce, had long been associated with the ASSA. He was a founding member of the Association and served as one of the select nominating committee at the first meeting in October 1865. He became a director in 1867 and served as vice president from 1869 on.

The Merger Proposal

When Gilman received Peirce's letter proposing the merger, he immediately recognized its importance. To be sure he understood Peirce's difficult scrawl, he copied over the main body of the letter. It began by informing him that he was to be selected as the ASSA's next president. It then went on to explain that the secretary, Frank Sanborn, and the treasurer, Gamaliel Bradford, both wished to resign their positions. Peirce observed that they could hardly be blamed, in view of the meager salary offered the secretary and the inherent difficulty of fund raising. The onerous chore of replacing them might be obviated, however, if Gilman would just adopt Peirce's grand plan. Peirce wrote as follows:

> The present plan is to carry in [sic] the Association just as the Scientific Association is conducted, and which will so much reduce the expenses that no addition to the members will be required. I have, for a long time, favored this course—but I have now—new ideas which I propose to submit to you for consultation and adoption or rejection.
>
> I regard the true University as the only hopeful way of educating the people to the right method of organizing the republic. I see in the past history of the Universities and in the present state of Scientific Association, that there is a great disposition in men to convene for the study of difficult problems, and that when the conventions are properly organized thousands will assemble at the discussion. I wish, then, to see the Associations regulated by the University and become part of it. Now your University is the only one which is untrammelled so that it can undertake this great task, and your own ardent and energetic nature is peculiarly fit to be the leader of the undertaking. I would have meetings held in Baltimore, in the rooms of your university. But I would have the subjects

---

[6.] Quoted in Hawkins, *Pioneer*, 77. See also ch. I and pp. 21–22 of that volume.
[7.] *Ibid.*, 34, 39, 44.

of discussion distributed through a large part of the year. Notice should be given for instance that financial matters will be discussed in one week—mathematical in another, chemical in another—etc. The members should be accommodated in rooms, some for single men, others for families—with one common table—and at a moderate expense.

It would be expedient perhaps to have the title of Chancellor of the University substituted for that of President—to have presidents of sections etc.—but the Chancellor to preside over the whole.

A secretary will be more than ever required to see to the arrangement of the meetings—but he should be a paid officer of the University. The fees of members will go to the University, and the members will be in fact pupils of the University. It will take a little time for scientists to understand the new system—but when understood, I feel sure that it will be found to be the way to fill the country with the needed light. We will see the substitution of a continuous debate—for the short and unsatisfactory discussions, to which we are now restricted.

I hope you will think the matter over and rest assured that I am wedded to no one feature of the plan—but earnestly desire to do what I may to awaken the country to the importance of some better way to stimulate thoughts and obtain the advantages of the interchange of ideas.

Your very sincere friend,
Benjamin Peirce[8]

Peirce's bold and rather presumptuous proposal was ambiguous at several points. His concern for the title of the presiding officer is puzzling. His reference to sessions on mathematics and chemistry seems to indicate that he hoped to see not only the ASSA, but also the American Association for the Advancement of Science merged with the university, yet this is not spelled out at all. The details of the proposal became clearer in the following months as Peirce, Gilman, and Sanborn exchanged thoughts and explored the practical ramifications of the merger. It was clear from the beginning, however, that Peirce envisioned a university that would absorb and "regulate," as he said, groups of citizen-scholars like the ASSA, and thus enhance communication among inquirers and insure the soundness of their views. The university would incorporate into its structure a means of regular

---

[8.] Peirce to Gilman, Cambridge, 10 June 1878, Gilman Papers, Johns Hopkins University. The original letter is quoted here; Gilman's copy errs in several inconsequential ways. The proposal to merge with Johns Hopkins was never mentioned in the *Journal of Social Science*, though there were vague references to the general idea of merging with some university. The Bernards in their *Origins of American Sociology* were unaware of the proposal but circumstantial evidence in their account might well lead one to expect that there was more in Gilman's presidency of the ASSA than meets the eye. I first saw portions of the correspondence between Peirce, Gilman, and Sanborn in an early draft of Betty P. Broadhurst's "Social Thought, Social Practice and Social Work Education: Sanborn, Ely, Warner, Richmond" (D.S.W. dissertation, Columbia University, 1971). I am grateful for her generosity.

interaction with adult non-academic inquirers, who presumably would contribute their knowledge and experience to the university, as teachers do, but whose relationship to the university nonetheless would be that of "pupils." The ultimate goal was to educate the people to the "right method of organizing the republic."

In retrospect, Peirce's proposal seems odd and unrealistic. What did the ASSA have to offer Johns Hopkins? It is tempting not even to take the proposal seriously, knowing as we do today that the ASSA and its style of social inquiry were soon plunged in obscurity, while Hopkins was soon to become the keystone in the modern American university system. Perhaps the strangest feature of the plan, from the vantage point of the present, is Peirce's unspoken assumption that much vital social inquiry would continue to take place outside the walls of the university and that the university, therefore, would suffer if it were not in contact with non-academic inquirers. It was essential to Peirce's plan that the university exercise a regulatory influence over social thinkers, but he seems not to have anticipated that social inquiry would become virtually the exclusive province of academic professionals.

Certainly Frank Sanborn did not see the proposal as a step toward exclusiveness. He regarded it as a means by which qualified amateurs might gain access to the university, rather than a means by which the university might subordinate, control, and finally exclude amateurs. After Peirce's death in 1880, Sanborn (whose editorial memory is not to be taken at face value) recalled the merger proposal in a way that stressed its democratizing rather than its professionalizing aspect. "Professor Peirce's conception of the American Social Science Association," said Sanborn, "was this,—that it should be a *university for the people*,—combining those who can contribute anything original in social science into a temporary academical senate, to meet for some weeks ... debate questions ... give out information. ..."[9] According to Sanborn, Peirce wished to see the universities become like their medieval predecessors, "camps for intellectual combat" rather than "stagnant endowments for the perpetuation of ancient use and error." Quoting Peirce, probably from memory, Sanborn said: " 'the old impromptu university died out: but our free republic is the place, and this is the time to revive it.' "[10]

Sanborn probably exaggerated the democratic emphasis of Peirce's merger proposal; certainly Peirce's own words, and his career as a member of the Lazzaroni, suggest that greater stress should be placed upon the regulatory and professionalizing aspects of the plan than

---

[9] *JSS*, 12 (Dec. 1880), xi. Italics in the original.
[10] *Ibid.*, 14 (Nov. 1881), 28.

Sanborn was willing to give. But regardless of the precise balance of motives in Peirce's mind, it is clear that he and Sanborn both thought that Johns Hopkins stood to gain from a connection with the ASSA and its nonprofessional inquirers into the nature of man and society. It is this assumption which gives the merger proposal its archaic ring. Within the next two decades it would become increasingly obvious that the amateur social thinker could not survive in competition for authority and legitimacy with the professional academic social scientist. The university and its professional inquirers would soon possess such authority that such a merger would seem positively to detract from the university's position. But the future emergence of professional social science was not obvious in 1878.

As a young university president and nominally a professor of political geography, Gilman was in a much better position than Peirce or Sanborn to judge trends in the development both of the university and of social inquiry. Had he dismissed Peirce's proposal out of hand, we would be justified in regarding the project as merely an ambitious pipe dream on the part of Peirce and Sanborn, a product of their remoteness from the latest developments in social thought and higher education. But Gilman did not dismiss the project; rather, it appears that he seriously considered Peirce's proposal. Only in retrospect, if at all, can it be said that the proposed merger was a practical impossibility.

Gilman wrote Peirce [11] his initial reaction to the proposal on 9 July 1878:

> You seem to me to outline the *ideal* of a scientific congress, & to be quite just in your conception of the inherent tendency of men to come together that they may learn from one another. If the great universities of our country would only assume the direction of scientific assemblies, & the existing associations would consent to come under the leadership of the universities, science would be efficiently advanced & diffused. If Harvard, for example, would assume the General Assoc. for Advt. of Science or the Nat. Acad. [National Academy of Science] or both; Yale the Oriental & Philological Associations; or either; Amherst or Dartmouth, the Amer. Inst. of Instruction & so on, —then I think the Baltimoreans would perhaps not be presumptuous in offering to the Social Science Association a home. Perhaps even without waiting for other institutions to act, a beginning might be made in Baltimore. I have my doubts whether even for this, the country is quite ripe, —& before speaking officially I ought to

---

[11.] Gilman first responded to Peirce's letter of 10 June 1878 with a brief note explaining that the closing work of the academic year prevented him from writing an adequate reply to "your very interesting suggestion." Gilman was on his way to Cambridge and promised to see Peirce within the week: Gilman to Peirce, Baltimore, 16 June 1878, B. Peirce Papers, Houghton Library, Harvard University. They were unable to meet, however, so Gilman wrote Peirce the letter of 9 July.

consult our professors & trustees. Probably the best way to shape public opinion, if it is thought best to go forward in this direction, would be to have one prolonged session in Baltimore or elsewhere, with abundant plans for discussions & lecturers, & see what results came. I am sure that there will be a cordial response on the part of the University there to any progressive proposal which is reasonable, & I need hardly assure you of my own earnest desire to advance the interests of the Amer. Soc. Sci. Association & to respond promptly and heartily to any suggestions coming from you by whose counsel & encouragement, I have been so greatly benefitted.[12]

Peirce took this promising but cautious response to mean that Gilman had adopted the scheme. "Your letter gave me immense satisfaction," wrote Peirce. "I regard the battle as gained and that nothing remains to but to [sic] act with caution and prudence."[13]

At the October meeting of the Executive Committee of the Social Science Association Peirce announced, apparently for the first time in public, his plan for the absorption of the ASSA by Johns Hopkins. As secretary, Sanborn wrote a long letter to Gilman making the formal offer of union. According to Sanborn, Peirce told the committee that the time had come when social science could best be advanced by such a connection with a single university. Papers and debates could become part of the university's regular work and be published at university expense, he noted. The Executive Committee received the proposition "very favorably," and Sanborn expressed his own confidence that the Association as a whole would approve of "a feasible working-plan, by which the promotion of Social Science should be made one of the objects of a University so liberally organized as yours. . . . In that case you would probably have *carte blanche* to arrange the details as might be most agreeable to your own theory of Education and of public discussion. . . ."[14] Sanborn reminded Gilman that a new treasurer was needed immediately to replace Gamaliel Bradford. Sanborn also reasserted his own intention to resign, but by implication he did not rule out a paid position for himself with the university, and he volunteered to continue as secretary on the present basis if needed to carry out the "transformation" of the Association.

His letter closed with the first of several gross misjudgments of the heavy burden borne by Gilman as a university president, and of the

---

[12.] Gilman to Peirce, Amherst, Mass., 9 July 1878, B. Peirce Papers, Houghton Library, Harvard University. Italics in the original.

[13.] Peirce to Gilman, Cambridge, 9 Aug. 1878, Gilman Papers, Johns Hopkins University. Peirce at the same time proposed another meeting which probably took place in Northampton, Mass., in mid-August 1878, of which I find no record.

[14.] Sanborn to Gilman, Concord, Mass., 17 Oct. 1878, Gilman Papers, Johns Hopkins University.

modest place (if any) that the ASSA would occupy in Johns Hopkins University. Sanborn asked Gilman to "set forth in some detail what your plan of operations would be," in time for the next meeting of the executive committee nine days hence. The demand was unrealistic. It suggests that Sanborn had little appreciation of the sensitivity of the proposed merger or of the demands made upon a modern university president's time.[15]

Gilman's response to this peremptory request reflects a growing, though not complete, disenchantment with the project. The idea of merger, or perhaps a somewhat less intimate form of connection between the ASSA and Johns Hopkins, seems not to have been totally rejected by Gilman until sometime in 1880 or later, after he had served for a time as president of the Association. But one major objection was clear to him in October 1878 when he replied to Sanborn's formal offer of union:

> In respect to the connection of the Association with this or any other University,—I think it would be well to consider the two-fold obligations,—investigation & agitation. A *University* should promote study, research, the accumulation of experience, the publication of results. Such work should be steady, quiet, prolonged, attracting but little, passing attention. The Association should endeavor to act upon the public, by meetings, addresses, newspaper-reports, & other modes of awakening attention to possible and necessary reforms. I doubt whether it would be wise to merge the functions of the Association, so far as agitation is concerned, in the University: but it does seem to me that any university, & this of Baltimore in particular should encourage heartily and in many ways scientific studies of social questions.[16]

From this point on, Gilman took care to keep the ASSA at arm's length, without ever quite rejecting the idea of some kind of connection with the Association. In lieu of a direct connection with one university, he proposed that the ASSA meet from time to time in various universities, "to enlist scientific help in various departments of activity." Johns Hopkins would of course welcome such a meeting, but if the ASSA were to meet in Baltimore he thought it preferable that someone other than himself—perhaps Peirce—be president, so that the university could appropriately "show honor and extend hospitality."

[15.] Later, when Gilman was slow to send his presidential address to Sanborn for publication, Sanborn sent him an angry letter complaining that "these are the things that make my secretaryship such an irksome task. Everybody makes my convenience wait on his...." Sanborn to Gilman, Concord, 6 Nov. 1879, Gilman Papers, Johns Hopkins University.

[16.] Gilman to Sanborn, Baltimore, 24 Oct. 1878, Gilman Papers, Johns Hopkins University. Italics in the original.

Gilman made it very clear that he would not be disappointed if the presidency went to someone else. In effect he had rejected the details of Peirce's proposal, though not the general principle.

> You observe that I have no plan to propose looking toward changes in the work of the association. If any affiliation should be established between the Johns Hopkins University and the society, the steps will probably be gradual. Thus far I have not consulted those on whose judgment I must rely among our Trustees and Professors for I was not quite clear enough to formulate any proposition.[17]

Gilman assured Sanborn that Hopkins would, in one way or another, advance the field of social science. "In the original plan of this institution," said Gilman, "it was expected that marked attention would be given to social science." At the "proper time," these plans, like many others, would be carried out. "Already something has been done in historical and political science, and social science comes naturally on."

Though he may not have realized it at the time, Gilman's most revealing comment was his recommendation for the office of treasurer. He suggested Henry W. Farnam, "a graduate of Yale who has just returned from Germany, where he took the degree of Doctor of *Staatswissenschaften*." The potency of the German doctorate was just beginning to be felt. In the next two decades its authority would gradually eclipse the gentlemen-scholars of the ASSA and their view of man and society. Farnam had virtues appropriate to both the old regime and the new. His father, Gilman took care to point out, was "a wealthy citizen of New Haven, well known as a benefactor of Yale College." Well-connected, well-educated, and wealthy too, he occurred to Gilman as a man who might contribute "time, learning and perhaps money to promote an object in which he is deeply interested." Indeed, Gilman went on, "if our institution were to look for a young professor in that branch, he is one of the first of whom I should think."[18] If men like Farnam were available—professional teachers and researchers with prestigious German degrees—what need was there for the ASSA?

Sanborn did his best to assure Gilman that the reformist stance of the Association would not become an albatross around the university's neck. "As to what you say about the two-fold character of our work," wrote Sanborn, "we agree. I do not see how the agitation we sometimes

---

[17.] *Ibid.*

[18.] *Ibid.* The treasurer's job was not offered to Farnam because he was deemed too young to work with the men who had money to give. The executive council acknowledged, however, that he would make an excellent secretary, especially if employed at Johns Hopkins. He taught at Yale and served as secretary of the ASSA Finance Department in 1882-83.

have carried on could find a proper place in your university, or any other." Having shown practical reform out the front door, Sanborn called it in the back.

> But if the work of investigation and exposition, such as properly belongs to a great university, were there undertaken, in Social Science as in the Physical Sciences, the other work, of agitation and practical illustration (as in the administration of charities and the application of economic principles) could be left to State and city governments, and to special associations or departments of our association, to carry forward according to the demand of the times. I have supposed that this portion of our task could best be performed by local organizations... and it would very much aid the efficiency of these local, and to some extent, temporary agencies, if there were a single recognized center of Social Science research, such as your University might perhaps become if the scheme of Prof. Peirce could be carried out.[19]

Sanborn reported that Peirce was "a little disappointed at your suggestion of dividing the Social Science work among several Universities,—hoping that you would see an opportunity to make your own foremost, at a very long interval, in a matter of this kind."[20] But Gilman was not persuaded. Though he seems not to have completely rejected a connection with the Association, he was not prepared to leap into an intimate relationship in 1878. At the end of the year Gilman attended a meeting of the executive council and discussed the merger plan with Sanborn privately. Following their discussion, Sanborn wrote to Peirce that it was "as you inferred,—that he was doubtful of his power to carry out such a plan as we talked about, and therefore hoped it would not be pressed at present."[21]

It should not be thought that Gilman grew cool to the idea of merger because Johns Hopkins already possessed a full faculty of social scientists. Far from it. The university was only two years old in 1878; among the first six full-time professors at Hopkins, who still constituted the whole permanent staff in 1878, not one could be called a social scientist. At the beginning the nearest thing to a regular appointment in the general area of social science was held by Austin Scott, George Bancroft's secretary, who commuted from Washington, D.C., to deliver lectures on history. Thomas M. Cooley held a lectureship in jurisprudence, but only on a visiting basis. Of the original twenty-one "fellows" who performed some teaching duties, only three could be called social

---

[19.] Sanborn to Gilman, Concord, 5 Dec. 1878, Gilman Papers, Johns Hopkins University.
[20.] *Ibid.*
[21.] Sanborn to "My Dear Sir" [Peirce], Concord, 2 Jan. 1879, B. Peirce Papers, Houghton Library, Harvard University.

scientists. Among these were the economist Henry Carter Adams, and Herbert Baxter Adams, peripatetic organizer and "germ theory" historian.[22]

When Peirce wrote his initial proposal to Gilman in June 1878, the university was just about to award its first four Ph.D.'s. One of these went to Henry C. Adams in political economy, but he had already done his course work in Germany and there was no regular offering in political economy at Johns Hopkins until the spring term of 1879. Then, remarkably, the subject was taught by the versatile Herbert Baxter Adams. Henry C. Adams returned to lecture in political economy in 1880, but there was no regular appointment in the field until autumn of the following year, when Richard T. Ely made his first appearance at Hopkins. Herbert Baxter Adams taught history from the beginning, and in fact drew more students than Austin Scott; but he was not offered a regular appointment as associate until 1881. Thus even history and political economy, the two oldest and best-established fields in the general area of man and society, were only modestly represented at Hopkins when Peirce, Sanborn, and Gilman talked about a merger with the ASSA. Psychology was introduced by G. Stanley Hall in January 1882, and he was the only "social scientist" to be given full professorial rank before the 1890's. To be sure, the low rank of the social scientists at Hopkins merely reflects their youthfulness, rather than a policy decision to downgrade the social sciences: during the seventies and eighties Johns Hopkins offered full professorships to Francis A. Walker in economics, Herman Von Holst and Henry Adams in history, and Thomas M. Cooley in jurisprudence, but none of these academic stars could be lured to Baltimore.[23]

It is not surprising, then, that Gilman was willing to become president of the ASSA even if he was unwilling to commit himself to a merger. As president he could gradually assess the possibilities and pitfalls of Peirce's merger project and try his hand at controlling a Boston-bred organization from the remote hinterlands of Baltimore. He must have been skeptical from the beginning that the ASSA could become a functioning part of the university, but he was not willing to let the opportunity go by the board without a trial. In a few years, perhaps even before the end of the eighties, the Association's speculative and undisciplined style of inquiry would seem distinctly out of place in a university. But in 1878 Gilman still wanted to keep the fish on the hook.

I have found no record of Gilman's private estimate of the ASSA during his two years as president, but his attitude is reflected in his serious efforts to reorganize and revitalize the Association. Under his

[22.] Hawkins, *Pioneer*, 55–56, 82–83.
[23.] See *ibid.*, 55–56, 169–173, 193 and *passim*.

leadership a number of constitutional reforms were adopted, with the stated purpose of "increasing the efficiency and reducing the expenses of this Association."[24] The chairman of the committee that formulated the new plan was Francis Wayland, Jr., son of the antebellum moral philosopher and dean of Yale Law School. The plan called for the Association to give up the office it had long maintained in Pemberton Square, in the shadow of the Massachusetts legislature, and to hold major meetings in cities other than Boston, thus ending the monopoly on the selection of officers that Massachusetts members had always enjoyed. The Executive Committee was streamlined, and the preparation of papers and debates for the annual meeting was supervised more carefully than ever before. The main effect of the reforms was to nationalize the Association (or at least detach it from its birthplace) and shift its orientation slightly, away from eclectic reform toward a more disciplined kind of inquiry.

Gilman's vigorous efforts to reform the Association suggest at once his faith in its potential and his concern for its failings. In a published 1880 review of the recent numbers of the *Journal of Social Science*, he observed that "the papers thus collected are of uneven quality—and some of them are too diffuse in style to be permanently valuable—but on the whole, they are suggestive, instructive, and comprehensive."[25]

In his 1880 presidential address Gilman spoke as if he still hoped the Association might be transformed in such a way as to make it a vital center of social inquiry and perhaps a fit partner for Johns Hopkins. "The aim of this Association," he declared, is to "study . . . the conditions which tend to make a perfect state of society where 'each is for all and all is for each,' and . . . [to discover] those laws of cooperation which will secure to every individual his highest development." But this progessive sentiment was not to be understood as advocacy of direct engagement in politics, or practical reform. The Association, Gilman emphasized, "is not a society for the promotion of reform, nor an assembly whose object is charity; but its object is the promotion of science, the ascertainment of principles and laws."[26] Gilman would have liked to see the Association become what it really was not, and probably could never become: an institution which—like the university—would stand aloof from the buzzing confusion of society, and yet influence society toward higher goals by indirect means.

In his last presidential address Gilman recommended a program that would have genuinely transformed the Association had it ever

---

[24.] *JSS*, 10 (Dec. 1879), v.
[25.] Unidentified newspaper clipping (presumably written by Gilman) dated Apr. 1880 in Box 1, Gilman Papers, Johns Hopkins University.
[26.] *JSS*, 12 (Dec. 1880), xxiii, xxii (abstract furnished by Gilman).

been carried out. He suggested that the ASSA's chief function, like that of any scientific society, should be to publish papers not publishable elsewhere. In a modest way, the ASSA was already performing this function with the *Journal of Social Science,* but Gilman wanted to go one step further: instead of passively waiting for individual researchers to submit their unrelated inquiries and speculations, he proposed that the ASSA take upon itself the responsibility for *organizing* and *sponsoring* inquiry. The Association, said Gilman, should "organize plans of investigation and research." It should produce progress reports on vital fields of concern and publish regular reviews of current literature. Ideally, Gilman went on, the Association ought to initiate inquiries and supply money to support them.[27]

If the Association had assumed these functions, it might very well have become a fit partner for Johns Hopkins. In any case, it would have been fit for survival in a modern society. Even if it were not integrated into the structure of a single university, it might have offered academic inquirers a viable framework within which to organize disciplines on a national basis. The German counterpart of the ASSA, the *Verein für Sozialpolitik,* did in fact organize and sponsor scholarly inquiry: no less a scholar than Max Weber benefited from its largesse in the 1880's and 1890's.[28] Moreover, Gilman wanted learned societies to provide an alternative haven for men of intellect, one even purer in its dedication to research than the university.[29] For the ASSA to assume such a role would have required major alterations in its structure, operations, and style of inquiry, but Gilman evidently thought the transformation possible.

Gilman's statements and actions from 1878 to 1880 demonstrate that he took the merger proposal seriously and still looked upon the ASSA as a viable means of organizing social inquiry, if only it could be suitably invigorated. It is highly significant that he did so, for Gilman was the single most important figure in the professionalization of the American academic world. No one contributed more than he did to the establishment of the modern academician as a research specialist, certified and judged by an organized disciplinary community.

Gilman's respectful, if ambivalent, attitude toward the ASSA as late as 1880 signifies that the processes impelling professionalization had not yet proceeded far enough to strip the ASSA of its authority or destroy its credibility as a fountainhead of sound social opinion. No one

---

[27] *Ibid.,* xxiii-xxiv.
[28] Oberschall, *Empirical Social Research in Germany,* ch. II and *passim.*
[29] See Gilman's address before the American Philosophical Society, "The Alliance of Universities and Learned Societies," reprint dated 15 Mar. 1880 in notebook entitled "Various Speeches and Articles," Vol. IV, Gilman Papers, Johns Hopkins University.

could be a better judge of this question of timing than Gilman. The Association was challengeable, and certain of its shortcomings were evident. But the university and its professionally organized social scientists had not yet monopolized authority in questions of man and society, even in the eyes of the man who did more than any other individual to create the modern university and the professional academician.

Within a few years, however, Gilman would change his mind. In 1880 he stood on the very edge of the transition from amateur to professional social science; by 1884, when he encouraged the formation of the American Historical Association, he stood on the other side. Once on the other side, out of touch with the premises that had made the ASSA seem a sensible way to organize social inquiry, he began to see it in a new light, as the frivolous gathering of dilettantes, cranks, and charlatans that it has appeared to be to all later generations. By 1890, when his good friend Andrew D. White suggested a rendevous at Saratoga for the annual ASSA meeting, Gilman turned him down. "It is too soon to promise to go to Saratoga," wrote Gilman. "I went last year & we had a very dull time. . . ."[30]

Gilman's Rejection of the Merger

Why did Gilman finally reject the merger proposal and embark on a course of action that ultimately doomed the ASSA? In answering this vital question several things must be remembered: men are motivated by factors of which they are not fully conscious; they are conscious of more than they say; they say more than they write; and of what they write only a fragment is preserved. It is the crudest of historical fallacies to suppose that the written record discloses all the considerations that shaped men's action. To explain human motivation, conjecture is not only permissible, but preferable to a literal-minded adherence to surviving documents alone. This is not to deny the risks of conjecture, but only to insist that the apparently risk-free strategy of documentary literalism is not so trustworthy as it appears. And of course conjectural attribution of motives must be controlled, for it must be compatible with the evidence.

The most prominent reason for Gilman's rejection of the merger requires no conjecture: he stated explicitly in his letter of 24 October his feeling that the Association and the University had divergent functions—"agitation" on the one hand and "investigation" on the other. He acknowledged the legitimacy of both functions but felt that it would be incongruous to mix the two in a single institution. This

---

[30.] Gilman to White, Northeast Harbor, Maine, 17 July 1890, A. D. White Papers, Cornell University.

perhaps is what he meant by his curious statment that he "was not quite clear enough to formulate any proposition" for his faculty or trustees. He simply could not see where to begin, so strong was his sense of the incongruity of the two functions.

Thirteen years earlier the founders of the ASSA—Gilman among them—had built their Association on the implicit premise that investigation and agitation worked hand in glove. Who could better agitate and carry out reform than the person who had investigated the facts? The loose structure of the Association, its undisciplined style of inquiry, its rosy optimism—all its essential features, in fact—were built on the assumption that rational minds, once possessed of the facts, would quickly reach a consensus about what needed to be done. This assumption had permitted the founders to ignore any distinction between agitation and investigation.

Yet by 1878 Gilman distinguished sharply between the two tasks and declared them incompatible. Why? Much had happened in the intervening years to compel men to recognize that the critical problem was not the collection of facts, but their interpretation—and that decent men might arrive at different interpretations. The Republican reform consensus of the war years had long since broken down; the depression of the early 1870's and the great railroad strikes of 1877 had inaugurated a period of rising class tension and social controversy that would not reach its peak until the 1890's. Consensus could no longer be taken for granted; competition among men and ideas for positions of authority was growing sharper all the time.

As president of a young university dependent on industrialists and businessmen for financial assistance, Gilman may have wanted to exclude the agitative function simply for the sake of prudence. Certainly it was safer for the university to investigate than to agitate. Gilman's sense of the incongruity of the merger proposal might then be resolved into a simple case of his timidity (supposing that he still wanted reform but was afraid to support it) or conservatism (if his conversion was sincere). In support of this thesis one might point to Sanborn's comment at the end of the merger negotiations: Gilman, he said, "was doubtful of his power to carry out such a plan," a phrase that suggests timidity. Against the thesis, however, one must remember Gilman's long identification with the Association and its respectable, essentially moderate, if not conservative, brand of reformism. It had not in the past and would not in the future do much to incur the wrath of conservatives. Some of Gilman's regular faculty, like Richard T. Ely, would do much more.

Still, many readers will be content to attribute Gilman's rejection of the ASSA primarily to timid conservatism. Indeed, the whole tendency of late nineteenth century intellectual development toward a more objective, "value-free" style of inquiry and a more disciplined, professional mode of organization has often been cast in this same framework of an intellectual retreat in the face of conservative intimidation.[31] This line of explanation is, I believe, profoundly misleading. It treats the shift from advocacy to objectivity and from amateur to professional social thought as a superficial phenomenon, lacking any deep social or intellectual roots. On the contrary, I believe that these changes were provoked by profound alterations in the conditions of authority and explanation in an interdependent, urban-industrial society. Earlier chapters of this study, focusing on the intellectual consequences of interdependence and the rise of a movement to establish authority, permit a fuller explanation of Gilman's decision.

Gilman rejected the merger not because of simple conservative caution, but because he recognized that of the two functions, investigation and agitation, only the former could be professionalized. Hence only the investigative function should be incorporated into the university, for only it could benefit from what the university had to offer—an institutional framework within which to organize professional communities of competent inquirers. Only such communities could provide a safe haven for sound opinion, and only in the university could such communities flourish. To later generations the university would display other faces and offer other opportunities, but to Gilman's generation—the generation that laid the groundwork of the American university system—a prime task of the university was to establish authority. Other goals could be subordinated to that end.

To understand that the agitative function could not be professionalized, Gilman needed to see only two things: that sound opinion required the support of an organized community of inquiry, and that communities of inquiry could not be organized in extremely controversial fields. The experience of the gentry class in the late nineteenth century taught these two lessons. The most elementary premise of the movement to establish authority was that sound opinion could not defend itself, but required institutional support. And the most effective form of institutional support was an organized community of the competent that would discredit charlatans, insulate serious thinkers from the distractions of a mass public, and permit the strongest to rise above idiosyncrasy and speak with the sanction of the whole community.

[31.] This is the dominant theme of explanation in Furner's *Advocacy and Objectivity*. See also J. P. Nettl, "Ideas, Intellectuals, and Structures of Dissent," in *On Intellectuals: Theoretical Studies, Case Studies*, ed. P. Rieff (Garden City, N.Y.: Doubleday, 1970), 57–134.

But in extremely controversial fields no community can form. The *sine qua non* of a community of the competent is an implicit consensus about the criteria of competence. The members of such a community must perceive even their bitterest rivals inside the community as fundamentally competent, no matter how grudging the admission. To do otherwise is to dissolve the bonds of community, for this very special kind of community includes *only* the competent. To perceive one's opponents as truly incompetent is to exclude them from the community.

Gilman knew, I surmise, that professional authority is an exotic phenomenon that thrives only at a point midway between the self-evident and the impossibly controversial. At the first extreme the layman has no use for the expert; at the second there are parties, prophets, and cults, but no professional experts, for consensus is impossible even among those with some claim to competence.

The point of Gilman's distinction between investigation and agitation was not to condemn agitation, but simply to observe its inherently controversial nature and declare it unfit for professionalization.[32] The founders of the ASSA in 1865 had not believed practical reform to be an inherently controversial activity: they assumed that rational inquirers would achieve a consensus not only about the basic principles of human affairs but also about practical reform measures. By 1878 this assumption was beginning to seem naive. Gilman recognized that abstract inquiry into the essential nature of man and society was sufficiently removed from the fray that it might yield a professional consensus; however, practical agitation was not so removed. If authority in the field of man and society was to be established at all, the friends of authority would have to settle for half a loaf—they would have to form rigorous communities devoted to investigation and interpretation, leaving agitation to fend for itself.

If Gilman understood (articulately or not) all that I suggest he did, then his action ought to reflect that understanding. It does. His analysis of the situation, as I reconstruct it, left him only two options: reject the ASSA utterly, or reform it. He tried first to reform it; failing that, he

---

[32.] Gilman's distinction between agitation and investigation is not simply an early version of the debate of the 1960's over "value-free" social science. The terms of the argument were quite different in the late nineteenth century. In the debate of the 1960's, the advocative style was consistently propounded by the left; in Gilman's era, the style was shared by right and left and was attacked by both. Prime examples of late nineteenth century "advocates" were David Wells, Edward Atkinson, and William Graham Sumner—all ASSA members and all conservative in the eyes of an "advocate" of the left such as Richard T. Ely. Opponents of advocacy included liberals such as Thorstein Veblen as well as conservatives such as James Laurence Laughlin. Opposition to advocacy, or "agitation" as Gilman put it, was not simply a reflection of conservatism; it was more closely correlated with a relativistic view of human affairs.

rejected it. The vital reform was to redefine "social science" as a principally investigative enterprise that stopped short of agitation. Only that reform would encourage a broader consensus about the criteria of competence in the field and thus make social science suitable for inclusion as a professional discipline (or cluster of disciplines, as it turned out) in the university. This is exactly what Gilman tried to do during his two years as president of the ASSA, and it is why the Association closed the lobbying office it had long maintained near the Massachusetts State House. This is the significance of Gilman's last address when he declared (wishfully and contrary to fact) that this "is not a society for the promotion of reform, nor an assembly whose object is charity; but its object is the promotion of science, the ascertainment of principles and laws."

But the Association would not so easily abandon the broader definition of "social science" upon which it was founded. An activist conception of "social science" was integral to the Association and could not be snipped off like a loose thread. Three interrelated factors obstructed Gilman's efforts to redefine the Association's mission: its role as "headquarters" of the movement to establish authority, its strongly individualistic bent, and its domination by men of affairs.

The leaders of the ASSA were practical men of affairs—men like Frank Sanborn, Massachusetts Inspector of Charities; David Wells, Commissioner of Revenue; Andrew White, Ambassador to Germany; William T. Harris, Commissioner of Education. They cut commanding figures and spoke with the authority of personal achievement. The audience fell silent and newspaper reporters took out pad and pen when one of these men strode to the lectern, not because he was a certified member of the ASSA, but because of his unique and direct experience in the affairs of state. His authority was concrete and idiosyncratic: it did not stem from membership or rank within a community of inquiry, or from familiarity with the internal communications of such a community (its "body of literature"). His opinions seemed sound because of who he was and what he had done.

How was Gilman to persuade men like this that their variety of authority was inferior, in the long run, to that of an organized community of inquiry? Or that theory benefited by being severed from action? Or that the scholar's indirect and vicarious knowledge of society was superior to their own direct experience? It was a hopeless task.

Not only were ASSA leaders wedded to an idiosyncratic principle of authority; in addition, their whole view of human affairs was profoundly individualistic. Theirs was an individualism not only of ideals and values (what ought to be) but also of method and perception (what actually is). In their eyes the universe of human affairs was populated by free moral agents, beings who were ordinarily autonomous, masters of their own fate, causes of their own life experience. The few interrelationships that clustered these basically free atoms together to form society and its institutions were loose and contractual bonds of choice, not at all like the heavy chains of cause and effect that bound natural phenomena together in a seamless web of action and reaction.

This universe of human affairs did not invite a rigorous, disciplined, collective style of inquiry. Why go to the trouble of organizing vast projects of empirical research into past and present human behavior, if the future is in the hands of free agents whose behavior is by definition unpredictable? If the cause of each man's behavior is his own conscious will, and if all social phenomena are explicable as aggregations of this same cause, then how could there be any determinative laws or principles to discover? In a universe of human affairs like this one, the intuitive insight of a single gifted thinker was just as likely to yield sound knowledge as the work of a whole regiment of scholars.

In short, most ASSA leaders continued to attribute to individuals a degree of causal potency that obscured historical, cultural, social, and economic determinants of human behavior. They were transitional figures whose minds were geared to a society still in the early stages of interdependence. Their presumption of individual autonomy permitted them to feel content with explanations that the twentieth century would regard as superficial and formalistic. Already they were losing ground to Spencerian positivism, which, whatever its excesses, at least avoided the idealists' exaggeration of individual autonomy and acknowledged that causation might lie outside conscious mind.

The transitional nature of the Association is nowhere more apparent than in its departmental structure. As a "headquarters" of sorts for the movement to establish authority, it had seemed natural in 1865 for the Association to divide the work of "social science" between departments of jurisprudence, health, education, and finance. By so structuring itself, the Association invited the leadership and participation of lawyers, doctors, and college educators—the most highly trained secular members of that mid-nineteenth-century gentry class whose authority it set out to defend.

The goal of the founders of the ASSA was not to sponsor the formation of new communities of inquiry in social science, but to extend the competence of the traditional professional communities of which they themselves were members. By strengthening the professions they would strengthen the backbone of the whole gentry class. "Social science" to them was not an autonomous branch of knowledge, but the scientific underpinning of the esoteric advice that professional men had always dispensed to the lay public. By enriching the intellectual foundations of law, medicine, and pedagogy, and throwing in a little political economy (their only concession to novelty), they hoped to rejuvenate the gentry class, enabling its leading members once again to assert their mastery over human affairs. From at least the time of the Renaissance forward, Western culture had allocated all questions surpassing common sense and custom to one or more of the classic professions; the founders of the ASSA merely took for granted the continued viability of that venerable division of professional labor.[33]

No wonder, then, that Peirce's merger proposal and Gilman's attempts to redefine social science fell through. Although they may not have recognized the full import of what they were doing, the measures they proposed implied nothing less than the creation of a new division of professional labor. Instead of extending the foundations of law, medicine, and general pedagogy, Peirce and Gilman were carving out of these old fields a new subject area which they proposed to assign to full-time practitioners. Instead of enhancing the existing professions and refurbishing their public image of omnicompetence, Peirce and Gilman were proposing measures that diminished their authority and set limits to their competence. To redefine "social science" as a university-based, research-oriented enterprise, with its own community of full-time practitioners, was to make social science virtually irrelevant to the professional men who founded and sustained the ASSA.

Not only did Gilman's rejection of the ASSA signal the breakdown of the conventional division of professional labor, but in addition it inaugurated a new phase in the movement to establish authority. By abandoning the ASSA and undertaking a new scheme of university-based organization for social inquirers, Gilman continued the essential task of the movement to establish authority but divorced the movement from the gentry class that had launched it. The effort to establish secure institutional foundations for authority continued and indeed gained momentum under university auspices after 1880. But the success of the effort now meant the obsolescence rather than the preservation of the nineteenth-century gentry. Doctors, lawyers, and ministers would of course continue to enjoy elite status, but they would no longer claim to be masters of all esoteric knowledge. New professions and quasi-professions—the social sciences among them—would rise to demand equal stature. Growth and internal specialization led in the twentieth century to a far more complex professional elite, too large

---

[33]. William J. Bouwsma, for example, has recently argued that "the culture of Renaissance humanism, especially in its earlier stages, was largely a creation of lawyers and notaries": see his "Lawyers and Early Modern Culture," *American Historical Review*, 78 (Apr. 1973), 308.

and heterogeneous for its members to feel any vivid sense of a common identity. Without that proud sense of a common professional role to knit its leading members together, the gentry could not survive; as a consequence, genteel culture was relegated to the fringes of American society. Professional authority thrived, but the opinions it sanctioned would seem increasingly alien and even threatening to those who first set out to establish authority.

Why did Gilman—who was, after all, a founder and longtime supporter of the ASSA—conclude that its weaknesses were irremediable? Not, I think, because he or the Association had changed since 1865, but because the larger social and intellectual environment had changed. In an increasingly interdependent society, idiosyncratic opinion was more vulnerable, the presumption of individual autonomy was less plausible, and the omnicompetent role of the classic professions was less credible than had been the case even a decade earlier. In the last analysis, Gilman rejected the Association because he sensed that neither its institutional foundations nor the presuppositions of its leading members were suited to the defense and cultivation of sound opinion in a densely interdependent universe of human affairs. If American society could have been structurally frozen at an early stage of interdependence, then perhaps "social science" might even today be an auxiliary function performed by doctors, lawyers, and unspecialized educators "moonlighting" from their regular work. But society became vastly more interdependent, and social science therefore had to become something quite different.

CHAPTER VIII

The Founding of the Historical and Economic Associations, 1884-85

The next three chapters introduce no new stages of explanation, but follow leads previously established to a logical conclusion. This anticlimactic turn of the narrative is dramaturgically regrettable but analytically necessary and historically authentic. From our own perspective in the twentieth century Gilman's rejection of the ASSA marks a momentous transition from amateur to professional social science. It is only proper for us in retrospect to attach a larger meaning to events than contemporaries were able to see. But decisive moments often are not recognized as such by contemporaries; this one passed without notice, and it is important to see why. The present chapter traces the close connections between the ASSA and the new Historical and Economic Associations and examines the complex pattern of similarities and differences among the three organizations. The next chapter briefly describes the activities of the ASSA in the 1880's and 1890's, when it was no longer a prime fountainhead of sound social opinion but still retained a high degree of public respect and attention. Chapter X traces the demise of the Association at the end of the century, when it was taken over by last-ditch defenders of gentry authority and put to trivial uses.

Formation of the American Historical Association

Scholars heretofore have failed to notice that the story of the founding of the American Historical Association properly begins with Daniel C. Gilman's 1880 presidential address to the American Social Science Association.[1] Gilman recommended that the ASSA form a sixth department devoted to history on the grounds that "sociology is based on

---

[1.] See, for example, John Higham's otherwise fine account, in Higham *et al.*, *History*.

history."[2] Gilman was leaving the presidency as he spoke, and no action was taken on his proposal. Had it been carried out, and had he thrown his weight behind it, it is entirely possible that at least the initial steps toward the professional organization of history might have occurred within the framework of the Social Science Association. One can only speculate, but if historians and economists had organized within the ASSA the subsequent pattern of disciplinary specialization might have been different and the ASSA might have survived as an institution, though probably not without major modifications.

Herbert Baxter Adams is justly remembered as the organizer of the Historical Association, but it was Gilman who provided the original inspiration and the milieu in which Adams's organizing talents flourished. Gilman enlisted his faculty members and graduate students in a continuing campaign to insinuate themselves into existing scholarly organizations in every field, thus spreading Johns Hopkins's influence far and wide. Where the existing communities of inquiry were disorganized or merely loose aggregations, like the ASSA, Hopkins men worked to organize them and give them formal structure. Gilman's own efforts to strengthen the Social Science Association were duplicated by his colleagues' efforts in a host of lesser societies. Special attention was given to journals and publication programs, the media of interaction by which a community of inquiry is knit together.

In this prosaic task of infiltrating scholarly institutions and constructing new ones, Gilman found in Herbert B. Adams an apt collaborator. Two years before the founding of the AHA Adams, in a letter to Gilman, reported on the far-flung activities of his students and himself. The conspiratorial air is characteristic of his correspondence with Gilman:

> It will be a good thing for the University if we can ally with us the Historical Societies and *quasi* historians in all the seaboard States. Gould did well for us in Philadelphia. Jameson is pretty well established with the New York set. I have now a good chance in Rhode Island for Foster wants me to read a paper before that State Society. Wisconsin has treated us very kindly, presenting books, making me a member of their Society; and Wilhelm and John Johnson have just been taken into the Maryland

[2] *JSS*, 12 (Dec. 1880), xxiv. Among Gilman's papers there is a Boston newspaper clipping dated April 1880 which observes that "the Social Science Association is not complete without a section devoted to history, that is to say, to a study of the origin and growth of social institutions. It happens that in this country historical societies are chiefly made up of local antiquaries. They do good service, but there is need of an organization by which historical students may be brought together for the discussion of broader questions as they are now taken up in the light of scientific method." The author was probably Gilman. See Box 1, Gilman Papers, Johns Hopkins University.

Historical. Ingle is on the Md. Epis. Libr. Committee [*sic* —Maryland or Methodist Episcopal Library?] and will hold that little cloister-fort, with the forces it represents. I have never begun to realize until this year the importance of corporate influences, of associations of men and money.[3]

Adams kept a map on which the location of each Hopkins graduate was marked with a pin, like a display of colonial outposts.[4] Woodrow Wilson, one of Adams's early graduate students at Hopkins, described him as "a great Captain of Industry" and "a disciple of Machiavelli, as he himself declares."[5]

The organizing fever that swept through American society in the last decades of the nineteenth century reached a peculiar intensity at Johns Hopkins. The idea of the American Historical Association was a product of this fever and the specific institutional concerns cultivated by Gilman. John Franklin Jameson, as a graduate student at Hopkins in 1882, recorded in his diary his own ambition to establish "an American Historical Assoc. or Congress, at whose annual meetings professors and others might meet, compare notes, get hints and stir up popular interest. . . . One can see the good it might do by observing how much interest its greater prototypes the Scientific, Philological and Social Science [Associations] excite every summer." Jameson thought at first that even a young and unknown man like himself might launch such a project, but a few months later (in May 1883) he laconically noted that his mentor, Herbert B. Adams, "is proposing to work around to the Am. Hist. Ass. I planned."[6]

When President Andrew D. White instructed historian Moses Coit Tyler to represent Cornell at the founding of the AHA, Tyler replied with a trace of indignation that it had been his idea to begin with. As early as 1880, he said, he had discussed such an association with Justin Winsor and afterwards with Herbert B. Adams and Charles Kendall Adams. According to Jameson's later recollection, however, "Moses Coit Tyler publicly stated that the first suggestion of such an organization had come to him from President Daniel C. Gilman, who pointed to the value accruing from the meetings of such bodies as the American Oriental Society and the American Association for the Advancement

---

[3.] H. B. Adams to D. C. Gilman, Amherst, Mass., 3 July 1882, in W. Stull Holt, ed., *Historical Scholarship in the United States, 1876–1901: As Revealed in the Correspondence of Herbert B. Adams,* The Johns Hopkins Studies in Historical and Political Science, LVI, no. 4 (Baltimore: Johns Hopkins University Press, 1938), 55. For Hopkins's role in creating and sponsoring the media of interaction within scholarly communities, see Hawkins, *Pioneer,* 73–76, 107–113.

[4.] Holt, *Historical Scholarship in the United States,* 10.

[5.] Quoted by Hawkins, *Pioneer,* 175.

[6.] Elizabeth Donnan and Leo F. Stock, eds., *An Historian's World: Selections from the Correspondence of John Franklin Jameson,* Memoirs of the American Philosophical Society, XLII (Philadelphia: American Philosophical Society, 1956), 36.

of Science."[7] Clearly the historical association was an idea whose time had come, but to the extent that any one man can be thought of as the originator of the idea, Gilman, the ex-president of the Social Science Association, is the man.

The AHA was not the first fruit of Gilman's campaign to organize American scholarship on a more professional basis. In 1883 the Modern Language Association had been formed under the leadership of Professor A. Marshall Elliott, who taught philology and romance languages at Johns Hopkins. Elliott had been regarded as a marginal scholar and was in danger of losing his job until he led in the creation of the MLA and became secretary of the new organization. A year later he was promoted.[8]

In planning the Historical Association Herbert B. Adams consulted closely with Gilman. It was at Gilman's suggestion that Adams invited Andrew D. White to preside at the opening session and become the first president of the AHA. "I have just received your letter," wrote Adams to Gilman, "and have written President White, urging him to be present at our Saratoga convention of teachers and friends of History and to read his article on the Ethical Value of Historical Studies. . . ."

> The convention will be a success in a quiet way. Mr. Winsor, whom I visited in Cambridge last week, and young Channing (Professor Torrey's protégé) will represent Harvard in the flesh. Emerton is coming with his newly imported German Professor, who is to represent German Literature *as German History* at Harvard College. He has worked on the Monumenta with Waitz and will strengthen historical work at Harvard. . . . Besides the new blood of Harvard, we shall have at the convention Levermore to represent *young* Yale and the Johns Hopkins. I am very proud of this delegate. Dr. Austin Scott has offered a 'constitutional' paper. C. K. Adams, Moses Coit Tyler, and their disciples will be on hand. Young Columbia, I think is well disposed, although Burgess was thinking of a convention in New York City. But Saratoga will win. C. K. Adams has been asked to give the 'send off,' but I have no doubt he would like to see President White preside at the meeting. Would not that course permit an ethical 'inaugural?' We shall form a very happy family and have a very good time.[9]

---

[7] M.C.T. [Tyler] to A.D.W. [White], Ithaca, N.Y., 13 Aug. 1884, A. D. White Papers, Cornell University; John Franklin Jameson, "The American Historical Association, 1884–1904," *American Historical Review*, 15 (Oct. 1909), 4. The Oriental Society (of which Sanborn and Ralph Waldo Emerson were also members) may have been as important a model in Gilman's mind as the ASSA. Gilman in the 1860's was secretary of the society.

[8] Hawkins, *Pioneer*, 161.

[9] H. B. Adams to D. C. Gilman, Amherst, Mass., 8 Aug. 1884, in Holt, *Historical Scholarship in the United States*, 71. Italics in the original. See also H. B. Adams to A. D. White, Amherst, Mass., 8 and 13 Aug. 1884, in A. D. White Papers, Cornell University.

Aside from the clear fact that the organization of professional associations was a deliberate policy at Hopkins, we cannot be sure what advice Gilman gave Herbert B. Adams about the formation of a historical association. But if Gilman had still favored a historical section within the ASSA in 1884, as he had in 1880, it is hardly likely that Adams would have defied his wish.

When the "teachers and friends" of history met under the auspices of the Social Science Association in a small parlor of the United States Hotel in Saratoga on 9 September 1884, only one of them (besides Frank Sanborn) spoke out firmly in favor of organizing a historical section within the old Association. John Eaton, president of the Social Science Association and U.S. Commissioner of Education, urged a close bond between historians and the ASSA on grounds that are perhaps deceptively attractive today. According to Herbert B. Adams, Eaton warned that "the tendency of scholarship in this country was toward excessive specialization. He thought students should seek larger relations than their own field of work afforded." Eaton promised the historians "perfect independence" in their section and reminded them of the prestige of the old organization and the benefits of being associated with inquirers in fields other than their own.[10]

But Eaton's appeal fell on deaf ears. There was no real likelihood that the historians would organize within the Social Science Association. Although the question was debated at the September meeting, one suspects that the decision had already been made. The call for a convention, sent out in June and signed by Eaton and Sanborn as well as the historians Herbert B. Adams, Charles Kendall Adams, and Moses Coit Tyler, spoke not of a historical section of the ASSA, but of "an American Historical Association." The notices that appeared in the *Nation,* the Springfield *Republican,* the *Independent,* and other journals carried the same implication.[11]

Charles Kendall Adams of the University of Michigan regarded the decision to meet under the auspices of the ASSA as "merely a prudential measure." The two most influential scholars in attendance, President Andrew D. White of Cornell and President Francis A. Walker of the Massachusetts Institute of Technology, both politely but firmly declined to support Eaton. (Gilman was conspicuous by his absence.) Walker, who one year later would become the first president of the Economic Association, agreed that "it is much safer for a new association, which has not tried its strength, to start in connection with an

---

[10.] *Papers of the American Historical Association,* ed. Herbert Baxter Adams (New York: G. P. Putnam's Sons, 1885), I, 12–13. These and following quotations from the charter meeting come from Adams's summary and are presumably paraphrases rather than direct quotations.

[11.] *Ibid.,* 5–6, 7–10.

older and stronger body; but if the strength of the American Historical Association is already well assured in point of numbers and in moneyed contributions, immediate independence might prove a safe policy." White struck a conciliatory note, suggesting that the AHA call its next annual meeting at Saratoga so as to coincide with the meeting of the Social Science Association. The principal consideration, he thought, was that "the membership of the American Historical Association would soon be as large as that of the body under whose auspices we were now assembled."[12]

The president of the Massachusetts Historical Society, Charles Deane, came close to the heart of the matter when he observed that affiliation with the ASSA would make sense only if the new group was formed largely from the membership of the old Association. That was not the case.[13] The Historical Association was not in any literal sense an offshoot of the ASSA, for there was very little overlap of membership. Of 189 AHA members in 1884, only 24 had been members of the ASSA the year before.[14]

What were the differences between the AHA and the ASSA that justified historians in organizing separately? In truth, there was little on the face of things to distinguish the new association from its venerable predecessor. Although the Historical Association did not draw to its meetings the same individuals, it did draw much the same social mix of academicians, gentleman-scholars, and men of affairs that had sustained the ASSA for the preceding two decades. The ASSA had followed an open membership policy. In form, the historians departed sharply from this policy by making membership subject to the approval of the executive council; but in practice this restrictive provision was not employed to ban "amateurs." Indeed, of the two chief sources of AHA membership, college teachers of history and members of the state and local historical societies, the latter provided the larger number of recruits. Of the original 189 members of the Historical Association, only about a third were academic men.

To a casual observer the AHA might not have seemed much more academic than the ASSA, for the old Association had often been led by academic men. But in fact the AHA, unlike the ASSA, was decisively oriented to the needs and ways of the professional scholar. Academic men were a minority of the general membership of the AHA, but they

---

[12] *Ibid.*, 12, 15, 14. See also the remarks of Clarence W. Bowen (p. 14), who mentioned that "a professor of history in New York City" (probably John W. Burgess) decided not to come to Saratoga when he mistakenly heard that the historians were to organize as a section of the ASSA.

[13] *Ibid.*, 12.

[14] Author's calculations. Another eleven men had been ASSA members at one time or another in the sample years 1869, 1874, or 1878. These thirty-five men comprised roughly 10 percent of 1883 ASSA membership.

comprised about half of the group that attended the charter meeting and accounted for six of the nine original officers.[15] In spirit, it was their association, and they knew it. The men who attended ASSA meetings typically knew each other socially and often were related by blood or marriage. They came together primarily as gentlemen and concerned citizens; their connection to a university, if any, was incidental to their common membership in the gentry class. This pattern was reversed in the Historical Association. The driving force behind its formation and operation was the college teacher of history, whose connections with the gentry, if any, were incidental to his professorial occupation. Most members of the ASSA had a basis for convivial interaction quite apart from any shared interests in a particular occupation or field of knowledge, but the members of the AHA seldom had any other reason for associating.

From the day of its founding the Historical Association was embarked on a more professional and more academic course than its eclectic predecessor, but its novel character was not immediately conspicuous. College teachers of history were too few, too insecure financially, and perhaps too weak even in terms of established intellectual authority to stand alone. They needed not only financial support but also intellectual recognition from other elements of society. They found their practical allies among their intellectual competitors, the non-academic men who wrote the histories of states, villages, or heroic ancestors. These amateur historians, working in state and local historical societies, left an intellectual heritage that is very uneven, but perhaps worthy of more respect than professional historians commonly give to it today. In the long run, the amateurs could not compete with the trained academic professionals for intellectual authority; nevertheless, for several decades after 1884 the professional academicians maintained an uneasy alliance with the amateurs, unwilling or unable to strike out on their own. Though the creation of the AHA meant the ultimate triumph of the cosmopolitan professional over the more provincial amateur, that triumph was not complete in the 1880's or 1890's.[16]

[15.] Adams, ed., *Papers of the AHA*, I, 21–23. A full list of 1884 members is given at pp. 40–44. The first officers of the AHA were A. D. White, president; Justin Winsor and Charles Kendall Adams, vice-presidents; H. B. Adams, secretary; Clarence Winthrop Bowen, treasurer; and William B. Weeden, Charles Deane, Moses Coit Tyler, and Ephraim Emerton, members of the Executive Council. Bowen (who held a Ph.D.), Weeden, and Deane are the officers I have classified as non-academic. Winsor was librarian at Harvard.

[16.] On the local amateurs, see David D. Van Tassel, *Recording America's Past: An Interpretation of the Development of Historical Studies in America, 1607–1884* (Chicago: University of Chicago Press, 1960), esp. the last chapter. For evidence of friction between H. B. Adams and J. F. Jameson over the continuing role of non-academic

The most obvious difference between the ASSA and the AHA was the narrower subject matter of the latter and the different political complexion that followed from that change in subject. No one could expect historical investigations to yield direct guidelines for reform action. But the reformism so typical of the old Social Science Association was by no means absent from the new AHA. Many historians, including their first president, Andrew D. White, regarded themselves as reformers by virtue of their indirect teaching influence over the next generation of political leaders.[17] In the long run it would make a great deal of difference that the Historical Association declared its goal to be merely "the promotion of historical studies," whereas the Social Science Association had aimed to conquer every social evil from crime to pauperism.[18] Still, in the 1880's whatever political differences existed between the two groups were incidental.

John Eaton tried to persuade the historians that the principal feature distinguishing their association from the ASSA was its excessive specialization. Later students have adopted the same theme, portraying the ASSA as a victim of specialization, gradually robbed of purpose by its narrower offspring.[19] Certainly history is a narrower subject than social science, but it is important to recognize that the ASSA did not really represent a unified conception of "social science" as that term is understood today. Nor is "history" properly understood as a fragment once embraced by the ASSA's version of "social science." Instead, as has been shown in previous chapters, the ASSA represented an altogether different conception of "social science" based on a division of labor designed to accommodate the classic professions in an enterprise devoted as much to practical reform as to investigation.

The formation of the AHA signaled not only the triumph of specialization but, more important, the transition to a new division of labor in social inquiry—one which made inquirers independent of, and largely irrelevant to, the classic professions. The implicit logic of the modern division of labor is a question for further investigation, but it is readily

men in the AHA, see Donnan and Stock, eds., *Historian's World*, 46, 47n, 54. Of the 1888 meeting Jameson said, "I am a little inclined to think the thing is getting into the hands of elderly swells who dabble in history, whereas at first it was run by young teachers, which I think made it more interesting" (47n).

[17] For an illuminating discussion of the reformist implications of historical scholarship in this period, see Robert L. Church, "Development of the Social Sciences as Academic Disciplines at Harvard," ch. II. Andrew D. White, who pioneered in the development of history as a modern scholarly discipline, was motivated by two thoughts: "first, how effective history might be made in bringing young men into fruitful trains of thought regarding present politics; and, secondly, how real an influence an earnest teacher might thus exercise upon his country." *Autobiography of Andrew D. White* (New York: Century, 1922), I, 256. See also 255, 262.

[18] Adams, ed., *Papers of the AHA*, I, 20–21.

[19] Bernard, *Origins of American Sociology*, 527–718.

seen to be a division suitable for inquirers whose first concern is to understand and explain the world rather than to change it. The ASSA departments constituted operational fields of action for the diagnosis and cure of society's ills; the modern disciplinary boundaries have practically no relevance to social problems and instead serve simply to allocate the raw materials of scholarship by assigning to each group of specialists a field of factual data in which its members enjoy first rights of exploitation.[20] It is a division of labor among investigators, not agitators.

The most profound difference between the American Historical Association and its predecessor is that the AHA was the formal expression of a genuine community of inquiry. The ASSA was never more than an aggregate of individuals; in contrast, the AHA represented the beginnings of a genuine community, solidaristic enough to insulate its members from public opinion, to submit their idiosyncrasies to communal discipline, and to perform the other authority-enhancing functions adumbrated by Joseph Henry and the Lazzaroni forty years earlier. Justin Winsor, one of the older historians present at the charter meeting, saw that this was the essential novelty of the new association: "We are drawn together because we believe there is a new spirit of research abroad,—a spirit which emulates the laboratory work of the naturalists, using that word in its broadest sense." Contrary to the common interpretation, these words do not mean that Winsor advocated a naturalistic or positivistic philosophy. Instead, what he was talking about was a pattern of social affiliation, an institutional arrangement that had reached its fullest development among natural scientists. What he had in mind was, in fact, the community of inquiry. As his next words made clear, it was this new mode of organizing inquirers that historians were borrowing from natural scientists, not the technical methods or ideas of science *per se:*

> This spirit requires for its sustenance mutual recognition and suggestion among its devotees. We can deduce encouragement and experience stimulation by this sort of personal contact. Scholars and students can no longer afford to live isolated. They must come together to derive that zest which arises from personal acquaintance, to submit idiosyncrasies to the contact of their fellows, and they come from the convocation healthier and more circumspect.[21]

---

[20] To the founders of the ASSA, the subdivisions of social science had a functional bearing on practical reform action: new discoveries in social science were to be propagated by the Education Department and formulated as statute law by the Jurisprudence Department. See ASSA, *Constitution, Address,* 15–16.

[21] Adams, ed., *Papers of the AHA,* 1, 11. Of course some historians also tried to adopt scientific ideas, Herbert B. Adams's "germ theory" being the most notable example and the most spectacular failure. See W. Stull Holt, "The Idea of Scientific History in America," *Journal of the History of Ideas,* 1 (June 1940), 352–362; Higham et al., *History,* 89–103.

Young historians and other specialized scholars could have organized their closely knit communities of inquiry under the umbrella of the ASSA, but it was not an attractive option. The Association was shaped by too many non-academic interests. It was too closely linked to the professions of law and medicine. It was too securely anchored in the granite of New England for men whose outlook and career opportunities were national in scope. It was too deeply interwoven with an older and more vivacious kind of community, the gentry class, whose members paid attention to one's family name and valued certain intangible graces at least as highly as a command of the pertinent literature. It was dominated by an imposing generation of men who had not read Gumplowicz or Knies or Schmoller, perhaps, but who had earned a sense of self-assurance by defeating the slave power and holding the reins of government. The ASSA at once intimidated men of the younger generation and also invited their contempt by leaving its doors open to "Saratoga cranks and hangers-on," as Herbert Baxter Adams called them.[22] It was not a genuine community of inquiry and could not readily have been converted into one.

Formation of the American Economic Association

Historians rejected the ASSA as a framework within which to work, but they were eager to meet under its protective wing in order to enhance the legitimacy of their independent organization. Economists also wanted a dignified host for their coming-out ceremonies. Officially the American Economic Association was founded under the auspices of the American Historical Association. But since the AHA was meeting in Saratoga concurrently with the ASSA in 1885, the economists enjoyed the informal sponsorship of both the new and the old associations. The Economic Association did not ask for a send-off from the Social Science Association—indeed, the old Association is nowhere mentioned in the official records of the first meeting of the AEA. Nonetheless, the economists benefited (just as the historians had a year earlier) by organizing in the shadow of the "mother of associations," as Sanborn liked to call it. Several of the young economists who came to Saratoga to found their own organization were included in the regular program of the ASSA. Edmund J. James delivered two papers, Edward W. Bemis read one, and Henry Carter Adams took part in a discussion of social science instruction.[23]

[22.] Adams to A. D. White, Baltimore, 6 Feb. 1887, A. D. White Papers, Cornell University.
[23.] At the preceding ASSA meeting in 1884 the program had included papers by Henry C. Adams and Francis A. Walker, two of the "new" economists, and a paper by one of their opponents, Edward Atkinson.

In retrospect, it is obvious that the Historical Association represented a step away from the Social Science Association toward a more professional and more academic mode of organizing social inquiry. This is not so obvious in the case of the Economic Association, but I think it no less true. The AEA and the AHA both came into being for essentially the same reason: to give formal, tangible representation to communities of inquiry developing primarily within the university. This interpretation conflicts with the conventional story of the founding of the AEA. According to the conventional story, the AEA was founded as a reform organization, committed just as firmly to agitation as to investigation. I believe the conventional interpretation exaggerates the strength of the commitment to reform. The relative priority of agitation and investigation was a controversial question in the early years of the AEA, but the question was so quickly resolved in favor of investigation that the episode constitutes a proof, rather than a counter-instance, of Gilman's conviction that only investigation could be professionalized.

The reformist image of the AEA derives from its chief organizer, Richard T. Ely, and the "platform" or statement of principles that he wrote for the association. Ely unquestionably regarded the organization as an instrument of reform (though, as we shall see, he also regarded it as an instrument of professionalization). Among the founders, he later recalled, "there was a striving for righteousness and perhaps here and there might have been one who felt a certain kinship with the old Hebrew prophets."[24]

Although Ely's platform was never made binding on individual members, by adopting it the AEA seemingly assumed a posture no less reformist than that of the ASSA—a posture, in fact, that seemed to differ only by being more vigorous and outspoken. The platform challenged the doctrine of laissez-faire, extolled the state as a means of progress, and called upon church and state as well as scientific investigators to heal the breach between capital and labor. Moreover, the platform had the practical effect of excluding from the association some of the most eminent economists in the country, thus splitting the existing community of economic inquirers—a most unprofessional thing to do. The Economic Association further departed from the professional model set by the AHA by opening its doors to all who would pay for membership, just as the old Social Science Association had done. Among its first members there were many, like Lyman

[24] Richard T. Ely, *Ground Under Our Feet: An Autobiography* (New York: Macmillan, 1938), 143. See also Benjamin G. Rader, *The Academic Mind and Reform: The Influence of Richard T. Ely in American Life* (Lexington: University of Kentucky Press, 1966), 35.

Abbott and Washington Gladden, who were far more interested in reform than in scholarship.[25]

The apparent similarity of the Economic Association to the ASSA was stressed by Frank Sanborn in his annual address at Saratoga. He observed that the ASSA had always been closely associated with a distrust of laissez-faire. It had long advocated the expansion of public power in the form of boards of charity and health, regulatory agencies, the establishment of parks and museums, and so on. The proposition that economics was an ethical science had always been the message of Sanborn's Department of Social Economy. Paraphrasing Edward Bemis, one of the "new" economists, Sanborn said: "Methinks this expresses very well what our association has been doing in its broader field and with more miscellaneous activity, for the last twenty years. To learn patiently what *is*—to promote diligently what *should* be,—this is the double duty of all the social sciences, of which political economy is one."[26]

Sanborn was not alone in thinking that the AEA had much in common with the Social Science Association. Before the formation of the Economic Association, when young professor Edmund J. James discussed the need for such an organization with a number of the finest academic economists in the country, several of them told him that it would be superfluous: "I found a general agreement that possibly such an association would do useful work, but in some cases also the view that the American Social Science Association practically performed the only available function of such an organization." James, however, wanted an organization that would have "quite a different attitude toward our economic problems from that which was characteristic even of the Social Science Association, broad and liberal as that was."[27]

The idea of a society of economic thinkers was "in the air" in the 1880's. A conservative group of New England economics teachers and non-academic writers who called themselves the Political Economy

[25.] The basic documents are contained in *Publications of the American Economic Association*, I, *Report of the Organization of the American Economic Association*, by Richard T. Ely, secretary (Baltimore: John Murphy, 1886) hereafter cited as *Report of the Organization of the AEA*. The original platform is given on pp. 6–7; a diluted version was adopted as article III of the constitution (35–36). Charles H. Hopkins counted thirteen prominent leaders of the social gospel among the AEA's 182 original members: *The Rise of the Social Gospel in American Protestantism, 1865–1915* (New Haven: Yale University Press, 1940), 117. On religionists in the AEA, see also Coats, "First Two Decades of the AEA," 562n.

[26.] Franklin B. Sanborn, "The Social Sciences: Their Growth and Future," *JSS*, 21 (Sept. 1886), 1–12, quotation from p. 6. Italics in the original. Sanborn went on, however, to urge the necessity of combining private initiative with state action whenever possible.

[27.] Edmund J. James, "Anniversary Meeting ... December 28, 1909," *Publications of the AEA*, 3rd ser., 11 (1910), 109.

Club began organizing in 1883. As the word "club" suggests, they made no serious attempt to embrace thinkers of all schools, or of all regions of the country.[28] The first concrete proposal that had some hope of incorporating all of the new German-trained economists (and thus some hope of being truly national) came from Edmund James and Simon Nelson Patten. James was already teaching at the University of Pennsylvania, where Patten soon joined him. In May 1885, after many conversations with economic thinkers around the country, the two young scholars formulated a plan for a "Society for the Study of National Economy."[29] The organization they proposed was modeled explicitly on the Verein für Sozialpolitik, of which their German mentor, Professor Johannes Conrad, had been a leading member since its inception. The Verein was founded in 1872 under the leadership of the younger historical school of economists to overthrow the Volkswirtschaftlicher Kongress, which since 1858 had guided German public opinion toward free trade, economic unification, and classical liberalism in general. The Verein, composed of government officials, journalists, and businessmen as well as professors, took an active political stance against economic individualism and in favor of state action on behalf of the lower classes. Its members were jokingly called "socialists of the chair," though it was really not a radical but a reformist organization. James and Patten would have liked to create an organization bearing much the same relation to the ASSA that the Verein bore to the Kongress. Their aim was to undermine the doctrine of laissez-faire and the radically individualistic conception of social reality of which it was a part.[30]

In consultation with Patten, James drew up a rather wordy and rambling constitution that leaned heavily on the two men's few years of

---

[28] Francis A. Walker and Richard T. Ely even toyed with the idea of "asserting ourselves aggressively" in the Political Economy Club, rather than organizing a separate organization: see A. W. Coats, "Political Economy Club," 624–637, quotation from 625.

[29] Daniel M. Fox, *The Discovery of Abundance: Simon N. Patten and the Transformation of Social Theory* (Ithaca: Cornell University Press, 1967), 37.

[30] James recalled that Conrad once urged his American students to create an organization at home similar to the Verein in order to influence the direction of social legislation: *ibid.*, 108. On the Verein see Eugen von Phillipovich, "The Verein für Sozialpolitik," *Quarterly Journal of Economics*, 5 (Jan. 1891), 220–237; John H. Gray, "The German Economic Association," *Annals*, 1 (1890–91), 515–522; Small, *Origins of Sociology*, ch. XV; Anthony Oberschall, *Empirical Social Research in Germany;* J. A. Schumpeter, *History of Economic Analysis*, ed. Elizabeth Boody Schumpeter (New York: Oxford University Press, 1954), 803–807; Herbst, *German Historical School in American Scholarship*, 144–147, 169 and *passim;* Abraham Ascher, "Professors as Propagandists: The Politics of the Kathedersozialisten," *Journal of Central European Affairs*, 23 (Oct. 1963), 282–302; and Fritz Ringer, *The Decline of the German Mandarins: The German Academic Community 1890–1933* (Cambridge: Harvard University Press, 1969), 146–151.

experience in Germany. They proposed an organization that would encourage investigation and support publication but which also would adopt a forthright political stand against "the widespread view that our economic problems will solve themselves." The organization was to have a "platform" beginning with this proposition: "The state is a positive factor in material production and has legitimate claims to a share of the product. The public interest can be best served by the state's appropriating and applying this share to promote public ends."[31] This was stiff medicine. In the context of practical American politics in the 1880's, the platform of the Society for the Study of National Economy seemed radically statist. It advocated a degree of political centralization and government involvement in the economy that went far beyond anything Frank Sanborn or most other ASSA members had in mind, and exceeded even the desires of most of the young German-trained economists.

The platform put forward by James and Patten probably could not have won broad support among American economists under any circumstances, but the immediate cause for its failure was the tactical maneuvering of Richard T. Ely. A few days after James and Patten began circulating their constitution in May 1885, Ely began sending out his own proposal for an "American Economic Association." His was less radical than the James and Patten plan, and it was drawn up in a much more precise and compact manner. It contained a platform which declared the state to be "an educational and ethical agency whose positive aid is an indispensable condition of human progress," continuing: "While we recognize the necessity of individual initiative in industrial life, we hold that the doctrine of laissez faire is unsafe in politics and unsound in morals; and that it suggests an inadequate explanation of the relations between the state and the citizens."[32] This sentence is

[31.] The constitution of James's and Patten's society is conveniently reprinted in Ely's *Ground Under Our Feet*, 296–299. Quotation from 296.

[32.] *Ibid.*, 136. Ely claimed that when James and Patten sent out their draft constitution, he "held back until it became absolutely certain that success could not be achieved along that line" (*ibid.*, 135). This explanation has been generally accepted by historians, but Daniel M. Fox has shown that Ely's response was more aggressive than he pretended: see *Discovery of Abundance*, 37–39. Ely's aggressiveness was even greater than Fox suggests. It appears that James and Patten were simply left at the starting gate by Ely, in whom an unusual capacity for self-advertisement was coupled with extraordinary drive and energy. As late as 15 June 1885 James and Patten had sent their platform to only five men, according to a postscript written at the bottom of a letter from James to A. D. White, Philadelphia, 15 June 1885, A. D. White Papers, Cornell University. The five men were White, Ely, John Bates Clark, Francis A. Walker, and Alexander Johnston. Six days *before* James and Patten sent White their proposal, Ely had sent White the platform of the AEA: see Ely to White, Baltimore, 9 June 1885, White Papers. Six weeks later, Ely's project had proceeded so far that notices of the forthcoming AEA meeting were appearing in prominent journals like the *Christian Union*.

often quoted as a characterization not just of Ely's original intentions for the AEA, but of the organization itself. This is most ironic because the sentence was too radical for most of the charter members of the Association. They rejected it; the statement of principles incorporated into the constitution does not include it.[33]

In planning the formation of the AEA, Ely had the enthusiastic assistance of Herbert Baxter Adams, fresh from his success with the Historical Association a year earlier. As Ely later observed, "that sort of thing was in the air at the Johns Hopkins and was encouraged by the authorities."[34] Ely kept Gilman posted on the plans for the first meeting and expressed the hope that the Association would bring credit to Hopkins.[35]

There is no reason to believe that Ely even considered the possibility of organizing within the ASSA. Given his reform goals, however, it is by no means obvious why this possibility was ignored. Edmund James, whose political sentiments at this time were at least as radical as Ely's, became secretary of Sanborn's Department of Social Economy in 1885 and remained in that post until he founded his own organization, the American Academy of Political and Social Science, in 1890. The secretary of the ASSA Finance Department since 1883 had been Henry Carter Adams, a Hopkins man. He remained in that post until 1887. The old guard of the Association included free-traders like David A. Wells, whom Ely regarded as an enemy; but there were others, like Sanborn and Andrew D. White, who could be expected to be sympathetic even if not in complete agreement with Ely's ideas. Indeed, White acted as Ely's sponsor on several occasions. Certainly there was no stark opposition between Ely and the ASSA on the legitimacy of the reform impulse, nor was there any unbridgeable chasm between them on the place of ethical considerations in economic thought. The men of Sanborn's generation who dominated the ASSA in the 1880's gave the individualistic tenets of classical liberalism more credence than Ely and

---

[33] See note 43, below.

[34] Ely, *Ground Under Our Feet*, 135, 137. Ely dedicated his autobiography to Daniel Coit Gilman, "creative genius in the field of education" (v).

[35] R. T. Ely to D. C. Gilman, Buchanan, Va., 11 July 1885, D. C. Gilman Papers, Johns Hopkins University: "I send you with this the [one word unclear—"revised"?] platform of the proposed American Economic Association.... At present the task before me is to organize the various elements in substantial harmony into a vigorous association. This is difficult and the attempt to form even a small organization has brought me to appreciate—if you will allow me to say it—better than ever before the work which has been done at the Johns Hopkins.

"I should be very much obliged for any suggestions in this matter. The prospects seem to be very favorable for an influential movement which will help in the diffusion of a sound Christian political economy. At the same time I trust it may benefit the Johns Hopkins."

his generation ever could; however, the ASSA already had a pronounced tilt in the direction of state intervention, the very direction in which Ely and his young colleagues wished to go. In terms of political attitudes and goals, the ASSA had much in common with what the young economists wanted.

But there were other considerations that made the ASSA unsuitable for the new economists. Most important, Ely and his young colleagues had many goals that were not political, but professional. The potential political compatibility between the new economists and the ASSA is really beside the point because the economists wanted not only to reform society, but also to give institutional structure to a developing community of inquiry that was rooted deeply in the academic world. The ASSA might have been an appropriate vehicle for their political aspirations, but for their professional goals it was not adequate.

For all his reforming zeal, even Ely was at the same time interested in creating a professional structure that would enhance interaction among inquirers and simultaneously insulate them from the general public. Many years later Ely recalled that there were two goals motivating the founders of the AEA. The first, to undermine the theory of laissez-faire, is the only one remembered in the conventional story of the founding. But there was another: "The second aspect, and the one on which we were all in complete agreement was the necessity of uniting in order to secure complete freedom of discussion; a freedom untrammeled by any restrictions whatsoever. It was this second point that came, in final analysis and after much debate, to be accepted as the foundation stone of our association."[36]

The cry for academic freedom is, of course, the exposed cutting edge of the professional academician's drive for internal solidarity and insulation from the general public. As Ely well knew, men who could agree on nothing else, not even the evils of laissez-faire, could unite on the superiority of professional to non-professional opinion. The "foundation stone of our association," then, as Ely described it, was the effort to define and articulate a community of the competent whose members would judge each other but tolerate no judgment from outside the community. In other respects, too, the professionalizing motives of Ely and his colleagues are quite obvious. In his initial statement at the first meeting of the AEA, Ely began by stressing the need for an organization to encourage research, publish monographs, and facilitate communication among inquirers. In addition to these professional services, "one aim of our association should be the educa-

[36.] Ely, *Ground Under Our Feet*, 132–133.

tion of public opinion in regard to economic questions and economic literature. In no other science is there so much quackery and it must be our province to expose it and bring it into merited contempt."[37]

It was the crusade against laissez-faire that gave the AEA the appearance of a reform organization rather than a professional one. But even that crusade can be seen in a professionalizing light, as a classic battle against charlatanism and "quackery." In denouncing pure laissez-faire theory the new economists were not boldly attacking a still-strong citadel of orthodoxy, as they pretended, but merely capitalizing on a shift of mood that was already underway and awaited articulation. This was not a pitched battle between evenly matched factions, but a mopping-up operation. By choosing laissez-faire as the central issue, Ely threw down the gauntlet not to other academicians, but to businessmen. The "let alone" principle was still a vigorous element in the commonsense economic thinking of the man in the street, but among academicians William Graham Sumner was unique for his uncompromising stand. Although Sumner had support from two prominent non-academic thinkers, David Wells and Edward Atkinson, they were not as extreme as he was, and they conformed even less closely to Ely's stereotype of the laissez-faire theorist. Other academicians who opposed the new historical school—such as Dunbar, Taussig, and Laughlin of Harvard, Hadley of Yale, and even Simon Newcomb, astronomer and occasional lecturer on economics at Harvard and Hopkins—were far from being pure laissez-faire theorists. They were offended as much by the contentious and imperialistic spirit as by the letter of Ely's platform. In "restoring ethics" to political economy Ely and his colleagues were not innovating, but reaffirming the traditional preoccupations of moral philosophy in American academic life. In doing so they acted in harmony with what was very nearly a consensus among *academic* economic thinkers in America. As Ely said, the AEA platform "had an inclusive as well as exclusive aim," and the former counted for much more than the latter.[38]

---

[37] Ely, *Report of the Organization of the AEA*, 14–15.

[38] Quotation from Ely, "The American Economic Association, 1885–1909," *Publications of the AEA*, 3rd ser., 11 (1910), 52. Ely's biographer, Benjamin G. Rader, agrees that Ely deliberately formulated a "straw man" that conformed neither to the classical economists nor to Ely's contemporary opponents: see Rader, *Academic Mind and Reform*, 42. The new and the old economists engaged in public debate in 1886 in the pages of *Science* magazine, published by a company of which Daniel Coit Gilman was the president. The articles were published as a book, Richard T. Ely *et al.*, *Science Economic Discussion* . . . (New York: The Science Co., 1886) which is the best compact source for the character of the conflict between the two schools. Ely freely admitted at the time that the crusade against laissez-faire served the practical function of advertisement: "it is not easy to arouse interest in an association which professes nothing. This proposed economic association has been greeted with enthusiasm precisely because it is not colorless, precisely because it stands for something." Ely, *Report of the Organization of the AEA*, 19.

The crusade against laissez-faire committed the AEA in principle to liberal reform, but only in a very general and abstract way that was acceptable to the great majority of academic economists. Moreover, in practice it was a commitment that rapidly faded into insignificance and soon was overwhelmed by more strictly professional concerns. From the very beginning, Ely and his most zealous colleagues were on the defensive—not against laissez-faire conservatives, but against advocates of undiluted professionalism. After Ely's opening remarks at the charter meeting of the AEA, one after another of the economists present stood up to urge a more moderate statement of principles. Only one participant, Andrew D. White, a long-time member and future president of the Social Science Association, explicitly defended Ely's platform as it stood.[39] E. R. A. Seligman, a German-trained scholar of Ely's generation, spoke against Ely. He declared that "the reaction against the principle of free competition had perhaps been somewhat exaggerated by certain economists." He doubted that there was valid scientific sanction either for or against government intervention, as a general principle. "He did not wish that any economist who worked in the true scientific spirit should deem himself precluded from joining the association through fear of any imagined Katheder-Socialistic tendencies."[40] Professor E. Benjamin Andrews of Brown University (who, a decade later, as the president of Brown, would be involved in an important academic freedom case) declared himself willing to drop the statement of principles, if necessary, so as to "exclude no earnest economist from the association."[41]

A large majority of the men attending the charter session of the AEA thought Ely's platform went too far. Accordingly, they appointed a committee of five to further consider, and in fact dilute, the statement of principles. Not only did they moderate the statement, but they also added, at Herbert B. Adams's suggestion, a proviso that the statement was only "a general indication of the views and purposes of those who founded the American Economic Association . . . not to be regarded as binding on individual members."[42] Ely was not repudiated, but he hardly carried the day.

The platform finally adopted by the charter AEA members was innocuous.[43] On the subject of reform it was utterly bland and evasive.

---

[39] Ely, *Report of the Organization of the AEA*, 24.
[40] *Ibid.*, 27-28.
[41] *Ibid.*, 28.
[42] *Ibid.*, 29, 35-36.
[43] The statement of principles that was finally adopted reads as follows:

"1. We regard the state as an agency whose positive assistance is one of the indispensable conditions of human progress.

2. We believe that political economy as a science is still in an early stage of its development. While we appreciate the work of former economists, we look not so

Its most forceful statement—not about reform at all—was a gratuitous and rather frivolous insult to economic thinkers not of the historical school.[44] Though still irritating, its substance was not enough to provoke much serious resistance.

But the damage had already been done by newspaper coverage. A number of eminent conservative scholars felt obliged to boycott the new association as a matter of principle. Arthur Twining Hadley of Yale, one of the holdouts, observed that "the worst elements of the platform have been omitted, but Art. I still retains its character, and gives the public the decisive impression of the whole. The newspapers continue to treat the Association as standing on its old platform, and consider this as simply an abbreviation."[45] It was not the substance of the adopted platform, but the belligerent spirit of Ely and his circle that inspired the boycott. The very existence of a platform was obnoxious to many, and for good reason, because Ely had used it to exclude from a community of the competent men whose competence was not seriously open to question. Simon Newcomb, perhaps Ely's bitterest enemy, stated the situation quite accurately: in Ely's view the AEA was "intended to be a sort of church, requiring for admission to its full communion a renunciation of ancient errors and an adhesion to the supposed new creed."[46]

Even after being watered down and deprived of its creed-like character, Ely's undeservedly famous platform was able to survive only two years. By April 1887 E. R. A. Seligman and other leading figures in the AEA were pushing hard for its repudiation. In December the AEA Council voted to drop the statement of principles altogether; that

---

much to speculation as to the historical and statistical study of actual conditions of economic life for the satisfactory accomplishment of that development.

3. We hold that the conflict of labor and capital has brought into prominence a vast number of social problems, whose solution requires the united efforts, each in its own sphere, of the church, of the state, and of science.

4. In the study of the industrial and commercial policy of governments we take no partisan attitude. We believe in a progressive development of economic conditions, which must be met by a corresponding development of legislative policy."

[44.] See Article 2 of the principles. The contest between the "new" and the "old" economists in America was a weak echo of the famous Methodenstreit which broke out in Europe in 1883. See Ringer, *Decline of the German Mandarins*, 152; Small, *Origins of Sociology;* Schumpeter, *History of Economic Analysis*, 800–824; R. S. Howey, *Rise of the Marginal Utility School, 1870–1889* (Lawrence: Universityof Kansas Press, 1960); and R. D. Collison Black, A. W. Coats, Craufurd D. W. Goodwin, eds., *The Marginal Revolution in Economics: Interpretation and Evaluation* (Durham, N.C.: Duke University Press, 1973). The latter volume also appeared as a special issue of *History of Political Economy*, 4 (Fall 1972).

[45.] Quoted by Furner, *Advocacy and Objectivity*, 79.

[46.] Newcomb, review of R. T. Ely, *Outlines of Economics*, in *Journal of Political Economy*, 3 (Dec. 1894), 106, quoted by Coats, "First Two Decades of the AEA," 558. See also 558n and 564–566 for other expressions of opposition to the platform.

decision was ratified by the whole association at its next meeting in 1888. The most significant of the holdouts, Charles Dunbar of Harvard, accepted a formal invitation to join in the same year and was elevated to the presidency in 1892. William Graham Sumner, the platform's chief target from the beginning, never joined despite cautious feelers from the association. Dunbar's rise to the presidency coincided with Ely's fall from the post of secretary in a dispute triggered by the latter's desire to associate the AEA with the Chautauqua summer camp sessions. Ely temporarily dropped out of the Association in the 1890's, becoming president in 1893 of a midwestern organization that firmly linked investigation with agitation, the American Institute of Christian Sociology.[47]

The conclusion is inescapable, I think, that the AEA's commitment to reform was extremely superficial. In fact, that commitment reduces to the irascible personality of one man, Richard T. Ely, who soon found himself almost totally isolated within the association. Ely's ouster in 1892 was a formal repudiation of reformist zeal, but it merely indicated symbolically what in substance had been true from the beginning: the AEA, like the AHA, was a professional organization.[48] It was oriented toward the occupational needs of academic men and was the concrete expression of a community of inquiry. Its members defined themselves primarily as investigators, and collectively the community would war-

[47] Furner, *Advocacy and Objectivity*, 115-124; Coats, "First Two Decades of the AEA," 559, 563-567; and Rader, *Academic Mind and Reform*, 121.

[48] Furner, *Advocacy and Objectivity*, 115-119. Furner's interpretation differs from mine. She believes that initially not only Ely and his small circle, but also *"the AEA* [i.e., the association itself] was committed to advocacy, no matter how the fine print of its platform might read. . . . True to Ely's plan, the AEA became a haven for reformers" (p. 76, emphasis added). Of course she does not deny the plain fact that two years after the founding Ely was practically isolated and the platform abandoned. To account for his sudden demise on the heels of seeming triumph, she tells a dramatic story of bitter warfare between two whole schools of economists (one of them encamped in the AEA), followed by a period of elaborate compromises and doctrinal concessions, resulting finally in a grand convergence in 1887. My own view is that when two "schools" divide and reunite as quickly as this, we are justified in treating the whole matter as epiphenomenal. Many issues divided individual American economists, of course, and the rather inexact parallel between the Ely-Newcomb debate of 1884 and the Schmoller-Menger debate of 1883 in Europe gave the whole affair an aura of gravity. No doubt the issues were momentous, but American economists simply were not polarized into two evenly matched, clearly defined schools. The failure of the participants in the *Science* magazine debate of 1886 to arrive at any clear statement of the issues dividing them suggests to my mind not concession and compromise, but simply recognition that from the beginning the contest had been ephemeral, largely the product of newspaper attention and the abrasive personalities of Ely and Newcomb, both notorious hotheads. By the same token, Ely's abandonment by even close colleagues in 1887 signifies no profound change in the AEA, but simply exasperation with Ely personally. Furner's marvelously full account of the AEA's early years supplies plenty of evidence for this view: see *ibid.*, 59-60, 62, 70-72, 93-94, 106, 115-119.

rant its members' performance only in the role of investigator, not agitator. As individuals, economists and historians were free to pursue whatever reforms they wished, but the communities to which they belonged could certify their competence only as investigators. Just as Gilman had seen in 1878, the price of professional authority was a degree of irrelevance.

In its structural arrangements the Economic Association was even more decisively oriented to the academic world than the Historical Association. Although anyone could become an ordinary member of the AEA, only members of the Council were entitled to vote for officers. Of the twenty-four initial members of the Council, all but six were university professors. Of the first officers of the association, all were academicians. The control of the professoriate could hardly have been more complete at the outset. In succeeding years, however, as the AEA became more "professional" in the sense of embracing competent inquirers of all viewpoints, it became less academic in the sense that it admitted to positions of leadership a larger number of men with business experience in lieu of scholarly credentials. From 1894 on the AEA actively solicited membership applications from businessmen and other non-academic persons. The possibility of electing a businessman as president was seriously considered in the late 1890's.[49]

The Economic Association, like the Historical Association, was at first unable or unwilling to exclude amateurs from membership or even from positions of leadership. No wonder, then, that contemporaries failed to see any striking difference between the new associations and the old Social Science Association. The ASSA did not topple over and die the moment its younger rivals came onto the scene; rather, it retained a considerable aura of authority for at least a decade, even though its roots had been cut off. The new associations were embarked on an irreversible course of professional development, but their essential novelty was not readily visible to contemporaries. What distinguished them from the ASSA and insured their eventual triumph over the old Association was the connection of each with a cohesive community of inquiry based in the university. Without such

---

[49.] Coats, "First Two Decades of the AEA," 571. The first officers of the AEA were, in addition to President Francis A. Walker and Secretary Ely: Henry C. Adams, Edmund J. James, and John B. Clark, vice-presidents, and Edwin R. A. Seligman, treasurer. The six council members who were not professors were Albert Shaw, journalist, who earned his Ph.D. under Ely; the Reverend Dr. G. M. Steele, principal of Wesleyan Academy; the Reverend Dr. Lyman Abbott, editor of the *Christian Union;* the Reverend Dr. Washington Gladden, a leader of the social gospel; the Honorable Eugene Schuyler; and the Honorable Carroll D. Wright, U.S. Commissioner of Labor and future president of the ASSA. See Ely, *Report of the Organization of the AEA,* 40–41.

institutional support as these communities provided, ideas and values could not be sure of survival in the competitive intellectual world of an interdependent society. The advantages of institutionally established authority finally proved overwhelming, but only after a transitional period in the 1880's and 90's during which older modes of authority competed with professional social science on nearly equal terms.

CHAPTER IX

From Social Science to Sociology: The 1880's and 1890's

Many intelligent observers continued to regard the ASSA as a viable center of social inquiry and an authoritative voice in questions of social policy for at least a decade after the creation of the Historical and Economic Associations. Foremost among these men was Andrew Dickson White, founding president of Cornell and president of the ASSA from 1888 to 1891. Far from being discouraged by the formation of the two specialist associations, in the 1880's the ASSA cherished high hopes of introducing its unique mode of social inquiry and even its peculiar division of labor into the university. But by the mid-nineties the rise of academic sociology confronted the Association with a more direct threat than had been posed by either historians or economists. The Association's version of social science then began to seem hopelessly eclectic and diffuse, in comparison to the more systematic work of university-based communities of inquiry.

Before the critical challenge from sociology in the 1890's, the ASSA possessed two trump cards that made it, for all its failings, an attractive forum for serious social thinkers. First, its orientation was strongly interdisciplinary. Its scheme of specialties (represented by the departments) ultimately proved too cumbersome to survive, but at least the Association was sensitive to the danger of specialization and tried to hold in one focus all the fragments of knowledge about man and society. Its second advantage was its voluntaristic bias. Even during Spencer's heyday, most serious social thinkers in America were fearful of the fatalistic implications of his necessitarian doctrine and eager to find an alternative to it. The ASSA, with its heavy infusion of transcendental idealism, supplied an alternative. Its members found comfort in each other's confidence that men were free and fully able to shape society to their will.

Sociologists in the 1890's deprived the ASSA of its trump cards by promising an analysis of society that would be both interdisciplinary and voluntaristic—and, more than that, systematic. It is often forgotten today that the first professional sociologists initially advertised their discipline as the Queen of the Social Sciences. It was to be not just another branch of knowledge, but a unifying synthesis of all the specialized branches of knowledge about man and society. The sociologists also promised to do justice to the voluntaristic principle, of which the ASSA had long seemed to be the lone defender. Sociologists not only admitted a degree of human freedom and therefore the possibility of social reform, *contra* Spencer, but they also promised—by recognizing remote causation—to be better reformers than the "social scientists" of the Social Science Association. Reformers in an interdependent social universe faced the constant danger of mistaking superficial symptoms for real, underlying causes. Only the most intensive inquiry by disciplined investigators would enable reformers to distinguish between symptom and cause, dependent and independent variable.

Sociology never really made good its claim to be the interdisciplinary discipline, nor did it succeed in reuniting social theory with reform practice. Though it was at first a haven for reformers and came closer than any other academic discipline to proving Gilman wrong about the incompatibility of agitation and investigation, in the long run sociology has contributed far more to the analysis of society than to the practical solution of social problems.[1] But it did offer the twentieth century an interpretation of human affairs that was more voluntaristic than Spencer's, and more realistic in its perception of interdependence than that espoused by the ASSA. The new view of man and society seemed plausible to a public whose younger members, at least, had learned to be skeptical of proximate causal attribution and distrustful of opinion not sanctioned by a community of inquiry.

Social Science at Cornell

The changing stature of the Social Science Association and the plausibility of its mode of inquiry can be measured by the attitudes

[1.] In its early years the American sociology profession included men and women who were just as reformist and practical-minded as Frank Sanborn. As late as 1948 Harry Elmer Barnes could say, "Probably the largest group of sociologists are what are usually called 'social economists' or 'practical sociologists,' namely those chiefly interested in social work and amelioration." Barnes included in this majority such diverse persons as Jane Addams, Graham Taylor, Mary E. Richmond, E. T. Devine, Samuel M. Lindsay, Robert Woods, Paul V. Kellogg, J. L. Gillin, and even Robert MacIver. Barnes also noted that "the emphasis has been progressively shifted from amelioration to prevention, though the 'uplift' attitude is still strong in many quarters." Barnes, ed., *An Introduction to the History of Sociology* (Chicago: University of Chicago Press, 1948), 741. See also C. Wright Mills, "The Professional Ideology of Social Pathologists," *American Journal of Sociology*, 49 (Sept. 1943), 165–180. After World War II sociology moved in a distinctly theoretical direction under the leadership of men such as Talcott Parsons.

toward it held by key figures such as Daniel C. Gilman and Andrew D. White. As we have seen, Gilman's disenchantment with the ASSA dates from the early 1880's; however, White retained a high opinion of the Association until at least 1890. That these two men should have devoted their time and energy to the Association is weighty testimony to its importance in the eyes of contemporaries. Their connection with the Association is especially significant, since both were prominent pioneers in the development of precisely those institutions of university scholarship that ultimately rendered the ASSA obsolete.[2]

White was associated with the ASSA from its earliest years, and, like his good friend Gilman, he was especially interested in introducing social science into the university curriculum. When Gilman became president of the ASSA in 1879, he immediately called on White to give an address—previously given at Johns Hopkins—at the next ASSA meeting. The address that Gilman so admired was, as White described it, "a plea for higher education in political and social science and general jurisprudence, with suggestions of method, replies to objections... and statements as to the attempts made... in certain of the Universities of the old world." In a postscript White added: "The title of my address might be given as 'Education & Politics' or 'education in Political & Social Science' or 'Instruction for the Coming Rulers.'"[3] The last phrase goes to the heart of White's conception of social science.

White is justly regarded as one of the first modern teachers of history in the United States. In 1857, the same year that H. W. Torrey and Francis Lieber began comparable innovations at Harvard and Columbia, White at the University of Michigan dispensed with recitations in favor of the lecture system and began introducing students to outside readings and original sources.[4] Even before Ezra Cornell selected him to preside over the new school at Ithaca, White dreamed of creating a "truly great university" where there would be "a chance for instruction

---

[2.] Charles William Eliot of Harvard, the third of the three key leaders of the American university movement, was also an officer of the ASSA, though he never became president. He was chairman of the Education Department in 1873, director and vice-president from 1873 to 1876. He was appointed director again in 1903, but the Association had changed character by that time.

[3.] White to Gilman, Ithaca, N.Y., 11 Feb. 1879, Gilman Papers, Johns Hopkins University. Gilman wanted to publish the address as a pamphlet for distribution in Baltimore, but White protested that it should have a wider audience. He especially hoped that Gilman would delay printing it long enough for him to give the address again at the Union League Club in New York, where George Walker (Sanborn's brother-in-law) and other of the "best men" in the club were making arrangements for him to appear. "I feel that *there* is the place to strike," said White: "there are wealthy graduates in great numbers, of Harvard, Yale, Princeton, Union, Columbia, Amherst and all the rest, and I think that if the idea could be set going it would enure to the benefit of many of these colleges." See White to Gilman, Ithaca, N.Y., 11 and 13 Mar. 1879, Gilman Papers, Johns Hopkins University. Italics in the original.

[4.] Ruth Bordin, *Andrew Dickson White: Teacher of History*, Michigan Historical Collections, Bulletin No. 8, Aug. 1958 (Ann Arbor, 1958).

in moral Philosophy, History and Political Economy unwarped to suit present abuses in Politics and Religion."⁵ White's original plan of organization for Cornell called for a special department devoted to "History, Jurisprudence, Political and Social Science."⁶

His eagerness to adapt social science to the university curriculum was intensified in the 1860's, when his duties as a New York state senator made him especially conscious of the general dearth of facts about pressing social questions. The work of the Social Science Association in filling that dearth seemed especially valuable. Inspired by the meetings of the Association, he renewed his urgings to the trustees of Cornell to "establish a department, or at least a lectureship, which should do something toward training up young men for future usefulness in these fields, giving them a better knowledge of such subjects than legislators generally possess. . . . As soon as the funds of Cornell University permitted this plan was carried out. . . ."⁷

Although history and economics were well represented at Cornell from the beginning, White did not get his first teacher of "social science" until 1884. In that year (when he himself became president of the AHA, the first professional association in the social sciences) White chose as Cornell's first professor of "social science" none other than Frank Sanborn, secretary of the ASSA. In White's eyes the dividing line between amateur and professional social inquiry was still indistinct. The amateur social science of the ASSA was still good enough for the man who ranked second only to Gilman as academic professionalizer.

For several years White had been urging Sanborn to come to Cornell to give a course that would be neither "ethics" nor "economics," nor "what is broadly termed sociology," but "practical lectures on the treatment of the public dependents."⁸ In conjunction with the lectures there were to be visits to functioning institutions, such as prisons, asylums, and poorhouses. Sanborn conceived of social science quite simply as dealing with all the techniques of meliorative action. White shared this view, but added to it the indirection of training the "coming rulers" in lieu of taking direct action.

Sanborn taught his intensely practical variety of social science at Cornell from 1885 to 1888. There seems to have been no theoretical content to the course, except perhaps for the introductory lectures, which, as Sanborn described them, traced "historically the methods of dealing with crime, pauperism, insanity, preventable disease, public

---

⁵· From a letter soliciting support from Gerrit Smith, 1 Sept. 1862, in Carl L. Becker, *Cornell University: Founders and the Founding* (Ithaca, N.Y.: Cornell University Press, 1943), 156.

⁶· Rogers, *Andrew D. White and the Modern University*, 127–128.

⁷· White, "Instruction in Social Science," 1–2.

⁸· Sanborn, "Social Sciences: Their Growth and Future," 9. See also White's *Autobiography*, I, 378–379, and *JSS*, 28 (Oct. 1891), 2–3.

vice, etc., and showing how these evils and therefore the mode of meeting them has been affected by the changing conditions of modern civilization."[9] Sanborn was typical of his generation and of the ASSA in indiscriminately lumping together all the imperfections of society, from crime to cholera, as within the proper sphere of social science. He was typical also in the practical, untheoretical cast of his thinking. It was a popular course in social problems, highly current and topical—even exciting, in view of the field trips into dark prisons and ominous asylums, places few undergraduates had ever seen. The course regularly drew over forty students.[10]

In a typical semester Sanborn's students traveled to the Elmira State Reformatory, the Tompkins County Jail and the County Poorhouse, Willards Asylum, the Auburn State Prison, the Auburn Asylum for Insane Criminals, the Western House of Refuge at Rochester (for dependent and criminal juveniles), the Monroe County Insane Asylum at Rochester, and the New York State Asylum for Idiots at Syracuse. Each student was required to submit a written report, comparing each institution by capacity and number of inmates, age and sex of inmates, penal and educational modes employed, and quality of management. The students found Elmira Reformatory, run by an officer appointed by the governor for a five-year term (in fact, Zebulon R. Brockway, a friend of Sanborn), to be the best managed. The Tompkins County Poorhouse, where the superintendent was elected by the county, was found to be "very poorly managed."[11] Sanborn's students also wrote short essays such as "The Relation of Insanity to Crime as Illustrated by the Auburn Asylum and Prison," "Journalism in the Elmira Reformatory," and "The Duty of Aiding the Poor and the Best Method of Giving Relief."[12]

[9.] *JSS*, 21 (Sept. 1886), 10. A rough preliminary outline of the course appears in Sanborn to White, Boston, 7 Mar. 1885, A. D. White Papers, Cornell University. The course was to be organized around three main topics: "pauperism," "crime and penalty," and "insanity and its causes."

[10.] *JSS*, 21 (Sept. 1886), 11. White's secretary reported that "Mr. Sanborn's lecture room was well-filled this morning and all seemed pleased with his lecture...." Ernest W. Huffcut to White, Ithaca, N.Y., 10 Apr. 1885, A. D. White Papers, Cornell University. Three years later Sanborn wrote: "I am here now giving my usual spring lectures,—perhaps for the last time. My class is large and the interest seems to be growing. I enjoy Ithaca greatly always." Sanborn to Caroline H. Dall, Ithaca, N.Y., 11 Apr. 1888, Clarkson Collection (original in Massachusetts Historical Society).

[11.] See Andrew Strong White, "The Institutions visited by Professor Sanborn and his class in Social Science during the year 1887, and comparisons between them," a student essay dated June 1887, in Andrew Strong White and Claire Howard Papers, Cornell University.

[12.] Essays by Andrew Strong White with these titles are located in the collection named above. For additional fragmentary information on Sanborn's course, see Haskell, "Safe Havens for Sound Opinion: The American Social Science Association and the Professionalization of Social Thought in the United States, 1865–1909" (Ph.D. dissertation, Stanford University, 1973), 222–223.

The students seem to have enjoyed Sanborn, and he seems to have enjoyed giving the course, but the constraints of institutionalized scholarship did not come naturally to him. When Cornell originally asked him to give an eleven-week course, to conform to its routine calendar, he responded with assurances that his course could be concentrated in fewer weeks so as to take "less time and be more useful."[13] Sanborn returned to Concord frequently while giving the course. Once, late in his first year at Cornell, he cancelled nearly a week of classes with no explanation, saying only that his presence was required in Concord.[14] Sanborn never lost the contempt for academic pedantry that he developed as a young apprentice to the Transcendentalists. I have found no record of conflict with the Cornell faculty, but Sanborn's estimate of the academic man is epitomized by his later expression of disgust with "that effusive omniscience which is the mark of the college professor of the present era,—especially in New York, New Haven, and Harvard."[15] Sanborn could bring his rich fund of practical experience into the university and thus contribute to the legitimization of social science as a field of professional expertise, but he himself was no professional. He found it amusing to be addressed as "professor."[16]

By the late 1870's, if not before, Sanborn and White recognized a clear distinction between their kind of inquiry—eclectic, practical, and reformist—and "sociology," a field whose boundaries were still ephemeral but which was closely identified with the bold constructions of Comte and Spencer. But Sanborn and White were not at all apologetic about their own kind of inquiry. Sanborn as late as 1890 declared: "It is to discover and amend what is wrong in the habitual life of men that social science applies itself most usefully,—not to promulgate broad theories or insist on ambitious panaceas for every human ill, but to consider the ailment and apply the remedy patiently and repeatedly, as a mother heals the hurts of her children."[17]

Inspired by Sanborn's position at Cornell, the Social Science Association launched a campaign in the mid-1880's to introduce its variety of social science into colleges and universities across the country. In

[13.] Sanborn to White, Concord, Mass., 8 Nov. 1884, A. D. White Papers, Cornell University.
[14.] Sanborn to White, Concord, Mass., 19 May 1885, *ibid.*
[15] Sanborn was speaking of John W. Burgess's *The Middle Period*, which greatly offended him by its ill-treatment of John Brown. Sanborn thought Burgess knew no more about Brown than "Brown's Virginia preachers did of Christianity. I respect Rabbi Burgess as a gentle man; but as a Hebrew gentleman who preferred Barabbas for a fellow citizen." See Sanborn to [D. W.?] Wilder, Concord, 11 Apr. 1897, and Sanborn to F. G. Adams, Concord, 15 Apr. 1897, Clarkson Collection (both originals in Kansas State Historical Society).
[16.] F. B. Sanborn to Charles Dudley Warner, Ithaca, N.Y., 7 May 1885, Clarkson Collection (original in Trinity College, Hartford, Conn.).
[17.] Sanborn, "Work of Twenty-five Years," xlv-xlvi.

collaboration with several other ASSA officers, Sanborn in 1885 and 1886 formulated a comprehensive model curriculum that was based on the ASSA tradition of inquiry and even organized around the ASSA's five departments of Education, Health, Jurisprudence, Social Economy, and Finance.[18] In this project Sanborn had the assistance of Francis Wayland, Jr., William Torrey Harris, and Edmund J. James. Wayland, son of the antebellum moral philosopher, was dean of the Yale Law School, a leader in prison reform and the charity organization movement, and a one-time lieutenant governor of Connecticut. He followed Gilman in the presidency of the ASSA, 1880–82. William Torrey Harris was the founder of the *Journal of Speculative Philosophy* and the foremost proponent of Hegelian idealism in America. He and Sanborn were closely associated in the famous Concord School of Philosophy in the 1880's. Harris was chairman of the ASSA Education Department from 1878 to 1886 and became U.S. Commissioner of Education in 1889. Edmund J. James was the only one of the young German-trained scholars to play an important role in the ASSA. After taking part in the founding of the American Economic Association, the University of Pennsylvania professor became secretary of Sanborn's Social Economy Department, occupying that position from 1885 to 1889. In 1890 he organized the American Academy of Political and Social Science, which immediately absorbed the Philadelphia branch of the ASSA and almost merged with the parent organization a few years later. James later became president of the University of Illinois.[19]

In conjunction with the Sanborn-Wayland-Harris-James project to formulate a model curriculum, the Association conducted in 1886 a study of the current state of social science instruction in the nation's colleges and universities. Over a hundred schools responded to a questionnaire which asked the president of each institution to indicate whether, and in what year, students were instructed in the following topics:

1. Theory of property, real and personal
2. Production and distribution of wealth
3. Theory of government—national, state, and municipal
4. Public and private corporations
5. Punishment and reform of criminals
6. Prevention of vice (intemperance, prostitution, vagrancy, etc.)

[18.] For the model curriculum see *JSS*, 21 (Sept. 1886), 13–20.

[19.] On James see Richard A. Swanson, "Edmund J. James, 1855–1925: A 'Conservative Progressive' in American Higher Education" (Ph.D. dissertation, University of Illinois, 1966). See also *Annals*, 7 (1896), 78–86.

7. Public and private charities (care of poor, insane, blind, idiotic, deaf-mute, foundlings, orphans, etc.)
8. Sanitation of cities and private dwellings (water-supply, ventilation, drainage, epidemics, etc.)
9. Theory of public elementary education
10. Higher education (as furnishing the directive power of society)[20]

The questionnaire conveys in a brief space the main concerns of the ASSA model curriculum and suggests the great differences between that curriculum and modern social science.

Sanborn and those who collaborated with him in formulating the model curriculum undertook what appears in retrospect to be a futile task by trying to build university social science on the crumbling foundations of the ASSA. Especially interesting, in retrospect, was their struggle to transfer the ASSA's departmental division of labor to the university. The Association's departmental scheme of organization was geared to the twin assumptions that social science should be oriented almost exclusively to publicly visible problems, and that social science should be primarily an avocation for men of the professional class, organized in their classic specialties. The scheme was not well suited to the needs of men who made research and teaching a full-time vocation. The ASSA campaign on behalf of college-level social science instruction probably therefore could not have succeeded in perpetuating the Association's own special variety of social science, for that variety was inherently inappropriate for academicians. But the campaign could and did publicize social science in a more general sense, and thus helped it secure a legitimate place in the university curriculum. Once in the university, social science would be transformed to suit the occupational needs of those who practiced it.

In the eyes of Andrew D. White, whose opinions carried great weight in late nineteenth century academic circles, the ASSA report on social science instruction was "the most noteworthy contribution to the whole subject." He believed that "No more suggestive document has appeared in this country, and it ought to be brought to the notice of every College and University governing board and faculty in the country." White specifically hoped that colleges and universities would establish courses "in all the five main divisions of Social Science, recognized by this Association."[21]

[20.] The questionnaire is in *JSS*, 21 (Sept. 1886), 23–24; the responses are *ibid.*, xxiv-xlix, and *ibid.*, 22 (June 1887), 12–27. The project is discussed by Emily Talbot, secretary of the Education Department, *ibid.*, 7–11.

[21.] White, "Instruction in Social Science," 4, 14. White wanted all of the ASSA departments represented in the college curriculum, but he placed special stress on the Social Economy Department.

Popularizing Sound Opinion

From the beginning the Social Science Association had tried to shape public opinion. In the late 1880's and 1890's, as the rise of academic social science began to undermine some facets of the role that the Association had carved out for itself, the Association began to shift toward a self-conception that stressed the popularization, rather than initiation, of sound opinion. The shift of emphasis was easy to make, for education for "good citizenship" had always been close to the heart of the Association. In his 1886 presidential address Carroll D. Wright, first U.S. Commissioner of Labor, urged the Association to broaden its public appeal by promoting social science instruction not only in colleges, but also in secondary and Sunday schools.[22] Both Wright and Andrew D. White wanted to expand the Association's practice of occasionally publishing special pamphlets espousing the best ideas and theories about social problems—a practice advocated at the 1869 meeting by Horace Greeley.[23]

What was perhaps the only viable role left for the ASSA, after the emergence of professional social scientists, was stated quite clearly by Joseph Lee, secretary of the Social Economy Department in the mid-1890's. Lee argued that the proper role of the Association should be to stand midway between the academic social scientist and the legislator,

---

[22.] Carroll D. Wright, "Popular Instruction in Social Science," *JSS*, 22 (June 1887), 28–35. On Wright see James Lieby, *Carroll Wright and Labor Reform: The Origins of Labor Statistics* (Cambridge: Harvard University Press, 1960). Furner in *Advocacy and Objectivity*, 316–318, argues—erroneously, I believe—that Wright and John Eaton were at loggerheads with Sanborn and others over the question whether the ASSA should address itself to university men or to the general public. I find no evidence of such a debate or of any opposition to Sanborn's model curriculum project. Far from opposing social science in the university, Wright was eager to place it there: see his "Study of Statistics in Colleges," *Publications of the AEA*, 3 (Mar. 1888), 1–28. Instead of a dramatic conflict, it appears to me that the leaders of the Association simply drifted toward the role of popularization for lack of anything better. Sanborn was anything but a devotee of ivory tower social science, but to say, as Furner does, that the ASSA was traditionally dedicated to "communicating directly with the masses" (317–318) is greatly to oversimplify the organization's role.

[23.] Wright, "Popular Instruction in Social Science," 28, and White, "Instruction in Social Science," 16. White at the 1891 meeting expressed his conception of the Association's role as follows: "I remember that during the anti-slavery struggle that very eminent and noble man, Gerrit Smith, once spoke to me with admiration of the splendid work then accomplished by Henry Ward Beecher. Said Mr. Smith, 'I have been trying all my life to clutch the truth with both hands; but Beecher gets one arm around the truth and the other around humanity, and then forcibly but lovingly brings them together.' Is there not in this a statement of our own function? Other societies may have it for their purpose to make minute researches, more especially for scholars; others may still have it as their entire object to popularize the principal results of social science. Ours, it seems to me, should be, while promoting the strictest research, to lay hold at the same time of the public at large, and bring the results of our research to bear upon society around us." White, "Opening Remarks at the General Meeting of 1891," *JSS*, 28 (Oct. 1891), 26–27.

supplying a practical-mindedness rare among professors and a sense of perspective and statesmanship rare among politicians.

> College professors ... are concerned with social principles and phenomena from the purely academic standpoint ... [they] are not called upon to form opinions as to what ought to be done. ... But our Association is not academic in its purpose. It *is* our business to have an opinion as to what had better be done. As I understand it, our main object is to stand between the student and the politician, translating the ideas of the one into a policy for the other to pursue.[24]

The role Lee proposed for the ASSA is plausible and appealing even today, but it would have been extremely difficult to carry out in practice. What he had in mind went beyond popularization, but in fact the Association was unable to avoid that lesser role and the inferior status that went with it. If the real locus of original social inquiry and discovery was in the university, why should legislators or anyone else listen to the advice of the ASSA?

The period from the Great Railroad Strikes of 1877 to the collapse of the Populist movement in 1896 was one of rising social tension and political polarization. There was a growing public demand for social commentary in these years as the painful problems of an industrializing society were forced into the open and impressed upon the public mind by a twenty-year decline in prices, proliferation of squalid tenement districts, widespread labor violence, and spectacular cases of reaction such as the Haymarket incident of 1886. Areas of life that had heretofore possessed a quality of opaque permanence and tranquil necessity began to appear to many minds as problematical and subject, for better or worse, to intentional manipulation. The chief beneficiaries of this expanding market for insight into social problems were at first free-lance intellectuals like Henry George and Edward Bellamy. George's *Progress and Poverty* (1879) and Bellamy's *Looking Backward* (1888) demonstrated the existence of a vast market for social thought.[25]

The Social Science Association was able to capitalize on this market only indirectly, by serving as a forum for prominent social thinkers and their opponents. In the late 1880's and 1890's the Association frequently organized its meetings around a central topic of public concern. The ASSA Single Tax Debate of 1890 and the Free Silver Debate

[24] Joseph Lee, "The Argument for Trade Schools," *JSS*, 33 (Nov. 1895), 208–212. Italics in the original.

[25] *Progress and Poverty* sold over 1,000 copies in the first year, more than any other existing work in political economy: Charles A. Barker, *Henry George* (New York: Oxford University Press, 1955), 330. *Looking Backward* sold 200,000 copies by 1890 and was then selling at the rate of 10,000 per week: Daniel Aaron, *Men of Good Hope: A Story of American Progressives* (New York: Oxford University Press, 1961), 104.

of 1895 attracted considerable public attention. The controversial theories of Henry George had been discussed at the 1886 ASSA meeting, but George had then declined an invitation to attend.[26] A full-scale debate in which George participated was organized for the 1890 meeting by the Reverend John Graham Brooks, acting secretary of the Association during the temporary absence of Frank Sanborn.[27] Supporting George in the debate were William Lloyd Garrison II, Samuel B. Clarke, a lawyer, and James R. Carrett, Boston land-title expert. Vigorously opposing George were ASSA regulars Edward Atkinson and William Torrey Harris, together with young professors John Bates Clark of Smith and E. R. A. Seligman of Columbia. More moderate comments were made by Professors Thomas Davidson, Edmund J. James, and E. Benjamin Andrews.[28]

The chief battle of the single-tax debate took place between George and Seligman. George had long denounced the pretensions of academic economists and had conducted a running battle with such men as William Graham Sumner and Francis A. Walker during the 1880's. In response to George's charge that professors were allies of the ruling class, Seligman pointed to Arnold Toynbee in England, to the Verein für Sozialpolitik in Germany, and by implication to the AEA, to show that the worker had no better friend than the academician. Seligman condemned George's amateurism and gave the professional social scientist credit for the recent "awakening of interest on the part of the so-called higher classes toward the laborer" and the inauguration of "an era of practical reforms."[29] George was fully committed to the idea that political economy should be simple and understandable to the average citizen. As Charles Barker suggests, he never understood that the rise of a professional community of inquiry in the 1880's deprived him of the easy credibility he had enjoyed in 1879.[30]

The Free Silver Debate of 1895 was organized by the Finance Department, of which Professor Jeremiah W. Jenks of Cornell was chair-

[26.] *JSS*, 22 (June 1887), xiii, 98–161.

[27.] Sanborn traveled in Europe during the spring and summer of 1890 following the suicide of one of his sons. Brooks was an occasional lecturer in economics and social questions at Harvard and Chicago and became president of the ASSA in 1903–5: see James E. Mooney, *John Graham Brooks, Prophet of Social Justice: A Career Story* (Worcester, Mass.: Privately printed, 1968).

[28.] *JSS*, 27 (Oct. 1890), lx; Barker, *Henry George*, 565–566. The Single-Tax Debate was widely reported by the press. The record was printed in the *Journal of Social Science* and also published as a pamphlet which sold for thirty cents each or twenty-five cents each in lots of a hundred. By 1898 the supply was sold out and a second printing was considered: *JSS*, 27 (Oct. 1890), iv; 36 (Dec. 1898), 58.

[29.] *JSS*, 27 (Oct. 1890), 88.

[30.] Barker notes that George's *Science of Political Economy* (1897) fell flat because economic thought had become infinitely more complex than it once had been: *Henry George*, 582–583. For George's relations with other professional economists, see *ibid.*, 425–431 and 555–558.

man. Jenks was to be prime mover in the organization of the American Political Science Association in 1903. Academic participants were strangely absent from the Free Silver Debate, and the speakers were much less prominent than those who took part in the Single-Tax Debate. Nonetheless, the Association found a market for a special printing of the debate record.[31]

Other special sessions and debates organized by the Association in the 1880's and 1890's were not so elaborate or so well publicized as the Single-Tax and Free Silver debates, but they served a similar purpose. A series of papers on the labor question at the 1885 meeting aroused much interest.[32] At the 1887 meeting the Social Economy Department focused attention on profit-sharing plans.[33] The question of labor organization was discussed at the 1891 meeting by Samuel Gompers and Professor E. W. Bemis, among others.[34] Prompted by the furor attending the report of a select committee of the House of Lords, the Social Economy Department devoted its 1892 meeting to "the Sweating System in Europe and America."[35] At the 1896 meeting Booker T. Washington contributed to a debate on "Higher Education for Negroes," and Florence Kelley of Hull House took part in a debate on trade schools and industrial education.[36] Race relations and the question of higher education for Negroes were discussed again at the 1899 and 1901 meetings. Black representatives who opposed Washington's acquiescent stance took part in both discussions.[37] At the 1894 meeting the Social Economy Department presented a valuable compilation of reports from charitable institutions on the relief of the unemployed during the winter of 1893–94.[38]

In addition to debates and special sessions, which served to give the Association's meetings a focus and keep the Association itself in the public eye, there were other less conspicuous activities that were no less important. In the early 1880's, for example, William Torrey Harris generated a brief flurry of interest in child study. Inspired by an exchange between Charles Darwin and other scholars in the English

---

[31] The debate appears in *JSS*, 33 (Nov. 1895), 49–133.
[32] *Ibid.*, 21 (Sept. 1886), xxx–xxxiii.
[33] *Ibid.*, 23 (Nov. 1887), 21–67.
[34] *Ibid.*, 28 (Nov. 1891), 28–125.
[35] *Ibid.*, 30 (Oct. 1892), 59–146. See also *ibid.*, 31 (Jan. 1894), 63–69.
[36] Other contributors to the debate on Higher Education for Negroes included Heman Lincoln Wayland, a Baptist minister (and brother of Francis Wayland, Jr.); T. J. Morgan, former Indian Commissioner; Professor Silas X. Floyd of Atlanta University; and Hugh M. Browne. *Ibid.*, 34 (Nov. 1896), 68–97. The Trade School Debate appears *ibid.*, 29–67.
[37] See *ibid.*, 37 (Dec. 1899), 52–66; and 39 (Nov. 1901), 100–138.
[38] The reports cover New York, New Haven, Baltimore, Buffalo, Cleveland, Denver, Pittsburgh, Louisville, Chicago, Philadelphia, St. Louis, Syracuse, and Waterbury, Conn.: *ibid.*, 32 (Nov. 1894), 1–51.

journal *Mind,* the Education Department prepared a protocol to guide Association members in the observation of their own children. The protocol is remarkable for its primitiveness. A typical question is: "At what age did the baby exhibit consciousness, and in what manner?"—a question to which a standardized unambiguous answer cannot be given and which, moreover, invites the recipient rather than the researcher to define the terms of the question. Only thirty persons completed and returned the form. Not surprisingly, no clear generalizations could be reached on the basis of their responses.[39]

In the late 1880's the Association organized a substantial study of savings and loan associations, cooperative finance companies, life insurance, and like enterprises in America for presentation to the Paris Congress of Provident Institutions. Young academicians E. J. James, Henry Carter Adams, and Jeremiah W. Jenks took part in the work.[40] Frank Sanborn, who was a neighbor of J. Warren Bailey, one-time president of the National League of Building and Loan Associations, made the "cooperative banking" or "building association movement" his pet reform in the 1890's. The movement was first brought to the attention of the Association in 1874 by Josiah Quincy and had been an occasional subject of discussion in the late 1870's and 1880's. The hope of reformers lay in the presumption that these institutions helped heal the breach between capital and labor by encouraging thrift and enabling the poor to accumulate capital. Sanborn repeatedly attacked associations organized on a national basis because they subverted the reform motive: "localization of function is the very essence of this form of banking."[41]

Sociology versus Social Science

There is much to admire in the practical humanitarian reform tradition carried on by the ASSA in the late nineteenth century. Cer-

---

[39.] The protocol appears *ibid.,* 13 (Mar. 1881), 190–192. See also *ibid.,* 15 (Feb. 1882), 1–76; 17 (May 1883), 145–155. It is worth noting that Max Weber at about the same time was using protocols hardly less primitive than this one: see Oberschall, *Empirical Social Research in Germany 1848–1914,* 19–32. In passing it may be suggested that the modern protocol could only come into being when the researcher overcame inhibitions against treating the protocol recipient as a mere instrument of the researcher's will. The modern protocol elicits unambiguous, strictly comparable answers by short-circuiting the recipient's personal judgment and demanding an impersonal response—ideally, "yes" or "no." To formulate questions in such a way would have seemed insulting to Sanborn's generation: they would have been embarrassed, I think, to ask a question in such a way as to prevent the respondent from giving a whole, individualized response.

[40.] *JSS,* 24 (Apr. 1888), v–viii; 25 (Dec. 1888), 98–177; 26 (Feb. 1890), ix–xi.

[41.] Quotation from Sanborn, "Cooperative Banking in the United States (1873–1898)," *JSS,* 36 (Dec. 1898), 131. See also *JSS,* 27 (Oct. 1890), liii–lviii; *ibid.,* 31 (Jan. 1894), 45–48, 52–53. On Sanborn and Bailey, see *ibid.,* 39 (Nov. 1901), 161. See also Sanborn, "Our Progress in Social Economy Since 1874," *ibid.,* 35 (Dec. 1897), 51–53.

tainly it was a precedent available for the Progressive movement, even if the heritage was seldom acknowledged. In the words of a recent investigator, "scarcely a modern issue involving the individual in his relationship to society was not explored in a paper delivered before the American Social Science Association in the seventies and eighties."[42] Through the Social Science Association one can draw a valid, if thin, line of continuity linking the romantic reformers of the antebellum era, such as Wendell Phillips, Samuel Gridley Howe, and Sanborn himself, to such Progressive reformers as John Graham Brooks, Florence Kelley, and Jane Addams. But the continuity is much stronger in subject matter and areas of concern than it is in personnel. The Progressive generation owed much to Sanborn and his colleagues for their early, groping efforts to articulate and conceptualize the tendencies of an urban-industrial society. But neither the practical reformers nor the theoreticians of the Progressive generation adopted the ASSA as an institutional framework for their labors. For them it was only a place to give a spare paper or to meet a potential patron—or, worse, a gathering of cranks to be ignored.

Albion W. Small, head of the University of Chicago sociology department and founding editor of the *American Journal of Sociology*, dismissed the Social Science Association with the observation that "it represented humanitarian sentiment more distinctly than a desire for critical methodology."[43] Amos G. Warner, Stanford sociologist whose interests lay very close to social work, and who therefore might have been expected to be sympathetic, complained that "the action of the American and English Social Science Association has served to render the term social science unusable because so frequently misapplied and misunderstood."[44]

The younger generation's rejection of the ASSA took on an almost ceremonial character in 1894 with the appearance of Franklin H. Giddings at the annual ASSA meeting. Giddings had just been appointed to Columbia's professorship of sociology, the first university chair in this country to bear that name. Giddings and Albion Small share honors as the founding fathers of academic sociology in America.[45] Giddings's message to the ASSA was blunt: social science is

---

[42.] Bledstein, "Cultivation and Custom," 217.

[43.] Albion Small, "Fifty Years of Sociology," *American Journal of Sociology*, 21 (1915–16), 729.

[44.] Amos Warner, "Philanthropology in Educational Institutions," in his *Sociology in Institutions of Learning: Being a Report of the Seventh Section of the International Congress of Charities, Correction and Philanthropy, Chicago, June, 1893* (Baltimore: Johns Hopkins University Press, 1894), 81–82.

[45.] Harry Elmer Barnes called Giddings "probably the ablest sociologist that the United States has ever produced": *An Introduction to the History of Sociology*, abridged ed. (Chicago: University of Chicago Press, 1966), 199n. Small began teaching sociology at Chicago in 1892.

dead, and its heir is scientific sociology. Like E. L. Youmans twenty years before, Giddings could find no thread of coherence in the work of the Association, no grounds for deferring to its collective judgment. It was "unscientific." But unlike Youmans, Giddings acknowledged the legitimacy of the reforming impulse. The Association was unscientific not because its motives were philanthropic, but because its philanthropy was out of touch with reality.[46]

Giddings, who worked for a time on the Springfield *Republican,* for which Sanborn wrote weekly columns, rudely dismissed the kind of social science course Sanborn had taught at Cornell and for which the ASSA had campaigned for the past decade. Such courses, said Giddings, were hopelessly miscellaneous in subject matter and "resembled each other in nothing but name."

> Some of them have been statistical studies of population; others have dealt with the so-called labor question; others with defectives and delinquents, charity, punishment and reformation; and others still, with public health and sanitation. Indeed, they have collectively well-represented the broad inclusiveness of the term "social science" as it is used in the title of this Association. All that could be said with certainty about such university courses is that they were concerned with groups of social facts not otherwise covered by the courses in history, political economy, politics, and ethics.[47]

Among European scholars, said Giddings, it is understood that social inquiry belongs to the province of "sociology." And in America, "to the discerning," it has become clear that sociology must displace what has been called social science.

> To Comte, who coined "sociology" the name meant always a philosophical explanation of society as a whole. To Spencer, who made it pass as coin current, it has always meant an explanation of society in terms of evolutional theory. To the European sociologists . . . the word stands invariably for the original conceptions of Comte and Spencer,—conceptions, namely, of society as a concrete whole, and of its scientific explanation in term of natural causes. "Sociology," then, in the view of all these scholars, is the descriptive, historical, and explanatory science of society. It is not a study of some one special group of social facts: it examines the relations of

---

[46.] Giddings, like Youmans, was an evolutionary naturalist. See Clarence H. Northcott, "The Sociological Theories of Franklin Henry Giddings," *ibid.,* 180–201. It is essential, however, to recognize that Giddings assigned a much larger role to human volition than did Spencer.

[47.] Franklin H. Giddings, "The Relation of Sociology to Other Scientific Studies," *JSS,* 32 (Nov. 1894), 144.

all groups to each other and to the whole. It is not philanthropy: it is the scientific groundwork on which a true philanthropy must build.[48]

Why was "social science" of the ASSA variety a failure? Because, Giddings argued, the "social science" inquirer dealt with social phenomena at a superficial level and was not really in touch with reality. The ASSA's variety of "social science" imputed independence to variables that were really the dependent, last links in long chains of causation. In order to grasp reality securely, Giddings implied, one must plunge beneath the surface of events to the fundamental phenomena of society. One must leave behind the everyday world of proximate, readily visible social influences, and instead seek out the remote, obscure, elemental factors which underlie and influence everything else. This was the unique mission of sociology, according to Giddings—to be the *fundamental* social science, preliminary to and more general than all the other branches of knowledge about man and society. As the fundamental social science, it would also be the coordinating science. "If we want to know what relations economic, legal, and political phenomena bear to each other and to the social whole, we must ask what fundamental conditions of life in society they grow out of." In a turgid but important passage, Giddings tried to explain to ASSA members his feeling that a new dimension of explanation would unfold if scholars could only get below the surface of life to the elemental, causal factors:

> If, now, guided by this thought, we inquire what very fundamental phenomena of human society are under our observation, we shall see many things in a new light. We shall expect, perhaps, that a sociology which restricts itself to elements and first principles will prove to be highly abstruse, and much too general to afford helpful guidance in practical affairs. This expectation will be curiously disappointed. As a matter of fact, it so happens that the elementary phenomena of society are precisely those that give rise to the pressing questions of practical policy and philanthropy. Nearly all the practical problems that confront society grow out of the alternate aggregation and dispersion of population; the migration and intermingling of races and nationalities; the unequal development of economic and of social instincts in different individuals; the consequent appearance in the community of different standards of living; the consequent segregation also of the population into the enterprising, the industrious, the criminal, the pauper, the unfortunate, and the degenerate; and finally, out of that ceaseless interchange of thought and feeling whereby the members of a community come to feel the same desires and antagonism, to cherish the same ideas, and to act in concert for

[48.] *Ibid.*, 144–145.

common ends. That these things are the fruitful causes of unrest, of agitation, of interest in social questions, of philanthropy itself, has long been understood. What we have not perceived is that these things are in truth the very elements of social phenomena of every description.[49]

Sociology, then, by Giddings's definition, was a discipline specifically concerned with remote causation, with the realm of underlying factors upon which politics and philanthropy must depend. Part of the sociologist's duty was to convince the educated public that such a realm did exist and that wise policy could not be made without knowledge of it.

> Here, then, is the opportunity for sociology. It must enter upon that thorough-going, systematic study of these phenomena which will demonstrate to everybody that they are the fundamental, the elementary things,—that they are, in fact, the germ-plasm of society,—and that the study of them, so far from being a mere supplement to older sciences, and without logical relation to other inquiries, is the true co-ordination of all social sciences, because it is the groundwork on which all must build.[50]

Here was a challenge the ASSA could not safely ignore. For nearly three decades the ASSA had been an oasis of voluntarism and optimistic reformism in a desert of ever-bleaker deterministic extrapolations from Darwinian evolution and Spencerian positivism. Sanborn and his ASSA colleagues from the beginning had rejected Spencer's necessitarian doctrine and had insisted on human freedom and the feasibility of interventionary reform. Their commitment to idealism as a philosophy and to professionalism as a model of social decision-making carried them unscathed through a period when systematic social theory, dominated by Spencer, seemed to render humanitarian reform a futile illusion. In the 1870's the ASSA stood alone as the custodian of voluntarism. A harbinger of the new era came in 1883 with publication of Lester Ward's *Dynamic Sociology,* which, from a basically naturalistic point of view, reasserted the place of conscious mind in evolutionary progress. But by the 1890's the Association faced not individual challengers, but a rising profession that acknowledged the voluntaristic element in human affairs and yet also acknowledged interdependence and remote causation.

[49] *Ibid.,* 146–147. Giddings's conception of sociology as a fundamental science, underlying all the particular social sciences, was an alternative to Albion Small's conception of sociology as the coordinating science, overarching and uniting the others. Although the two notions of sociology were different in form, as Giddings observed, "in substance, they do not differ materially. Co-ordinating principles are fundamental or first principles, always" (*ibid.,* 146). Both conceptions served to justify a place in the university curriculum for sociology, a relative latecomer, and both served to deny legitimacy to the style of social inquiry associated with the ASSA.

[50] *Ibid.,* 148.

The voluntaristic bias of the Association had once been a source of strength, for most serious social thinkers in America were eager to reject the necessitarian aspect of Spencer's doctrine. But to insightful men and women of the 1890's, living as they did in the intensely interdependent surroundings of an urban-industrial society, the ASSA seemed to cling to voluntarism only at the expense of realism. Even the ordinary routines of life now taught that men were not, after all, autonomous masters of their own fate. Personal recollection taught that the texture of social experience had changed in the space of one lifetime. People sensed, even when they couldn't explain it, that they were bound together by ties far more extensive and compelling than had existed in a simple society of island communities and semi-subsistence farms. The universe of human affairs had changed. By the 1890's it had taken on a density that made the educated public receptive to an interpretation of man that portrayed him as partly free, but also constrained by a host of influences susceptible to empirical analysis.

The rise of professional sociology in the 1890's posed a more severe (because more direct) threat to the ASSA than had been the case with history or economics. But even more threatening was what sociologists stood for: namely, the growing recognition of interdependence. Sociologists took as their most basic premise the existence of a densely interdependent universe of human relations in which nothing, not even human volition, was autonomous. Everything was caused. As Lester Ward put it, "Every fact and every phenomenon is indissolubly linked to every other . . . and change is the result of some antecedent change and the occasion of some subsequent change."[51] To Ward and other sociologists this did not mean—as Spencer had seemed to imply—that man was impotent, always effect and never cause. On the contrary, as Giddings assured the ASSA in 1894, "Sociology will not be satisfied to accept as a sufficient interpretation of the social process an account in terms of physical causation only, or an account in terms of volition only. Physical causation and volition act and react upon each other."[52]

The task that Giddings set for himself and his discipline was the task of all social thinkers in the 1890's. Their goal was to rise above the nineteenth-century debate between positivism and idealism. They had to acknowledge that man's sense of freedom was no mere illusion— that, in choosing, *I* do become a genuine cause of my action. Yet they also had to acknowledge that man was part of nature, that human affairs could be understood in terms of cause and effect terminology,

---

[51.] Ward, "Sociology and Cosmology," *American Journal of Sociology,* 1 (Sept. 1895), 132.
[52] Giddings, "Relation of Sociology to Other Scientific Studies," 148.

and that the individual was in large measure a product of social circumstances.

Giddings's solution to the problem of the 1890's was not as elegant or as memorable as the work of such men as Weber and Durkheim, but it was perhaps more honest in its close approach to outright self-contradiction. In an early preview of his *Principles of Sociology* (1896) he tried courageously to fuse the positivist and idealist perspectives: "In the universe as known to science," said Giddings,

> there are no independent, unrelated, uncaused causes. By natural causation, therefore, the scientific man means a process in which every cause is itself an effect of antecedent causes; in which every action is at the same time a reaction. Nature [of which man is a part] is but the totality of related things, in which every change has been caused by antecedent change and will itself cause subsequent change, and in which, among all changes, there are relations of coexistence and sequence that are themselves unchanging.
>
> In this mighty but exquisite system man is indeed a variable, but not an independent variable. He is a function of innumerable variables. In a world of endless change he acts upon that world, but only because he is of that world. His volition is a true cause, but only because it is a true effect.[53]

Sociologists outside the naturalistic tradition were not as likely as Giddings to be enraptured by the spectacle of universal interdependence, but they too acknowledged that social interdependence was the root assumption of their profession. Albion Small, a man of deep religious convictions, began the first page of the first issue of the *American Journal of Sociology* with the observation that *"In our age the fact of human association is more obtrusive and relatively more influential than in any previous epoch.* Modern men are made aware in more ways than former generations that their lot is affected by the existence of other men."[54] Small saw the growing interdependence of men as the prime justification for his new journal and for the profession it represented.

He argued that men could not fail to be increasingly aware of their mutual dependence "as industries become diversified, as division of labor and competition become territorial and international, not less than individual, as occupations are more visibly affected by the actions of distant persons, as communication becomes accurate and rapid between men industrially related though geographically separate." Indeed, said Small, this rapid proliferation of dependencies had already produced in popular consciousness an irritated feeling that we

---

[53.] Giddings, "The Theory of Sociology," [special supplement] *Annals*, 5 (July 1894), 71.

[54.] Small, "The Era of Sociology," *American Journal of Sociology*, 1 (July 1895), 1. Italics in the original.

are all " *'meddlers one with another.'* " The common man did not need the sociologist to tell him that he was no longer immune to social influences; on the contrary, the sociologists' job was to answer "obtrusive questions about society that the ordinary man is proposing every hour."[55]

The increasing pressure and density of social relationships was, said Small, the source of a growing popular demand for social philosophy. If sociologists did not meet that demand, someone else would. "The millions have fragmentary knowledge of social relations, and they are trying to transmute that meager knowledge into social doctrine and policy." It would be better for society to rely on the "revised second thoughts" of professional sociologists than on the "hasty first thoughts" of the man in the street. "The conditions of human association are so involved," warned Small, "that it is no longer pardonable to increase present popular sensitiveness and irritation by theorizing about plans for accelerating the rate of human improvement, unless we have reduced all available pertinent facts about past and present human association to generalized knowledge, which shall indicate both direction and means of improvement."[56]

Small's words were self-serving but true. What undermined the Social Science Association most severely was the growing sense, first articulated by the positivists but taught also by concrete daily experience in an urban-industrial society, that human events were profoundly interrelated, that nothing moved independently, that all things were dependent variables shaped by factors lying hidden beneath the visible surface of life. The conviction that the social world was a realm of extensive mutual dependence in which there were few, if any, independent variables, and therefore precious little basis for explanation or effective intervention—it was this conviction which led men to believe sociologists and ignore the ASSA. If society were half as complex as Spencer's famous iron plate, which would not be flattened by a direct blow, then the reform and even the maintenance of society required the guidance of inquirers who focused not on the surface of life, but on the essential elements, the potentially independent variables upon which everything else depended.[57] Opinion not grounded

[55.] *Ibid.*, 1, 2, 6.
[56.] *Ibid.*, 3, 6, 8.
[57.] Spencer's classic statement from *The Study of Sociology* reads as follows: "You see that this wrought-iron plate is not quite flat; it sticks up a little here toward the left—cockles, as we say. How shall we flatten it? Obviously, you reply, by hitting down on the part that is prominent. Well, here is a hammer, and I give the plate a blow as you advise. Harder, you say. Still no effect. Another stroke: well, there is one, and another, and another. The prominence remains, you see: the evil is as great as ever—greater, indeed. But this in not all. Look at the warp which the plate has got near the opposite edge. Where it was flat

in this realm of remote causation could no longer be regarded as sound.

---

before it is now curved. A pretty bungle we have made of it. Instead of curing the original defect, we have produced a second. Had we asked an artisan practised in 'planishing', as it is called, he would have told us that no good was to be done, but only mischief, by hitting down on the projecting part. He would have taught us how to give variously-directed and specially-adjusted blows with a hammer elsewhere: so attacking the evil not by direct but by indirect actions. The required process is less simple than you thought. Even a sheet of metal is not to be successfully dealt with after those common-sense methods in which you have so much confidence. What, then, shall we say about a society? 'Do you think I am easier to be played on than a pipe?' asks Hamlet. Is humanity more readily straightened than an iron plate?" Quoted by Harry Elmer Barnes, *An Introduction to the History of Sociology* (1948 ed.), 133–134. See also Philip Abrams, *Origin of British Sociology*, 66–76. This conservative rendering of the implications of interdependence is one of the most striking and memorable passages in Spencer's work.

CHAPTER X

Professionalism Unhinged

In the late 1890's, when Frank Sanborn and his generation finally stepped aside, the American Social Science Association fell into the hands of men of narrow spirit and small intellect who were not primarily concerned either with understanding society or with improving it. The aspirations of the gentry class and its professionally trained members had always played a central role in the life of the Association. Now genteel professionalism became an obsession. What had been a cautious and responsible elitism degenerated into frivolous ambitions for a pseudo-aristocracy. The ASSA was put to use by men who sought to institutionalize on a nationwide basis the Mugwump ideal of the "best men." Their ambitions were partly realized in the creation of the ASSA's two legal heirs (still alive today), the National Institute of Arts and Letters and the National Institute of Social Science. Within a few years after the turn of the century the ASSA was a hollow shell; by 1909 it was dead in all but name.

Except for the ASSA's tangential role in the founding of the American Political Science Association in 1903, the last years of the old Association are largely irrelevant to the history of the social sciences. There is one respect, however, in which the ASSA's demise is both relevant and illuminating. The sad fate of the Association demonstrates the depth and continuity of its commitment to a defense of authority. That commitment became sour and petty as the Association neared death, but it had been a source of vitality and expansiveness at the founding forty-five years earlier.

Before it became clear that the traditional division of professional labor had broken down, the ASSA with its traditionally structured departments had seemed a useful "headquarters" for the movement to establish authority. As such, it attracted men of strong intellect and serious purpose. But in the later decades of the century men began to realize that doctors, lawyers, ministers, and general educators could no longer claim mastery over the entire realm of esoteric matters calling for professional advice. Daniel Gilman's rejection of the ASSA as a framework for social inquiry symbolizes this growing recognition that the professional class of an interdependent society would have to adopt a new division of labor. The professional elite would so expand in numbers and complexity of organization as to deprive its members of any vivid sense of common identity. The new elite would be more competent, but it would not possess the morale of the nineteenth-century gentry.

By the turn of the century the only people left to staff the ASSA's Departments were men who did not recognize, or did not wish to admit, the breakdown of the traditional division of professional labor. Seldom strong representatives of their professions, they were the last-ditch defenders of gentry culture. They cherished gentility and professional prestige more than mastery of esoteric knowledge.

Sanborn's Last Years in the Association

Even Frank Sanborn had begun to lose interest in the Social Science Association by the mid-1890's. From the beginning, of course, it had been only one among his many interests, and distractions multiplied in his later years. In 1878, when plans for a merger between the ASSA and Johns Hopkins reached a dead end, Sanborn immediately engaged Benjamin Peirce in another project which to his eclectic mind was oddly similar in purpose to the ASSA and certainly just as important. This was the Concord Summer School of Philosophy. Sanborn, as always, was the behind-the-scenes organizer of this heady gathering of idealists that convened annually in Concord from 1879 to 1888. The speakers included, in addition to Peirce and Sanborn, such diverse figures as Ralph Waldo Emerson, Noah Porter, and William James, but the star attractions were William Torrey Harris and Bronson Alcott, representing German and native American idealism, respectively.[1]

In addition to the Concord Summer School, Sanborn continued to occupy a variety of appointive positions with the Massachusetts Boards of Charities and Health; he wrote regularly for the Springfield *Republican* and occasionally for the Boston *Advertiser*, the *Atlantic*, *The New Englander*, and other journals; and at the same time he wrote a number of books, including his substantial biography of John Brown, published in 1885. Spare moments were filled with work for the National Conference of Charities and Correction, prison and asylum reform

---

[1.] On the Concord School, see Austin Warren, "The Concord School of Philosophy," *New England Quarterly*, 2 (Apr. 1929), 199–233; Henry A. Pochmann, *New England Transcendentalism and St. Louis Hegelianism: Phases in the History of American Idealism* (Philadelphia: Carl Schurz Memorial Foundation, 1948); Leidecker, *Yankee Teacher*, 357–372, 403–421.

groups, and a half-dozen or more social and cultural clubs and organizations such as the Free Religious Association, of which he was a director. In the nineties and after, Sanborn's correspondence reveals a growing absorption in Greek literature and a parallel decline of interest in social science.

Sanborn never gave up hope for the improvement of the Social Science Association, but he, like Giddings, sensed failure. He admitted in 1897 that "the first ten years of our Association (1865–75) were certainly more fruitful in visible results than the last ten years have been," continuing: "It must be plain to any close observer that the American Social Science Association (though never forfeiting its claim to a respectable place among those influences that have made for good in our national life since the Civil War) has not kept pace with the advancing needs of the country, and bears now a smaller part in its attempted mission than during its earlier period." With some bitterness, Sanborn quoted Petrarch: " 'Povera e nuda vai Philosophia,'— which in modern vernacular would read, 'Social science pays small dividends. . . .' "[2] Sanborn knew the sad fate of men who cling to an outworn paradigm after younger inquirers have moved on to a new one. It was perhaps a sense of isolation, a feeling that there was no one left who would understand, that prevented him from completing the volume of his memoirs which was to have dealt with his social science activities.

Sanborn generously blamed himself for the weakness of the Association, saying that a more energetic secretary like Henry Villard would have kept membership up and preserved the organization's original vigor. More to the point, Sanborn attributed the failure of the Association to its regional character and its inability to construct a strong foundation of local societies. Always fundamentally a democrat of the most optimistic kind, he complained that ASSA meetings had often lacked a "real representative quality." "Our society . . . might be charged, and possibly has been, with being a close corporation, proceeding in its selection of subjects and writers from personal and restricted views, and not opening to the great public that opportunity for discussion which is the best guarantee that truth will be elicited and recognized in our debates."[3] Far from believing that the ASSA had embraced members and opinions too promiscuously, he thought the

[2] *JSS*, 35 (Dec. 1897), 23, 27, 29.
[3] *Ibid.*, 23, 27. Sanborn made these remarks at the 1897 meeting, at which the constitution was amended to make membership by invitation only, and at which he resigned as secretary—probably in protest. These developments are discussed below. It should be noted that his analysis of the causes for the Association's failure is probably colored by his efforts to resist the change in admission policy.

Association had not been sufficiently open to the general public. The new academic specialist associations would achieve success through a policy contrary to his.

The weaknesses of the Association were always cause and consequence of its empty treasury. After a period of comparative affluence in the 1880's, the Association returned in the early nineties to the same state of penury that had nearly killed it in the seventies. Both periods were, of course, eras of nationwide depression.[4] In times of trouble, the idea of merging with other associations (or, better yet, with a university) again became attractive. A zealous member of the American Association for the Advancement of Science, James A. Skilton, made overtures to the ASSA in 1895. This New York patent attorney hoped to combine all scientific social inquirers in a national Sociological Institution, modeled on the Smithsonian and financed by the federal government. As a first step he hoped that the ASSA would allow itself to be absorbed into Section I of the AAAS, devoted to "Social and Economic Science." There, as Skilton put it, "the scientific spirit and methods may more easily be made and kept dominant" than in the ASSA.[5]

Skilton's plan met with a chilly reception. Sanborn said:

> I hardly think we lose anything by not studying sociology in company with Mr. S. [Skilton] and his friends. But of the propriety of our uniting with the real students and teachers of our 'science' in universities and elsewhere, I have not the least doubt.... If we could sit down with... [A. D. White] and a few of his circle at Ithaca, I believe we could arrange a scheme for an excellent society to take our place in the near future, when we veterans retire to sleep on our laurels. But the new society will not meet at Saratoga.[6]

---

[4] In 1891 Sanborn complained to A. D. White that there was not enough money on hand to print another issue of the *Journal:* Sanborn to White, Concord, 27 June 1891, A. D. White Papers, Cornell University Library.

[5] J. A. Skilton to Frederick J. Kingsbury, New York City, 17 and 23 Sept. 1895, Kingsbury Papers, Concord Antiquarian Society. Section I of the AAAS, originally titled "Economic Science and Statistics," was formed in 1880–81 during a general reorganization of the AAAS. Edward Atkinson dominated its meetings in the 1880's. In the early 1890's the leadership of the section passed into the hands of Edmund J. James, Lester Ward, and Edward A. Ross. Ross resigned as section secretary in 1895, and leadership then passed into more conservative and less notable hands. Ross's resignation resulted from a battle in which Skilton (and, I presume, the young academicians) tried without success to change the name of the section to "Sociology." The title instead was changed to "Social and Economic Science." The official record of the meeting notes that "not everything said in the section is sound. Some wild monetary theories have been broached; some revolutionary socialistic schemes advocated, but the sound common sense of the majority of the members gives them a speedy quietus." *Science*, n.s., 2 (27 Sept. 1895), 406. In 1896 a bill to incorporate Skilton's "Sociological Institution" was actually introduced in Congress: *Science*, n.s., 4 (16 Oct. 1896), 560.

[6] Sanborn to F. J. Kingsbury, Concord, 8 Oct. 1895, Kingsbury Papers, Concord Antiquarian Society.

The Association openly advertised its willingness to be adopted by a university, but the only offer that even approached fruition came from what was indirectly its own offspring, now greatly expanded and transformed: the American Academy of Political and Social Science. Formed in 1890 by Edmund J. James, the Academy at its charter meeting absorbed the Philadelphia branch of the ASSA. Within a few years it built a membership list of 1,500 names and enjoyed an annual income of ten to twelve thousand dollars. Negotiations for a merger of the Academy with the Social Science Association went far enough in 1896 that Sanborn paid a visit to Philadelphia, and Roland P. Falkner, vice-president of the Academy, attended an ASSA council meeting in New York. The precise terms of the proposed merger are not known to me, but it is clear that the ASSA would have taken second place to the Academy in any joint organization.[7]

Under the leadership of Edmund J. James, who was one of the great academic entrepreneurs of the period, the American Academy quite simply beat the Social Science Association at its own game. Though led by academicians, the Academy warmly embraced both practical and theoretical inquirers. In its ranks businessmen, reformers, and academicians mingled amiably. Moreover, for a time it successfully held together in a single focus all the policy sciences, from history and economics to political science and sociology. Its success in attracting the support of the young academicians (James's editorial assistants were men such as Franklin H. Giddings and James Harvey Robinson) was directly related to its prodigious publishing program. The Academy published more in its first year than the ASSA published in five. The *Annals*, which came out every two months in the mid-nineties, attracted first-rate scholarly articles as well as important pieces from men of practical experience. Academic men were no doubt gratified to find book reviews and a steady stream of gossip about the academic world, including occasional biographies and bibliographies of prominent figures. In addition to the *Annals*, the Academy published numerous pamphlets, some of a scholarly nature, others designed to have an immediate impact on public opinion. The Academy seemed to unite effectively scholarly social inquiry of all kinds with a practical-minded, moderate reformism.

Money and energy explain the success of the Academy. Because it was oriented toward the academic social scientist, it drew upon the energies of ambitious men, the first to have careers in social science.

---

[7] The only account I have seen of the merger proposal is Sanborn to White, Concord, 12 Dec. 1896, A. D. White Papers, Cornell University. For Sanborn's visit to Philadelphia, see Sanborn to Benjamin Smith Lyman, Concord, 11 Dec. 1896, Clarkson Collection (original in Historical Society of Pennsylvania).

Social science for Sanborn and his generation was an interest and a pastime, not a career. With lifetime commitment came an improved ability to claim expertise in social phenomena and the related ability to attract financial support. The financing of the Academy in its early years is a mysterious matter, but it probably gained financially by making membership a matter of special invitation, and therefore prestige.[8] Unlike the ASSA, which admitted to equal membership all who paid the standard fee, the Academy—following the trend established by the AHA—restricted membership, admitting only persons nominated by a member and approved by the governing council.[9] Coupled with an aggressive promotion policy, membership as a kind of distinction could be a substantial source of income.

The proposed merger of the Academy and the Social Science Association provoked major disagreements within the Association in 1896–97. The record is not complete, but Sanborn, Frederick J. Kingsbury, and the Reverend Dr. Joseph Anderson favored union with the Philadelphia Academy. The decision went against them. The chief opponent of the merger seems to have been Francis Wayland, Jr. He, Kingsbury, and Oscar S. Straus formed a committee intended, as Sanborn said, "to meet the difficulty of raising funds, and thus make the union unnecessary."[10] In the end the Association retained its independence but adopted a selective membership policy like that of the Academy, presumably in hopes of improving its financial status.

With the new membership policy in 1897 came a new president, Simeon E. Baldwin of Yale Law School, the most conservative man ever to preside over the Social Science Association. Upon Baldwin's election, Sanborn resigned his post as secretary. The treasurer, Anson Phelps Stokes, resigned shortly thereafter. Although direct evidence is

---

[8.] The Harvard historian Albert Bushnell Hart in 1913 accused the Academy of carrying "paper members" on its rolls. He noted that the Academy had $30,000 a year to spend, while the combined income of the AHA, the AEA, and the American Political Science Association was $27,000. Simon Nelson Patten made a very brief response, stating only that Hart had omitted such things as institutional subscriptions. See Hart, "Standardization of the Accounts of Learned Societies," *Science*, n.s., 37 (10 Jan. 1913), 66–72, and Patten, "Facts about the Accounts of Learned Societies," *ibid.* (7 Mar. 1913), 371.

[9.] See the constitution of the Academy, *Handbook of the American Academy of Political and Social Science* (Philadelphia, 1891), printed as a supplement to, and bound with, the *Annals*, I.

[10.] Sanborn to White, Concord, 12 Dec. 1896, A. D. White Papers, Cornell University. Kingsbury, president of the Association from 1893 to 1896, was a wealthy Connecticut manufacturer and a member of the Yale Corporation. Anderson was a resident of Waterbury, Conn., and served as chairman of the Education Department in the late 1890's. Oscar S. Straus, diplomat and future Progressive governor of New York, had recently been named a director of the Association.

lacking, it can hardly be doubted that Sanborn and Stokes resigned because of the politics and policies of the new president.[11]

Sanborn remained active as chairman of the Department of Social Economy; nevertheless, his departure from the position of general secretary marks a major turning point in the history of the Association. It is not quite accurate to say that the Association became a conservative organization at this point, for some of its leaders in the early 1900's were Progressive liberals, politically as distant from Baldwin as Sanborn had been. The humanitarian reform tradition associated with the ASSA from its earliest years was diluted after 1897, but it never died out. The spirit of inquiry, however, was extinguished. At the same time, the impulse to defend authority, which had always been the common thread knitting together the diverse efforts of the Association, grew out of all proportion to other goals and motives.

Genteel Professionalism

The rise to the presidency of a conservative like Simeon E. Baldwin was not so dramatic a change as one might think, for the Association (and in some ways even Sanborn himself) had become increasingly conservative with advancing age and changes in the values of the society at large. Of course, a genteel and paternalistic conservatism had been implicit in the activities of the Association from the beginning. But there were changes in tone. In the anti-radical furor that followed the Haymarket incident, Sanborn, though never a nativist or immigration restrictionist, talked like many another outraged bourgeois about "that scum of European revolutions and conspiracies that has floated to this country on the wave of emigration." He criticized the trade unions and the Knights of Labor for "terrorizing poor men and subjugating rich and ambitious men."[12] In the 1890's Sanborn's views seem to have swung leftward again: though he could not bring himself to vote for Bryan, whom he thought the "better man" with the "better cause," neither would he vote for McKinley, whom he regarded as a tool of "dropsical wealth."[13]

---

[11] Concerning the resignations, see Sanborn to Kingsbury, Concord, 3 Aug. 1897, Kingsbury Papers, Concord Antiquarian Society, and Frederick S. Root to S. E. Baldwin, New Haven, 23 Nov. 1897, Baldwin Papers, Yale University.

[12] He went on, however, to note that the anarchists made more noise than their importance justified: *JSS*, 22 (June 1887), 105. Sanborn's tone had been more conciliatory after the labor violence of 1877, but he acknowledged the necessity of military force as a final weapon" against "riot and anarchy": *ibid.*, 9 (Jan. 1878), 8.

[13] Sanborn thought free silver a "foolish" issue but thought it "nonsense" to call Bryan's party "socialistic." Instead, he saw it as a continuation of the Jeffersonian and

But Sanborn probably was almost alone among ASSA members in offering even token support to Bryan. The decidedly conservative mood of the Association in the early 1890's is better represented by the words of Frederick J. Kingsbury, president from 1893 to 1896. Kingsbury thought the "most alarming feature" of the past quarter-century had been the rapid growth of antagonism to "everything sane people have been accustomed to consider necessary to the safety of society," including property, "respect for good order," and "loyal service in return for wages paid." Going "under the various names of Socialism, Anarchy, and Populism," claimed Kingsbury, "or under no name whatsoever . . . a vast number of men seem ready for acts of violence and any form of crime." This vast "outburst of lawlessness" sprang from Central Europe, but the contagion was carried by immigration to these shores, where already it "has succeeded in taking control of the machinery of one of our great political parties, and the end is not yet."[14]

No incident better illustrates the growing conservatism of the Association, relative to the rest of the nation, than the controversy over Zebulon R. Brockway, superintendent of the Elmira Reformatory. To Sanborn's generation Brockway was the very paragon of enlightened penology. Using the indeterminate sentence and an elaborate educational and grading system, Brockway made Elmira a model imitated by reformers throughout the world. In the words of Blake McKelvey, he was "the one vital genius" of applied reformatory penology in the late nineteenth century.[15]

Brockway's program included use of corporal punishment on intractable prisoners. When he was accused of cruelty by members of the New York Board of Charities in 1894, Sanborn, Francis Wayland, and

---

Lincolnian tradition: Sanborn to F. J. Kingsbury, Concord, 8 Oct. 1896, Kingsbury Papers, Concord Antiquarian Society. Sanborn's 1899 address, "Social Relations in the United States," *JSS*, 37 (Dec. 1899), 69–74, is perhaps the most eloquent of his career, and also the most radical and modern-sounding. It provoked a quite bitter exchange with Baldwin and others (*ibid.*, 75–77). Sanborn was a staunch anti-imperialist: see Daniel B. Schirmer, *Republic or Empire: American Resistance to the Philippine War* (Cambridge: Schenkman, 1972), 192, 242.

[14.] The 1896 presidential address from which this quotation is taken set a new low in quality for the ASSA: Kingsbury, "A Sociological Retrospect," *JSS*, 34 (Nov. 1896), 1–14. Actually a relatively enlightened manufacturer and a member of the Yale Corporation, Kingsbury had long been interested in profit-sharing plans, cooperative banking, and the like. In terms of career, mental acuity, and contemporary reputation, he is perhaps the least notable of the Association's presidents. It is tempting to speculate that financial largesse compensated for his other deficiencies. The Association always had a hard time surviving periods of depression like the mid-1890's.

[15.] McKelvey, *American Prisons*, 143. McKelvey rates Brockway very highly and gives little credit to his critics: *ibid.*, 144. A less favorable view is presented by Anthony M. Platt, *The Child Savers: The Invention of Delinquency* (Chicago: University of Chicago Press, 1969), ch. III.

other longtime prison reformers leaped to his defense. An extensive editorial campaign was organized and Charles Dudley Warner, coauthor with Mark Twain of *The Gilded Age* (1873), was invited to speak in Brockway's defense at the ASSA meeting. The harsh tone of Warner's address neatly draws the line between the sensibilities of the older generation and the new. Warner called Brockway's critics (including thick-skinned Theodore Roosevelt) "sentimentalists":

> For over thirty years in this country, from ten thousand pulpits and platforms, they have been preaching the gospel of moral-mush. They have incited discontent, they have stirred up hostile feelings between classes, they have talked always of rights, rights, rights, and very little of duties, they have taught that, whoever has a hardship, somebody else is responsible for it. Whether it is a crime or drunkenness or poverty or laziness, somebody is responsible, either society or some rich man, —a rich man who in nine cases out of ten in this country has worked his way by industry and thrift and ability....[16]

The words Warner used were too bitter for some of his friends, but he expressed a sense of frustration common to them all. Values that thirty years before had seemed radically humanitarian now seemed to the younger generation heartless and cruel. The growing conservatism of the ASSA was in large part a relative phenomenon caused not by any movement of opinion within the Association, but by the emergence in the society at large of a redefined humanitarianism that was more inclusive and less discriminating than that of the abolitionist era.

But by any standard of humanitarianism, new or old, Simeon E. Baldwin was remarkably conservative. Baldwin, an austere and puritanical man, was associated with Francis Wayland in the revival of Yale Law School in the early 1870's. Baldwin taught there for the remainder of a very active life that included service as chief justice of the Supreme Court of Errors of Connecticut and, at the age of seventy, two terms as governor of Connecticut in the midst of the Progressive era. From the early 1870's to 1893 he was counsel for the New York and New England Railroad, and he was considered for appointment to the U.S. Supreme Court in 1893. Grandson of a Federalist, son of a Whig turned Republican, Baldwin himself was a Mugwump who remained in the Democratic fold after the exodus of the eighties.[17]

[16.] Charles Dudley Warner, "The Elmira System," *JSS*, 32 (Nov. 1894), 63. The campaign in defense of Brockway is documented in the correspondence between Warner, Sanborn, and Wayland in the C. D. Warner Papers, Trinity College, Hartford, Conn. Sanborn reported that Warner's tirade against "sentimentalists" had "gravelled [Thomas Wentworth] Higginson, a little," and perhaps provoked Mrs. Charles Russell Lowell into an attack on Brockway: Sanborn to Warner, Concord, 19 Sept. 1894, Clarkson Collection (original in Trinity College).

[17.] See Frederick H. Jackson, *Simeon Eben Baldwin: Lawyer, Social Scientist, Statesman* (New York: Columbia University Press, 1955).

Baldwin was ruthless, tough-minded, and ideologically consistent. His conservatism was of the individualist, laissez-faire, states' rights variety. He condemned the Fourteenth Amendment as a "collectivist" document which undermined the sovereignty of the states. The chief dangers to society, in his eyes, came from the left: "the foe that threatens American institutions to-day is not absolutism, but anarchy; not the tyranny of a man, but a tyranny of the mob."[18] His radical views on penology set him sharply apart from the traditional humanitarianism of the ASSA. In 1900 he attacked the mainstay of late nineteenth century prison reform, the indeterminate sentence, on the grounds that the trial judge was best equipped to determine length of sentence.[19] While president of the ASSA from 1897 to 1899, he published several articles in newspapers and popular magazines advocating flogging as a punishment for minor first offenses. In the *Yale Law Journal* he proposed not only a return to the lash, but also castration as a punishment for rape. "It reforms his body if not his soul," the judge observed.[20]

For so hard-spirited a man to occupy the presidency of the Social Science Association seems mystery enough. But if Baldwin was at odds with the ASSA tradition of humanitarian reform, he was entirely at one with another tradition equally central to the life of the Association: namely, the defense of professional authority. As an organizer of professional institutions, Baldwin stands in a line of natural succession from Agassiz and Peirce to Gilman and White. A callous and irresponsible elitism was always the hazard to which the professionalizers were peculiarly susceptible; in Baldwin the hazard was at least partly realized.

Ever since the founding in 1865, side by side with the ASSA's reform projects and more or less abstract inquiries into the nature of society and man, there had been activities of a professionalizing character. The aspirations of professional men—some narrow and selfish, others in the best tradition of altruistic professional service—continually played a major role in the Association. The professional interest was

---

[18.] Simeon Eben Baldwin, "Absolute Power an American Institution," *JSS*, 35 (Dec. 1897), 19, 20.

[19.] Jackson, *Simeon Eben Baldwin*, 139.

[20.] Quotation from Baldwin, "Whipping and Castration as Punishments for Crime," *Yale Law Journal*, 8 (1898–99), 381. He went on to say, with no trace of disapproval, that feebleminded children, "progeny of a worthless stock," were already being castrated "in a quiet way" by "not a few" doctors in charge of almshouses and other public institutions: *ibid.*, 382. See also Jackson, *Simeon Eben Baldwin*, 139, and the following articles by Baldwin: "Flog the Kidnappers," *Leslie's Illustrated Weekly*, 92 (1901), 50–59; "The Restoration of Whipping as a Punishment for Crime," *Green Bag*, 8 (1901), 65–67; and "Corporal Punishments for Crime," *Medico-Legal Journal*, 17 (1899), 61–73. Baldwin's ruthlessness was shared by other leaders of the law: see Richard M. Brown, "Legal and Behavioral Perspectives on American Vigilantism," *Perspectives in American History*, 5 (1971), 140–144.

especially strong in the Departments of Health and Jurisprudence.

Men such as ex-president Woolsey of Yale, professors Langdell and Thayer of Harvard, and T. W. Dwight of Columbia Law School were working within the framework of the ASSA as early as 1870 to "consider the state of the science of jurisprudence in the United States, and to take proper action for enlarging the opportunities for instruction in that science in the universities of the country." [21] Papers such as "Legal Education and the Study of Jurisprudence in the West and North-West," delivered in 1876 by Professor W. G. Hammond of the University of Iowa, were not untypical.[22] The Jurisprudence Department was often led by Harvard law professors up through the mid-1870's, but from 1878 to the end of the century it was a stronghold of the Yale Law School, led by Dean Francis Wayland, Jr.[23]

When Baldwin made his first appearance before the ASSA in 1879, he came as the founder only a year earlier of the American Bar Association, the central institutional structure of the American legal profession. Baldwin created the ABA almost singlehandedly. Working through the State Bar Association of Connecticut, only three years old itself, Baldwin issued a call for a national association of lawyers to meet in Saratoga Springs, New York, on 21 August 1878. Baldwin selected the names to whom the call was sent, wrote the constitution which was adopted with only slight changes at the charter meeting, and was prime mover of the ABA for at least its first decade. The first project undertaken by the organization was intended to improve legal education and standardize bar admission requirements throughout the nation. Its official goal was "to advance the science of jurisprudence, promote the administration of justice and uniformity of legislation throughout the Union, uphold the honor of the profession of the law, and encourage cordial intercourse among members of the American Bar."[24]

The spirit of the Lazzaroni lived on in the Department of Health as well as Jurisprudence. From 1874 on, when the American Public Health Association was organized in conjunction with the ASSA annual meeting (the same meeting at which the National Conference of Charities and Correction was formed), there was an informal connection between the ASSA Health Department and doctors working in state boards of health and related activities.[25] Professional issues in-

[21] *JSS*, 3 (1871), 199.

[22] *Ibid.*, 8 (May 1876), 165–176.

[23] Wayland became secretary in 1878 and was chairman from 1879 to 1902.

[24] Quoted by Baldwin, "The Founding of the American Bar Association," *ABA Journal*, 3 (1917), 695. See also Jackson, *Simeon Eben Baldwin*, 79–81, and Francis Rawle, "How the Association Was Organized," *ABA Journal*, 14 (July 1928), 375–377.

[25] Sanborn occasionally claimed the APHA as one of the offspring of the ASSA, "mother of associations," but the connection was not nearly so close as that with the NCCC. He also claimed some maternal rights over the National Prison Association (1870) and the Association for the Protection of the Insane and the Prevention of

evitably drew attention at annual meetings. In 1877, for example, some members of the Health Department brought about the enactment in Massachusetts of a law abolishing coroners and replacing them with trained medical examiners. The idea was borrowed from the British Association for the Promotion of Social Science. The ASSA Health Department engaged a Boston lawyer to research the question and persuaded the governor to recommend abolition in his annual address. The result, in Boston: two physicians replaced nearly fifty coroners. "A few responsible and educated men," as Sanborn put it, "had taken the place of some hundreds of inefficient and ignorant officials."[26] In the same spirit, the Health Department in 1880 backed efforts in Massachusetts to establish a board of medical examiners, with the aim of "limiting quackery."[27]

All professional men, even at their best, walk a narrow line between self-serving greed and altruistic service to the community and to truth. The ASSA's professionally oriented activities were often beneficial by any standard. For example, at the 1883 meeting Theodore Bacon delivered a paper on "Professional Ethics" that warned of dangers inherent in the growing power of the legal profession.[28] In 1887 Frank Sanborn gave four lectures at the Boston University School of Medicine on "The Duties and Opportunities of the Medical Profession towards the Inmates of Public Institutions and in Regards to the Dependent and Delinquent Classes in General."[29]

But a more selfish variety of the professional interest began to dominate the ASSA in the late 1880's. Its influence became glaringly evident during an 1888 Health Department session devoted to medical education. The point of departure for the meeting was the urgent "need of legislative interference to prevent the medical schools from flooding the country with an uneducated rabble of incompetent pretenders."[30] The discussion soon developed into a general consideration of the needs of the professional class. The organizer of the session was the chairman of the Health Department, a New York City physician named H. Holbrook Curtis. Curtis, who was trained at Yale's Sheffield Scientific School, later became Simeon Baldwin's lieutenant in an effort to create under the auspices of the ASSA an American Institute which would recognize all the "best men" of the nation.

Insanity (1880):*JSS*, 14 (Nov. 1881), 33. Regarding the founding of the APHA, see *ibid.*, 6 (July 1874), 15; 7 (Sept. 1874), 210–250.

[26] *Ibid.*, 9 (Jan. 1878), 5–6.
[27] *Ibid.*, 11 (May 1880), xxiii.
[28] *Ibid.*, 17 (May 1883), 37–48.
[29] The syllabus appears *ibid.*, 28 (Oct. 1891), 20–22.
[30] *Ibid.*, 25 (Dec. 1888), 15. The papers and the discussions following them—often equally interesting—are at 15–58.

In view of the rapid growth of the medical profession (25,000 new doctors in the decade 1870–80), the high ratio of doctors to general population in the United States (higher than in Canada, England, France, or Germany), and the existence of medical schools whose standards were unquestionably inadequate, the participants in the discussion were practically unanimous in their support of state action to raise the standards of medical education. As one said, "Nowhere more than here is laissez-faire seen to be the most transparent nonsense."[31] A lone, cautious note of dissent was struck by Frank Sanborn, who suggested that the quality of the profession might best be improved by endowing basic research, like Pasteur's, rather than by changing requirements for admission.[32]

The participants in the discussion included ministers, educators, and lawyers, as well as doctors. They displayed a vivid awareness of their common cause as members of the traditional professional class. One of the discussants, the Reverend Dr. Joseph Anderson, observed that the three classic professions of law, medicine, and theology no longer dominated the field; new professions and semi-professions were emerging, and "the lines are not drawn as distinctly as before." But with the proliferation of specialities had come what he called "deprofessionalization," a watering-down of standards and status. "What is wanted in the ministry," said Anderson, "is wanted in the medical and legal professions; and it seems to me that we should cooperate with one another in that matter."[33]

President Sylvester F. Scovel of Wooster University (Ohio) argued that doctors needed a liberal education before receiving technical training so that they could properly fill their role as members of an aristocracy of esoteric knowledge:

> Better education is the watchword of these closing decades of the nineteenth century.... The old standard of trust cannot be maintained without that better education of professional men, which will preserve the relative advantage, heretofore a marked feature of civilization.... The masses demand leaders. The armies must have pioneers. There must be physical, mental, and moral, political and religious investigators. These have been and always will be the professions. On with them, then! society cries. Tell us which is the way out of this wilderness of suffering, and ignorance and vice, and official corruption and doubt.[34]

[31.] *Ibid.*, 45. Italics deleted. The statistics cited above are also from 48.
[32.] *Ibid.*, 25 (Dec. 1888), 26.
[33.] *Ibid.*, 21–22. Anderson later became chairman of the Education Department and helped Baldwin and Curtis organize the National Institute project.
[34.] Scovel, "The Value of a Liberal Education Antecedent to the Study of Medicine," *JSS*, 25 (Dec. 1888), 43.

Formation of the National Institute of Arts and Letters

When Simeon Baldwin became president in September 1897 there was precious little ground left for the Association to stand on. Its claim to be a center of original social inquiry was badly eroded by the rise of academic social science; as a popularizer it could hardly compete with the vigorous, mass-media assaults of the muckrakers; and with Baldwin at the helm it could hardly go any farther in a liberal humanitarian direction. Professionalism was the one area in which continuity of development was possible. The professional concerns that had always been the leitmotif of the Association's work were now carried on by men who had few, if any, balancing commitments. Unadulterated professionalism came to the fore.

Even before the leadership of the Association formally changed hands, H. Holbrook Curtis, who ten years earlier had organized the discussion of professionalism in medicine, persuaded Baldwin to use the ASSA—"mother of associations"—to create a National Institute of Arts and Letters. In conception the NIAL was modeled on the Institute of France: only the nation's most prominent men were to be offered membership.[35] Baldwin seems to have assumed that the NIAL would remain a section of the ASSA and thus bring dignity, new members, and new money to its venerable parent. Curtis, whose brainchild it was, had larger ambitions.

Rank-and-file ASSA members heard nothing of the project until 1899, and even Sanborn and other leaders probably did not know at first its full scope. But secret preparations were underway at the 1897 meeting. The Department of Education was renamed and its scope broadened to include "Education and Art." A committee was formed with Curtis at its head to select an insignia for the ASSA and "to consider what, if any, decoration should be given in recognition of past service or distinguished merit."[36] Never before had the Association worried about such things as insignia and decorations. In the following year it was cautiously mentioned that at a recent Council meeting "some plans and suggestions looking to the enlargement of the clientele and the extension of the influence of the Association were definitely shaped."[37]

Acting in what he described as "absolute secrecy," Curtis began contacting famous men as soon as Baldwin was installed in the presi-

---

[35.] Curtis mentions the French Institute as his model in a letter to Baldwin, Southampton, N.Y., 20 July 1899, enclosed with Curtis to C. D. Warner, Southampton, N.Y., 20 July 1899, Warner Papers, Trinity College, Hartford, Conn.

[36] *JSS*, 35 (Dec. 1897), x.

[37] *Ibid.*, 36 (Dec. 1898), 59.

dency. Calling himself Baldwin's "lieutenant," he periodically reported on the progress of his scheme:

> I take great pleasure in reporting to you that my conference with Edward Simmonds, [Augustus] St. Gaudens and [Charles Follen] McKim has aroused in them great enthusiasm, and they have suggested associating with themselves, [James Abbott McNeill] Whistler, of London and [John S.] Sargent, of Boston, to act as a sub-committee, to select such members of the professions they represent in painting, sculpture and architecture as they deem worthy of becoming members of a decorated chapter of the Social Science Association.[38]

Curtis, who stumped for McKinley in 1896, assured Baldwin that the men he contacted were "the most conservative, as well as the ablest exponents of their respective branches." Curtis hoped to have one hundred acceptances by the end of the year and to use that group as a nucleus to attract another one hundred to fill the Institute. He predicted a "revivication next September, which will surprise even our friend Sanborn. I see at once from my conversation with these men, that it will not be a question of whom we can get into the American Social Science [Association], but our difficulty will be in keeping people out."[39]

By the summer of 1898 Curtis could report that of thirty-eight invitations in the literary category there had been thirty-six favorable replies; fourteen of twenty in the musical group had already accepted; and he was then at work on the artists.[40] In addition to gathering his two hundred dignitaries for the NIAL, Curtis also recruited lesser men in what amounted to a general membership drive for the Social Science Association. He expected to add "some thousands" to the Association membership.[41] In reality, Curtis's high-pressure campaign increased membership from about 250 in 1897 to a little less than 400 in 1898 and just short of 900 a year later.

There was no pretense that these new members had competence or even interest in social science. What they had in common was high social standing and a wish to distinguish themselves by membership in an exclusive association. Curtis delegated the responsibility of organizing various occupational groups to prominent individuals: musical people to Rudolph Schirmer, publishers to Robert Russell, and artists

---

[38.] Curtis to Baldwin, New York City, 27 Sept. 1897, Baldwin Papers, Yale University.
[39.] *Ibid.* Curtis asked Reginald De Koven and Walter Damrosch to form a subcommittee of five to select the best composers. Marion F. Crawford was to organize the novelists.
[40.] Curtis to Baldwin, New York City, 1 June 1898, Baldwin Papers, Yale University.
[41.] Curtis to Baldwin, Southampton, N.Y., 20 July 1899, enclosure to Curtis to C. D. Warner, Southampton, N.Y., 20 July 1899, Warner Papers, Trinity College, Hartford, Conn.

to Hopkinson Smith. To Mr. Emile Bruguiere, "a very popular man in San Francisco," was given the task of organizing twenty-five of "the leading men of that place," apparently without regard to occupation. Baldwin agreed that the "personal qualities of 'clubability' and the likelihood of adhesion to the project, must be important considerations."[42] In his boundless ambition, Curtis seems to have dreamed of organizing *all* the "best men" of the nation: "I am enrolling in the Social Science Association such men as August Belmont, George Sheldon, and Robert Bacon of the J. Pierpont Morgan Company, who will constitute themselves a committee to invite some fifty bankers who are gentlemen to join the Association. This worked out so well among the architects that I propose to carry it through the different professions."[43]

Charles Dudley Warner presided at the first meeting of the National Institute, held in February 1899 in New York City, entirely apart from the Social Science Association. The immediate objective in Warner's mind was to organize literary men behind a drive for tighter copyright laws.[44] Among the fewer than twenty persons attending the meeting were William Dean Howells, St. Clair McKelway, Richard R. Bowker, Hamlin Garland, and Curtis.[45] In September, at the regular Social Science Association meeting, the NIAL made its official debut. Its host was the Department of Education and Art, chaired by the Reverend Dr. Joseph Anderson, who had been, like Curtis, a participant in the 1888 ASSA discussion of professionalism. Contention immediately developed between the Institute, whose members perhaps were doubtful of the calibre of their hosts, and the ASSA, whose old-time members were no doubt puzzled to find in their midst a brilliant assembly of gentlemen who had no interest whatsoever in social science, however broadly defined.

Shortly after taking over the Association, Baldwin and Curtis had set in motion the legal and political machinery necessary to win from Congress a federal charter similar to the one held by the American Historical Association.[46] The charter was approved on 1 February 1899. Baldwin deliberately drafted it so as to provide a firm legal basis not only for the five existing departments, "but [for] such other ones as

---

[42] Curtis to Baldwin, New York City, 28 Mar. 1898, Baldwin Papers, Yale University (rough of Baldwin's reply on the back).

[43] Curtis to C. D. Warner, New York City, 6 Oct. 1899, Warner Papers, Trinity College, Hartford, Conn.

[44] Warner to R. R. Bowker, Hartford, Conn., 12 Dec. 1899, Bowker Papers, New York Public Library.

[45] Warner to Baldwin, [New York City,] 13 Feb. 1899, Baldwin Papers, Yale University.

[46] Curtis to Baldwin, New York City, 21 Mar. 1898, *ibid.*

may from time to time be decided on." In using these terms he meant, among other things, to provide a convenient basis for the creation of the NIAL as a section of the Association.[47]

But Curtis and the leading members of the Institute wished to remain at arm's length—or even farther—from the ASSA. They needed a respectable corporate identity to legitimize their work and to blunt the charge of presumptuousness that might be leveled at individuals engaged in such an openly elitist scheme, but they also wanted autonomy. Even before the first meeting Curtis warned that "these men only joined the Social Science Association as a means to an end."[48] When Baldwin pressed for a clear affirmation that the Institute was a subordinate part of the Association, Curtis protested vigorously, reminding him that the invitations to the Institute members had specified that they would choose whether or not to remain a department of the Association. It would be very embarrassing, declared Curtis, if it turned out that the formation of the NIAL "was simply a trick of the American Social Science Association to obtain members for that Association."[49]

To smooth over differences between the two organizations, reciprocal membership arrangements were worked out, and Warner was even made president of both organizations when Baldwin retired in 1899. But there was no chance of reconciling men of such divergent purposes. The first joint meeting in 1899 was also the last. Although the NIAL dignitaries and hundreds of the lesser recruits were carried on the ASSA membership rolls as associate members for the next decade, few of them ever attended a meeting.

In 1904 the National Institute selected from among its own members an elite of thirty, later expanded to fifty, which was christened the American Academy of Arts and Letters. On its own terms, the Institute-Academy project was a roaring success. The members of the Academy included, besides those already mentioned, Samuel Clemens, Henry James, Theodore Roosevelt, John Hay, Edmund Clarence Stedman, Henry Adams, Charles Eliot Norton, Thomas R. Lounsbury, Thomas Bailey Aldrich, and nearly every other notable figure in the world of literature and art. The Arminian urge to make merit visible and concrete was strong in these years, even among men

---

[47.] Baldwin to Warner, New Haven, 4 Feb. 1899, Warner Papers, Trinity College, Hartford, Conn.

[48.] Curtis to Baldwin, New York City, 27 Nov. 1898, Baldwin Papers, Yale University. Robert Underwood Johnson, the second secretary of the NIAL, thought it important to note that the first members of the Institute were chosen "not by themselves, but by a committee of the Association." *Remembered Yesterdays* (Boston: Little, Brown, 1923), 440.

[49.] Curtis to Warner, New York City, 11 July 1899, Warner Papers, Trinity College. The question of the relationship between the two organizations was triggered by the ASSA's efforts to collect regular dues from Curtis's new recruits.

of insight. In tandem, the NIAL and the American Academy were the central headquarters of the men whom Henry May has dubbed "custodians of culture." They were the last twentieth-century defenders of a genteel tradition of moral idealism and literary culture, a tradition that was sincere, certainly, and decent in its way, but flawed by its Anglo-Saxon exclusiveness and smug rejection of all but the smiling aspects of life.[50]

That charming antinomian, William James, was the only one to turn down an Academy invitation. He had succumbed to the lure of the Institute, but balked at the idea of an even greater ascension into the Academy: "I am not informed that this Academy has any very definite work cut out for it of the sort in which I could bear a useful part; and it suggests *tant soit peu* the notion of an organization for the mere purpose of distinguishing certain individuals (with their own connivance) and enabling them to say to the world at large 'we are in and you are out.'"[51]

Decline

With the departure of the National Institute men, the Social Science Association was left with an inflated membership list which looked imposing but provided no revenue. Much worse, there was a confusion of role and purpose that was beyond remedy. Although each issue of the *Journal of Social Science* from 1900 to 1909 contains a few essays of reasonably high quality, the general level of contributions is decidedly low. Intellectual measures of worth were giving way to considerations of social standing and personal connection. The depths were reached in 1901 with a paper by Mrs. Orra Langhorne, of Lynchburg, Virginia, who spoke on "Domestic Service in the South"—that is, the difficulty of finding a good maid. This struck her as "the absorbing, apparently insoluble problem of the times."[52] To be sure, this was not a typical paper, even in the ASSA's worst years. Next to it appears a commentary on "Social Settlement and Educational Work in the Kentucky Mountains," more in the ASSA tradition.

The last three presidents of the Association—Oscar Straus (1901-3), John Graham Brooks (1903-5), and John Huston Finley (1905-9)–each tried without success to breathe life back into the organization. Brooks was a Unitarian minister who became a free-lance writer,

---

[50.] May, *End of American Innocence*, 78–79. See also Larzer Ziff, *The American 1890's: Life and Times of a Lost Generation* (New York: Viking, 1966), 345–346, and Johnson, *Remembered Yesterdays*, 439–452.

[51] W. James to Robert U. Johnson, written in 1905 and published in the *New York Times Book Review*, 77 (16 Apr. 1972), 36–37, by Jacques Barzun, current president of the National Institute of Arts and Letters, in a letter to the editor.

[52.] *JSS*, 39 (Nov. 1901), 169.

social critic, and gentleman-reformer. He first appeared before the Social Science Association in 1887 with a sympathetic study of labor organization; from 1889 to 1891 he taught the first social reform–oriented course in Harvard's Economics Department. When Frank Sanborn finally retired from the chairmanship of the Social Economy Department in 1901, Brooks took his place. As a prolific author and lecturer, and president of the National Consumer's League from its founding to 1915, Brooks's circle of friends is a "Who's Who" of the Progressive era, embracing everyone from the Fabians to William James, Richard T. Ely, and Theodore Roosevelt.[53]

Ironically, the last president of the ASSA, bastion of genteel amateurism, was John Huston Finley, one of the first men trained as a professional social scientist at the Johns Hopkins University. Richard T. Ely classed Finley with Woodrow Wilson and Albert Shaw as one of his three ablest students.[54] Although exposed to professional training, Finley's main interests did not lie in social science scholarship. Upon leaving Hopkins in 1889 he became secretary of the State Charities Aid Association of New York City and edited both its organ, *State Charities Record*, and the journal of the state Charity Organization Society, *The Charities Review: A Journal of Practical Sociology*. In the early nineties he was called back to his alma mater, Knox College in Illinois, as president. After leaving Knox and briefly teaching politics at Princeton he became president of the City College of New York in 1903. He later became State Commissioner of Education and an associate editor of the New York *Times*. He was active in charity work all his life and, like Sanborn, was a devotee of the classics.[55]

Despite the best efforts of Brooks and especially Finley, who cherished a genuine affection for the ASSA and its ways, the decline of the Association was rapid after the turn of the century. The Finance Department collapsed in 1901; the Health Department never met after 1906. After a modest membership drive in 1903–4, membership was for all practical purposes frozen.

[53.] See John Graham Brooks, "Labor Organizations: Their Political and Economic Service to Society," *JSS*, 23 (Nov. 1887), 68–75; Robert L. Church, "The Economists Study Society: Sociology at Harvard, 1891–1902," in Buck, ed., *Social Sciences at Harvard*, and Mooney, *John Graham Brooks*.
[54] Ely, *Ground Under Our Feet*, 104.
[55.] *Dictionary of American Biography*, XXIII, supplement II, ed. R. L. Schuyler (New York: Scribners, 1958), 185–186; Frank D. Watson, *The Charity Organization Movement in the United States* (New York: Macmillan, 1922), 236; and the following articles by Marvin E. Gettleman: "John H. Finley and the Academic Origins of Social Work, 1887–1892," *Studies in History and Society*, 2 (Fall 1969/Spring 1970), 13–26; "John H. Finley at CCNY: 1903–1913," *History of Education Quarterly*, 10 (Winter 1970), 423–439; and "Charity and Social Classes in the United States, 1874–1900," *American Journal of Economics and Sociology*, 22 (Apr. 1963), 313–329, and (July 1963), 417–426.

The death blow to the Social Science Association was the organization of the American Political Science Association in 1903. ASSA officers and ex-officers were very prominent among the founders of the new association. The chairman of the organizing committee was Professor Jeremiah Jenks of Cornell, who had chaired the ASSA Finance Department in the late 1890's. First secretary-treasurer of the new organization was Professor W. W. Willoughby of Johns Hopkins, who was secretary of the ASSA Social Economy Department in 1901–2. Ex-presidents of the ASSA Simeon E. Baldwin and Carroll D. Wright and future president J. H. Finley were active in the early stages of organization. Baldwin was one of the two initial vice-presidents of the APSA and became its seventh president. Andrew D. White did not take part in the organization of the new association, but he was made a member of the executive council.

Baldwin wanted to organize the APSA within the ASSA: "My own opinion is quite strongly that the Am Soc Sci Assn would furnish the best stock to graft on, and that a graft and not a new stock is the [desired?] thing *to start with*, at least."[56]

About a third of the men most active in the founding of the APSA were connected with the ASSA. What is more, the Political Science Association, like the ASSA, at first had a pronounced tilt toward men of affairs and nonacademic "amateurs." As late as 1912 it appears that as few as one of every five APSA members was a "professor and teacher." Academicians were outnumbered by lawyers and businessmen, who made up 35 percent of the individual members.[57]

In spite of the close connection in spirit and personnel between the ASSA and the Political Science Association—closer, really, than that between the ASSA and the AHA or AEA in the 1880's—Baldwin's recommendation seems to have aroused little enthusiasm. The ASSA did not seem suitable even as root stock to the political scientists.

[56.] Baldwin's rough reply, on the back of W. W. Willoughby to Baldwin, Baltimore, 6 May 1903, Baldwin Papers, Yale University. Underlining in the original. For the identity of the organizers of the APSA, see the above letter and J. W. Jenks to Baldwin, Ithaca, N.Y., 20 Jan. 1903, with enclosure dated "January 1903;" rough of Baldwin to Straus, New Haven, 27 Apr. 1903; Straus to Baldwin, New York, 4 May 1903; all in Baldwin Papers.

[57.] "Unclassified individuals" made up the remainder; there was a total of 1,180 individual members. These figures are based on data appearing in "The Teaching Personnel in American Political Science Departments: A Report of the Sub-Committee on Personnel of the Committee on Policy to the A.P.S.A.," *American Political Science Review*, 28 (1934), 729, cited by Somit and Tanenhaus, *Development of Political Science*, 55. The data are based on membership lists of uncertain reliability. The same data indicate that by 1932 "professors and teachers" constituted 52% of APSA individual members, and "lawyers, businessmen, etc." constituted only 8% of a total of 1,121. "Unclassified individuals" remain roughly constant—494 in 1912 and 451 in 1932. Institutional subscriptions rose from 156 in 1912 to 565 in 1932.

Instead, they organized themselves under the auspices of the Historical and Economic Associations.

With the departure of the political scientists the ASSA was left, practically speaking, with no viable connection to the academic world. The sociologists, who logically would seem the proper heirs of Frank Sanborn's Association, in practice owed less to it than did the historians, economists, or political scientists. When the American Sociological Society was organized in 1905, the ASSA was a hollow shell and had a reputation rather too conservative for a reform-minded discipline like sociology. The sociologists traced their lineage through the Economic Association. To be sure, the ASSA still had friends in academia, but they tended to be "philandropists," as Thorstein Veblen called them: "Captains of erudition" and others who liked to attend "gatherings of the well-to-do for convivial deliberation on the state of mankind at large."[58]

In 1908, for the first time since the early 1870's, the ASSA failed to hold its annual meeting. In 1909, in a spirit of desperation, President John H. Finley put together a purely ceremonial session so that the "mother of associations" would not go unrepresented at the twenty-fifth anniversary celebration of the AHA and the AEA. Frank Sanborn was unable to attend, but his letter recounting the early history of the Association was read for him. Thomas Wentworth Higginson and Caroline H. Dall, the only other living charter members, were too sick to travel.[59] Finley's remarks, addressed to an audience of professional historians and economists, many of whom knew nothing of the ASSA, were apologetic in tone:

> I asked for the American Social Science Association the privilege and honor of representation at this great associational festival,—not that I desired its president to be heard on any of the social, economic, or political questions of the day, but because I wished the noteworthy service of this most remarkable and distinguished institution to have filial remembrance; for she is the mother, the enfeebled mother, I regret to say, grandmother or aunt of most, if not all, of the associations now existent in the territory where she once dealt alone in her omniscient interest. She sits in old age, impoverished by the very activity, the highly specialized and splendid activity, of her learned and scientific children ... who have so intensively cultivated each its field of the once wide-stretching territory that nothing is left to her except to live off their fruits and in her own memories.[60]

[58.] Thorstein Veblen, *The Higher Learning in America: A Memorandum on the Conduct of Universities by Business Men* (New York: B. W. Huebsch, 1918), 255–256n.

[59.] Sanborn to I. F. Russell, Concord, 20 Dec. 1909, Finley Papers, New York Public Library.

[60.] *JSS*, 46 (Dec. 1909), 1.

The Social Science Association never again held a public meeting. In all but the legal aspect it was quite dead. Legally, however, it possessed a federal charter designed for flexibility by the expert hand of Judge Simeon E. Baldwin. On the strength of that charter the poor old "mother of associations" was made to give birth to one last palsied child in 1912. None other than H. Holbrook Curtis returned to the scene of his earlier triumphs to organize a twin for the NIAL, the National Institute of Social Sciences. John Huston Finley was made president, Frank Sanborn was called honorary president, and the list of vice-presidents and directors included many familiar names—Simeon Baldwin, A. D. White, Charles William Eliot, Seth Low, James B. Angell, and Oscar Straus, to name a few.[61]

Master promoter that he was, Curtis soon had so many members and so much money in the bank ($4,700) that the organization was compelled to find something to do—a detail the organizers had not previously faced in any clear way. To Finley, Curtis wrote: "The institute now has eight-hundred members, and unless we begin to show we are in earnest, we will soon be open to criticism."[62] In lieu of any better task, Curtis suggested that the Institute bestow Honorary Membership Medals on three South American mediators who had met recently in Niagara Falls in an effort to resolve a difficult diplomatic crisis involving Mexico and the United States.[63] From that day to this, the chief function of the NISS seems to have been the annual presentation of gold medals in recognition of outstanding achievements in "social science." Recent honorees have included General Maxwell D. Taylor, Francis Cardinal Spellman of New York, Frank Borman, the astronaut, and Mrs. Norman Chandler, president of the Los Angeles Symphony

---

[61.] The first treasurer was Henry P. Davison of J. P. Morgan and Company. The NISS was created on 21 Dec. 1912 at what was technically a "meeting" of the ASSA. A copy of the minutes of that meeting was furnished the author by Felicia Geffen, assistant to the president of the American Academy of Arts and Letters. To preserve the federal charter, ASSA officers were appointed and "meetings" periodically held as late as the 1920's. In 1926 Congress approved an amendment effectively transferring the ASSA charter to the NISS: see correspondence between Finley and Rosina Hahn, 21 and 25 Oct. 1926, Finley Papers, New York Public Library, and Mary C. Flynn to Executive Secretary, National Conference on Social Welfare, New York City, 16 Oct. 1965, copy furnished the author by Mrs. Mabel Davis of the National Conference on Social Welfare.

[62.] Curtis to Finley, New York City, 1 Oct. 1914, Finley Papers, New York Public Library.

[63.] *Ibid.* Curtis also was eager to start a journal: "I think that we can cull some very excellent articles from the Academy of Political Science, [presumably the *Political Science Quarterly*] and the Philadelphia Academy of Political and Social Science Journals [*Annals*]. It would certainly not take much time of any man who is accustomed to read these periodicals." (*Ibid.*) The result was the *Journal of the National Institute of the Social Sciences*, 1915 to date.

Association.[64] Even Frank Sanborn, no stickler for precision himself, would have thought the Institute's conception of social science remarkably broad.

[64.] Other honorees in 1965 and 1969 were James A. Perkins, president of Cornell University; Lester B. Pearson, former Canadian prime minister; the Reverend Theodore M. Hesburgh, president of Notre Dame; and Lady Barbara Ward Jackson, British economist. The current president of the Institute is Frank Pace, Jr. Emory R. Johnson, president of the NISS from 1918 to 1922, expressed the purposes of the organization as follows: "The officers of the Institute have deemed it wise so to develop the activities of the National Institute of the Social Sciences, that it shall be, on the whole, a conservative organization. It was felt by the officers that there was no need for a radical society that would break new paths in social theory, but that there was a real necessity for an organization whose publications should be essentially and candidly sane. I am not quite disposed to use the word 'conservative,' because that overstates the thought. I do not intend . . . to give the impression that the National Institute of Social Sciences is not open to the impress of new ideas. But its management has been careful not to give public expression to new ideas until they have been carefully weighed, and until a reasoned opinion can be obtained upon public questions of vital moment." Reprint from *Journal of the NISS,* sent to the author by Elizabeth Becker, secretary to Mr. Pace, 12 Feb. 1969.

CHAPTER XI

Conclusion: Explanation and Causal Attribution in Modern Society

Few people in the twentieth century have ever looked at the articles and reports published by the American Social Science Association; nor would anyone trying to solve current social problems profit much by doing so. Anyone who brushes off the dust and opens the yellowed pages of the *Journal of Social Science* today will feel, I think, that he has entered a subtly alien world of thought, a world full of the familiar phenomena of an urban-industrial society, but quaintly out of focus, as if seen through eyes accustomed to a different order of human existence. The chief argumentative burden of the preceding pages has been that this quaintness is not attributable merely to the ignorance or the frivolity of the Association's members, but is instead a measure of the magnitude of the great cultural divide that stands at the end of the nineteenth century. The sober effort of ASSA members to come to grips intellectually with a society racing toward interdependence failed, but their failure suggests how difficult was the task they faced and how great is the intellectual distance between their time and our own.

The failure of the ASSA was virtually inescapable. The organization was inherently transitional in nature because the crisis of intellectual authority that provoked its creation ultimately necessitated either its death or a transformation so extensive that its founders would no longer have recognized it. The ASSA did not survive, and probably could not have survived in its original form, because it was unable to adapt to two imperatives generated by the changing conditions of explanation in an urbanizing, industrializing society. The first of these imperatives was the necessity for highly disciplined, intensive communities of inquiry. The second was the necessity for a style of explanation that would acknowledge the interdependent quality of human affairs, being neither as voluntaristic as nineteenth-century idealism had been, nor as reductionistic as positivism.

The Necessity for Disciplined Communities of Inquiry

The founding of the Association in itself testifies to the development by 1865 of an unprecedentedly problematic view of human affairs among educated New Englanders, a recognition that common sense and customary knowledge were no longer enough. Moreover, the founding of the Association implicitly signified awareness that the individual inquirer—confined as he was to a single locality and able only imperfectly to see the whole of society—was unable to comprehend a social reality that was increasingly supra-local in character. The founders hoped that, by exchanging information and sharing viewpoints, inquirers might rise above idiosyncrasy and make the whole of society understandable again. The Association was a major step toward institutionalized inquiry in that it tried to overcome the shortsightedness that by 1865 had come to seem inescapable in isolated, independent inquiries.

But in virtually every other respect the Association was organized on the basis of assumptions that were still profoundly individualistic. Essentially, the Association hoped to overcome idiosyncrasy and advance social inquiry simply by providing independent inquirers with a central meeting place. It was taken for granted that there already were inquirers scattered about the country; that new inquirers would naturally emerge when the old ones passed on; that each could be trusted to have something worthwhile to say; that each using his own good judgment was perfectly competent to choose an appropriate subject area and investigate it in a sound manner; and that all these spontaneously initiated investigations would add up to a meaningful assessment of society and its problems.

These were optimistic assumptions. In sum, they meant that the Association made no effort to *organize* inquirers, other than grouping them by departments. The Association was built not on a community of inquiry, but on a shifting aggregate of independent inquirers. The result was chaotic eclecticism and lack of focus. The outcome would have been even more chaotic but for the fact that informally the Association was dominated by a circle of genteel New England intellectuals who constituted a community of sorts, one organized not primarily by functional and scholarly criteria but by social and familial ones. The standards they set, though interlaced with many non-intellectual considerations, were on the whole quite high. The members of this elite communicated with each other on a personal basis often enough that there was among them something approaching a coherent universe of discourse. And of course the ASSA, simply in its role as a clearinghouse, served to a limited extent to define issues, focus energies, and intensify interaction among inquirers.

But, in an increasingly interdependent society, truly authoritative opinion could come only from full-time inquirers organized in a highly disciplined community of inquiry, one whose members would police each other's work with ruthless intensity. Membership in a truly professional community could not be based overtly on charm, social standing, personal connection, good character, or perhaps even decency, but on demonstrated intellectual merit alone. Such a community would provide for its perpetuation across the generations by establishing more or less uniform training programs of a highly selective nature. Having no sufficient clientele among private individuals or business corporations, it would have to find a home in government, the church, or the university—the major embodiments of concern for the commonwealth in an age of individual wealth. It would not leave its members entirely free to select research subjects and methods idiosyncratically, but would forcibly inculcate at the apprenticeship level a division of labor and certain acceptable methodologies. Subtler mechanisms would operate to the same general effect among mature inquirers.

Most of all, a truly persuasive professional community would have to draw upon all the emotional energies of its members. To sustain it at the requisite level of intensity, members would have to invest in it their deepest hopes and ambitions—they would have to make it a career.[1] The rigorous demands of such a community on its individual members would be evident in its native genre, the monograph, with its burden of footnotes calculated to demonstrate wide acquaintance with fellow-workers, to leave a trail for followers, to acknowledge debts, declare loyalties, display alliances, cover flanks, condemn errors, harass deviants, and otherwise to restrain idiosyncrasy and invigorate community life. What would be classed as obsessive name-dropping and paranoid defensiveness in a natural community is normal in the communications of professional inquirers.

The younger generation of academic social scientists had numerous motives for creating truly professional bodies. The most obvious was access to employment. The same genteel elite that dominated the ASSA also controlled teaching positions in the nation's finest schools. The private correspondence of the first generation of professionals is punctuated with complaints about the difficulty of breaking into the establishment. By undermining the omnicompetent gentleman-

---

[1] Ely nostalgically recalled "tramping through the rain" with E. R. A. Seligman to the Associated Press office after the charter meeting of the AEA, "to see that we had such publicity as we both felt we deserved." The errand would have seemed vain to Frank Sanborn, who would have stayed on the veranda of the United States Hotel to keep dry. Ely, *Ground Under Our Feet,* 138.

scholar, the members of the younger generation with their German degrees made room for themselves.[2]

But the collectivization of inquirers and the systematization of inquiry were strategically valuable for the younger generation only because the resulting professional community promised a more trustworthy explanation of society than could be achieved by other means—trustworthy not only in their own eyes, but in the eyes of educated people generally. The independent inquirer had to be replaced by collective inquiry for fundamentally epistemological reasons: in a highly interdependent society the mind of the isolated investigator seemed powerless to encompass the whole, and therefore unable to comprehend even the part in its true nature, for the meaning of the part was defined by its place in the whole. The success of professionalization was not simply a matter of status-seeking strategy, but also the consequence of a profound change in the conditions of satisfactory explanation.

The man who most eloquently apotheosized the community of inquiry, the philosopher Charles Sanders Peirce, was, appropriately enough, the son of Benjamin Peirce, member of the Lazzaroni and leading figure in the ASSA. Perhaps because he was in some respects a social outcast himself, the younger Peirce keenly appreciated the social, consensual quality of all that passed for truth or ever would pass for truth among human beings. "No matter how strong and well-rooted in habit any rational conviction of ours may be," wrote Peirce, "we no sooner find that another equally well-informed person doubts it, than we begin to doubt it ourselves." Since we necessarily influence each other's opinions, "the problem becomes how to fix belief, not in the individual merely, but in the community." If the origin of doubt is social, then inquiry too must be a social process, for inquiry is the struggle to escape doubt and attain belief. If the origin of doubt and the escape from it through inquiry are social in character, then so is the intended goal of inquiry, truth. "The opinion which is fated to be ultimately agreed to by all who investigate, is what we mean by the truth, and the object represented in this opinion is the real. That is the way I would explain reality."[3]

[2.] After a conversation with Henry Adams in 1884, J. F. Jameson wrote in his diary: "I learned of some chances, though to a man not of distinguished family they are scarcely chances." See also Ely's desperate job-hunting pleas to A. D. White, Fredonia, N.Y., 7 Aug. 1880, and Lexington, Va., 10 June 1883, A. D. White Papers, Cornell University. The latter reads: "Most people in this situation seem to me to have influential uncles or brothers or [some?] [near?] friend who can make their desires known in a private way, but I have not. You are the only person of influence who ever befriended me."

[3.] Quotations from Mills, *Sociology and Pragmatism*, 159; Charles S. Peirce, "The Fixation of Belief," in *Chance, Love and Logic*, ed. M. R. Cohen (London: Kegan Paul, 1923), 20;

In an interdependent society, ideas, like men, are subjected to intense competition. Therefore doubt, as Peirce defined it, must always be rampant. The lone individual is powerless to attain confident belief, impotent to know the truth. His only grounds for confidence lay in what Peirce's good friend Francis E. Abbot (also a close associate of Frank Sanborn) called a "consensus of the competent"—an inherently collective test of truth.[4] As R. Jackson Wilson observes, the necessity of collective inquiry followed necessarily from Peirce's theory of truth: "The truth, and, hence, reality itself, was accessible only to a community of inquirers, infinite in number, and capable of carrying on inquiry for an infinitely long time."[5]

On the face of it, Peirce's theory risks radical skepticism. If truth is that which is *ultimately* agreed to by all competent inquirers (the yet to be born, as well as the living), then how as a practical matter are we *today* to know truth from falsehood? How are we to recognize sound opinion?

But in reality Peirce's theory erected a bulwark against skepticism. Even though Peirce ruled out for all practical purposes the very possibility of absolute certainty, he also promised an escape from radical uncertainty in the here and now. No one could claim to know the final, "true" opinion of a community that extended into the indefinite future; but the very existence of a community of inquiry was a guarantee against intellectual chaos, because the community's current best opinion was the closest approach to the truth that mankind could ever hope to achieve in practice. The question of which opinion to believe was, in principle, settled. Peirce of course would have resisted a literal identifi-

---

Charles S. Peirce, "How to Make Our Ideas Clear," *ibid.*, 57. My analysis follows that of C. Wright Mills, 159–160. As he points out, Peirce's words should not be mistaken for a cynical subjectivism. The stress is on the word "fated." Peirce was confident that the universe was so structured that the community of inquirers ultimately could not fail to apprehend what it really was. The two essays quoted here were first published in *Popular Science Monthly* in 1877–78, just before Benjamin Peirce wrote to Gilman to propose the merger of Johns Hopkins with the ASSA.

[4.] Persons, *Free Religion*, 31, 125–129. Persons argued, incorrectly I think, that Abbot's insistence on the consensus of the competent as opposed to individual conscience places him outside the experimentalist tradition of pragmatism. On Abbot's close relation to other pragmatists, see Wiener, *Evolution and the Founders of Pragmatism*.

[5.] Wilson, *In Quest of Community*, 44. The following paragraphs owe a great deal to Wilson's extraordinary essay on Peirce. To Durkheim as well as Peirce, the collectivization of inquiry and the death of the dilettante were moral imperatives in modern society. "It is not without reason that public sentiment reproves an ever more pronounced tendency on the part of dilettantes and even others to be taken up with an exclusively general culture and refuse to take part in occupational organization. That is because they are not sufficiently attached to society. . . . They have no cognizance of all the obligations their positions as social beings demand of them." In a footnote to this passage Durkheim noted that in the educational system, although specialization should not be begun prematurely, "it is necessary to get him [the student] to like the idea of circumscribed tasks and limited horizons." *Division of Labor*, 402.

cation of his ideal community with any concrete professional body. But the fact remains that his theory provided an elegant epistemological rationale for the professionalizing activities of his father, the Lazzaroni, and many other late nineteenth century friends of institutionalized authority.

The great functional advantage of the disciplined community of inquiry over unorganized individual inquirers is that the community, by its very existence, supplies mankind with indirect criteria of credibility and authority. It provides a practical way of recognizing (probable) sound opinion when that opinion cannot be directly checked against "the truth." Deprived of self-evident truth, unable to trust tradition, confronted with a chaos of idiosyncratic and subjective claims to truth, modern man has no other recourse, Peirce seems to have believed, than the consensus of the competent. Sound opinion becomes that opinion which wins the broadest and deepest support in the existing community of inquiry: there is, according to Peirce, no higher test of reality. The sound thinker is preferably one who is a certified member of the community; at least his views, if they are to be regarded as sound, must be compatible with the community's best opinion. The thinker's reputation and standing within the community provide even finer measures of his authoritativeness because presumably these indicators of professional status reflect the degree to which the thinker's peers accept his ideas and regard him as a reliable spokesman for the whole community. In general, the more highly integrated the community and the more vigorous its communal life, the more extensive the area of professional consensus and the more trustworthy its member's opinions. Neither Peirce nor anyone else would argue that mere conformity can earn a person high standing in the community—but neither, certainly, can idiosyncrasy.

If, as Peirce implies, dissociation from the community of inquiry diminishes a person's chances for valid knowledge, then it requires only a slight vulgarization of Peirce's view to argue that *familiarity* with the community is in itself a sign of valid knowledge. How easy it then becomes to recognize sound opinion! As pioneer sociologist Albion Small put it in 1924,

> One of the readiest ways of distinguishing the educated from the uneducated man is to probe into the acquaintance of the man in question with other workers in the field in which he claims knowledge. In my experience as editor of the *American Journal of Sociology* for twenty-seven years, I have found this test to be an almost invariably reliable criterion for deciding whether a man has anything to say that I can afford to spend my time considering.[6]

[6.] Small, *Origins of Sociology*, 7.

As at least a preliminary test of credibility, Small's criterion is a necessary article of faith in any modern, organized field of inquiry. Whether a man has "anything to say" is determined by asking how familiar he is with the work of other inquirers—whose own credibility, in turn, depends in large part upon *their* mutual familiarity. The patent circularity of this familial test of credibility would be amusing if we did not place so much trust in it and depend so heavily upon it. The danger that such self-justifying communities will drift into a truly circular scholasticism, utterly divorced from reality, is not to be dismissed lightly; but what better general criterion of intellectual authority can be found in the competitive mental world of an interdependent universe? Peirce's doctrine has some very discomforting implications, but it appears to explain the actual historical process of professionalization rather well, and to account also for the actual behavior of present-day professional inquirers.

It is not inconceivable that the highly disciplined communities of inquiry needed to authoritatively explain a complex society might have developed within the ASSA, thus insuring its survival as an institution. But it would not have been the same association. In order to incorporate such communities, the Association's departmental organization would almost certainly have had to change. Although the outcome need not have been exactly the division of labor that ultimately prevailed between History, Economics, Sociology and Political Science, it would have had to suit the needs of teacher-scholars better than the traditional ASSA division of labor between Departments of Education, Jurisprudence, Health, Finance, and Social Economy. Modern society would not be persuasively explained by genteel professional men "moonlighting" from their regular work, nor would academic specialists long tolerate a scheme of organization that subordinated their field of specialization to the extraneous categories of the classic professions.

The Necessity for a New Style of Explanation

The basic character of the Social Science Association was fixed at its founding by men whose habits of mind were formed at an early stage in the development of an interdependent, urban-industrial society. The same men remained in control of the Association through the 1890's. By that time a new generation of social thinkers, born into a far more interdependent society, had risen to maturity within the occupational framework supplied by the university. They rejected the ASSA because it was not adequate to their needs. It was too deeply rooted in the granite of New England; its genteel atmosphere at once offended and intimidated them; its tradition of practical reformism was an embarrassment to their academic orientation; and its organization structure, geared to the classic professions, seemed from their vantage point as teacher-researchers to be impossibly cumbersome.

In spite of all these obstacles, the ASSA could have survived in some modified form if the younger generation of inquirers had entered it, taken it over, and reorganized it to suit their needs. But they never even tried to do so because an intellectual chasm began opening between them and their elders in the 1880's. By the 1890's the chasm was unbridgeable. The point of division was a matter of causal attribution. At stake were two different generalized sets of expectations about where causation was likely to be found in human affairs. In the last analysis the younger generation rejected the ASSA because it was dominated by men whose style of explanation seemed superficial, and whose moral perspective was, for the same reason, beginning to seem suspect.

In the present analytical framework the term "superficial" acquires an exact meaning. The Association's style of inquiry seemed superficial to later inquirers in the literal sense that the elder generation located causation too close at hand, too near the surface of events. The ASSA saw causes where today we see symptoms. Its members exaggerated the autonomy of the individual and failed to notice external factors influencing behavior. Advancement in social theory in the late nineteenth century repeatedly took this form: what once had been seen as causes were shown to be symptomatic reflexes of some deeper cause; what once had been seen as a discrete area of inquiry was shown to be causally interwoven with other areas, thus requiring an expansion of the realm of inquiry; what once had been accepted as an adequate explanation was later seen as superficial, merely formalistic. In short, the Association was rejected by the younger generation because it typically imputed independence to variables which the younger inquirers perceived to be interdependent.

The heightened sensitivity of the younger generation of social thinkers to the multiplicity and mutuality of causal influences at work in society was produced by their everyday experience in a world moving rapidly toward interdependence. Although the alteration in perspective was triggered by objective changes in the material conditions of life, it was not correlated with those changes on a simple, one-to-one basis. Once people began to look for interdependence, they found it everywhere. In a few decades around the turn of the century illusions of local self-sufficiency and individual autonomy that in the past had been pierced only momentarily, by isolated thinkers, collapsed throughout much of literate society.

The alteration in perspective had a profound moral, as well as intellectual, dimension. Where a person locates causation determines not only his understanding of how things happen, but also his moral sensibility, for attribution of moral responsibility hinges absolutely upon a perception of causal potency. We hold a person responsible only for those evils to which he appears to be a contributing cause, and then only in the degree of his contribution. Likewise, we admire a person only for achievements which seem genuinely produced by him, and not the product of chance or privilege. By habitually locating causation outside man's intending, conscious mind, the younger generation of social thinkers adopted a moral perspective that made man appear neither as praiseworthy nor as blameworthy as he did to the founders of the ASSA.

Although the generation of the 1890's rejected the ASSA's style of explanation, it is essential to recognize that the younger generation also rejected the competing style of explanation against which the Association had struggled throughout the late nineteenth century. The ASSA, with its strong idealist and voluntarist emphases, had always stood in sharp opposition to those popular intellectual tendencies that went under the names of positivism, materialism, determinism, and naturalism. Though each of these "isms" has its own technical definition and to some extent its own historical career, they display more than a passing family resemblance and generally went together. Following the lead of H. Stuart Hughes, we may lump all of these patterns of thought together under the heading "positivism."[7]

Within the present framework of analysis the common theme of these various forms of positivism stands out clearly: it is the tendency to carry remote causal attribution to a radical extreme. The social theorists of the 1890's rejected the naive voluntarism of the ASSA because it was too superficial; but they also rejected the extremism of the positivists, who were so keen to avoid proximate causal attribution that they risked driving praise and blame out of the discussion of human affairs altogether.

Positivism can be understood as an overreaction to the breakdown of traditional habits of proximate causal attribution. During a period of growing social interdependence, when causation in everyday affairs appears to recede out of the observer's familiar milieu into more remote locations, it is to be expected that some observers will become so zealous in their pursuit of it that they overshoot the mark. After all, it is always difficult to say just how far the pursuit of causation should go,

---

[7.] I agree with J. D. Y. Peel, however, that Hughes's definition of positivism—"the whole tendency to discuss human behavior in terms of analogies drawn from natural science"—is too vague. Hughes, *Consciousness and Society*, 33–37, and Peel, *Herbert Spencer*, 237–239.

for in principle any prospective object of causal attribution can be viewed as the effect of some still more remote set of causes. Given an effect and asked for its cause, how does one decide when to stop climbing up the chain of intermediate causes and effects to name one of them "the cause"? We perform this intellectual act routinely, but with a confidence born principally of social reinforcement and convention; the rules governing it remain obscure.

Traditional conventions of causal attribution broke down in the nineteenth century. Sensitized to the dangers of proximate causal attribution by the objective changes underway in their social universe, the positivists leaped out of the frying pan of shallow formalism, into the fire of exclusively remote causal attribution. Some found the only true causal matrix of human affairs in the mode of production; others saw it in the blind instinct of reproduction and survival that propelled the evolutionary struggle for existence. Still others found real causation in chemical and biological determinants, or outside human society altogether in climate and geography. At the very least the positivists denied the autonomy of the individual conscious mind and treated it as a reflex, delayed or immediate, of external circumstances.

Positivistic social theories flourished in the last half of the nineteenth century not primarily because social thinkers wished to emulate the triumphs of natural science, but because the analogies and strategies of explanation they found in natural science meshed neatly with the principal cognitive lessons that contemporary social change impressed upon them: avoid proximate causal attribution; treat nothing as autonomous. The positivists' single most fertile source of inspiration and analogy, Charles Darwin's theory of evolution, expressed these lessons with special force.

Like Adam Smith and the classical economists before him, Darwin employed a strategy of explanation that stressed not individual agents and their conscious decisions, but an impersonal realm of collective and unintended consequences. The focal point of Smith's economic analysis had been the irrelevance of individual intentions to collective consequences that Mandeville succinctly expressed in the epigram, "private vices, public benefits." Darwin showed that the same basic disjunction between intentions and consequences held true on a cosmic scale, and by so doing he unlocked the secret of evolutionary development. Men and women copulate and secure food with no thought to the evolution of the species: though that grand process is the ultimate consequence of their acts, one could never guess it from an inspection of their intentions. Given the instincts of sex and hunger, plus random variation, Darwin was able to account for the development of species without reference to individual intentions, attributing effective causation instead to the selective pressures of the environment.[8]

In judging Darwin's significance for intellectual history, it is difficult to say which was most important: that his theory did not require God as a cause, or that it did not require individual intentions. Historians have been obsessed with the former implication of Darwin's theory, but they have ignored the latter, which was almost as novel and perhaps as unsettling to established views. In conjunction with the rapidly expanding theoretical structure of political economy, especially its marginalist branch, Darwin's theory posed a dramatic challenge to mankind's ancient predisposition to explain human affairs (excepting occasional providential interventions) as a simple sum of individual decisions and intended actions.

Positivistic styles of explanation were no doubt encouraged by the thought that a life process as grand as evolution was not attributable to individual intentions. But science may have imitated society in this case, for both Darwin and the generations that found his theory so revealing lived in a society which, by virtue of its growing interdependence, was to an increasing degree explicable only in terms of unintended consequences. Darwin's theory may have been easier for him to conceive and for others to believe because he and they lived in an age in which an individual's acts generated unplanned consequences far beyond his range of vision, and in which the individual himself was subjected to influences not intended by others. It was not mere adulation or imitation of science that drove social thinkers toward positivism in the middle and late nineteenth century. The expansion of the realm of unintended consequences in their own lives supplied them with a strong inducement to locate causation outside conscious mind.[9]

[8.] By way of contrast to Darwin's strategy of explanation consider the following passage from Frederick Bastiat, *Harmonies of Political Economy*, trans. Patrick James Stirling (London: John Murray, 1860), 159–160: "As regards the man who acts, Responsibility is the natural link which exists between the act and its consequences. It is a complete system of *inevitable* Rewards and Punishments which no man has invented, which acts with all the regularity of the great natural laws, and which may, consequently, be regarded as of Divine institution. The evident object of Responsibility is to restrain the number of hurtful actions and increase the number of such as are useful." Italics in the original. Bastiat's views are congruent with those of typical American moral philosophers before the Civil War, and he was popularized even after the war by prestigious A. L. Perry of Williams College. See D. H. Meyer, "American Moralists," 97–98, 283, 289. On the emergence of the problem of unintended consequences, see Schneider, ed., *Scottish Moralists on Human Nature and Society*, xxix-xlvii. On the novelty of "population thinking," see Ernst Mayr's introduction to Darwin, *On the Origin of Species* [A Facsimile of the First Edition] (New York: Atheneum, 1967).

[9.] What is said here about the appeal of Darwin is applicable also to the thought of Herbert Spencer. As Philip Abrams points out, his was a "brilliant sociology of unanticipated consequences." *Origins of British Sociology 1834–1914*, 70.

CONCLUSION

In its headlong pursuit of underlying causation, positivism constantly threatened to drain human society of causal potency altogether, leaving it a wasteland of secondary causes aimlessly reverberating from some primordial tremor. Since there could be no grounds for moral responsibility in a universe of human affairs devoid of causally potent actors, morally sensitive thinkers were understandably suspicious of positivism. The most formidable positivist for both Americans and the British was not Marx, but Herbert Spencer. As Philip Abrams says, serious social thinkers were not so much persuaded by Spencer, as provoked by him to formulate a credible alternative:

> Confronted with Spencer's conception of sociology many British intellectuals of the 1880's felt themselves drowning in their own thought world. Comte and LePlay were the straws they clutched at. Modern British sociology was built, more than anything else, as a defense against Spencer. It is in this sense that his influence is decisive.[10]

In one sense it is ironic that today we remember as "positivists" men like Spencer, who were insensitive to the moral dilemma inherent in a world without primary causal agents, and we categorize as rebels against positivism men like Dostoevsky, in whom the positivist vision of an amoral world without primary causes was most vivid and plausible. "How am I," asked Dostoevsky's underground man, "to set my mind at rest?"

> I exercise myself in reflection, and consequently with me every primary cause at once draws after itself another still more primary, and so on to infinity. That is just the essence of every sort of consciousness and reflection.... In consequence again of those accursed laws of consciousness, anger in me is subject to chemical disintegration.[11]

---

[10] *Ibid.*, 67. The passage is quoted with approval in Peel, *Herbert Spencer*, 238. R. Jackson Wilson observes that in America "among intellectuals of real stature . . . Spencer was more whipping boy than master." *In Quest of Community*, 155.

[11] Fyodor Dostoevsky, *Notes From Underground* (1864), excerpt from *Existentialism: From Dostoevsky to Sartre*, ed. Walter Kaufmann (New York: Meridian Books, 1957), 64–65. The whole passage reads as follows: "You know the direct, legitimate fruit of consciousness is inertia, that is, conscious sitting-with-the-hands-folded.... all 'direct' persons and men of action are active just because they are stupid and limited. How explain that? I will tell you: in consequence of their limitation they take immediate and secondary causes for primary ones, and in that way persuade themselves more quickly and easily than other people do that they have found an infallible foundation for their activity, and their minds are at ease and you know that is the chief thing. To begin to act, you know, you must first have your mind completely at ease and no trace of doubt left in it. Why, how am I, for example, to set my mind at rest? Where are my foundations? Where am I to get them from? I exercise myself in reflection, and consequently with me every primary cause at once draws after itself another still more primary, and so on to infinity. That is just the essence of every sort of consciousness and reflection.... I said that a man revenges himself because he sees justice in it. Therefore he has found a primary cause, that is, justice. And so he is at rest on all sides, and consequently he carries

The most memorable if not the most rigorous American protest against Spencerian reductionism was voiced by William James. It is fitting that James should have battled the enemies of the Social Science Association, for he was raised, intellectually speaking, in the same Concord drawing rooms that Frank Sanborn frequented; he was trained as a doctor in Harvard's Lawrence Scientific School, where he was a devoted pupil of Louis Agassiz; and he was a close friend of Benjamin Peirce's philosopher son, Charles. In the early 1880's, before he gave up "brass instrument" psychology for the philosopher's easychair, James published several attacks on positivism, of which "Great Men and Their Environment" (1880) deals most directly with the problem of causal attribution.[12] James's words supply the most striking illustration of the primacy of the question of causal attribution among late nineteenth century intellectuals.

James took as his text a passage from Spencer's *Study of Sociology* which declared that individuals, no matter how great, were not suitable objects for causal attribution:

> If not stopping at the explanation of social progress as due to the great man, we go back a step, and ask, Whence comes the great man? we find that the theory [of great men as causes] breaks down completely.... The origin of the great man is natural ... he must be classed with all other phenomena in the society that gave him birth as a product of its antecedents ... he is a *resultant*. ... Before he can remake his society, his society must make him. All those changes of which he is the proximate initiator have their chief causes in the generations he descended from. If there is to be anything like a real explanation of those changes, it must be sought in that aggregate of conditions out of which both he and they have arisen.[13]

The passage perfectly exemplified the inclination felt not only by Spencer, but also by many men who would not be called positivists, to see secondary and proximate causation where there once had been self-acting, causally potent creatures. To dispel that vision and to revitalize the individual, James brought to bear all his artistry and wit. "Can it be," he asked, "that Mr. Spencer holds the convergence of

---

out his revenge calmly and successfully, being persuaded that he is doing a just and honest thing. But I see no justice in it, I find no sort of virtue in it either, and consequently if I attempt to revenge myself, it is only out of spite.... In consequence again of those accursed laws of consciousness, anger in me is subject to chemical disintegration."

[12.] The others are "Reflex Action and Theism" (1881), "The Dilemma of Determinism" (1884), and "The Importance of Individuals," not published until 1890 but written earlier in response to criticism of "Great Men and Their Environment." All are contained in *The Will to Believe and Other Essays in Popular Philosophy* (1897; New York: Longman's, Green, 1912). This confrontation between James and Spencer is also discussed by Hofstadter, *Social Darwinism*, 132–135, and Boller, *American Thought in Transition*, 136–137. Their interpretation is quite different from mine.

[13.] Quoted in James, "Great Men and Their Environment," 233. Italics in the original.

CONCLUSION

sociological pressures to have so impinged on Stratford-upon-Avon about the 26th of April, 1564, that a W. Shakespeare, with all his mental peculiarities, had to be born there,—as the pressure of water outside a certain boat will cause a stream of a certain form to ooze into a particular leak?"[14] James relied as much on compelling rhetoric as logic to carry his argument. He boldly committed himself at the start to the view that historical development of the sort that makes the England of Queen Anne so different from the England of Queen Elizabeth can be explained adequately as "due to the accumulated influences of individuals, of their examples, their initiatives, and their decisions." But instead of arguing that difficult proposition, he tried to show that Spencer's argument inevitably boiled down to an absolutely rigid, fatalistic predestination, a method "identical with that of one who would invoke the zodiac to account for the fall of the sparrow."[15] By arguing an extreme case based on a straw man, James really failed to deal with the main problem—namely, that many people, perhaps without going as far as Spencer, found historical explanation in terms of "great men" or individuals of any kind no longer adequate.

The valuable part of James's essay lay in his concept of "cycles of operation." With this notion he tried to take account of the undeniable fact that satisfactory explanation does not consist of finding the most remote conceivable cause of a thing, or of displaying the totality of its relations with everything else in the universe.

> There are, in short, *different cycles of operation* in nature; different departments, so to speak, relatively independent of one another, so that what goes on at any moment in one may be compatible with almost any condition of things at the same time in the next. The mould on the biscuit in the storeroom of a man-of-war vegetates in absolute indifference to the nationality of the flag, the direction of the voyage, the weather, and the human dramas that may go on on board; and a mycologist may study it in complete abstraction from all these larger details. Only by so studying it, in fact, is there any chance of the mental concentration by which alone he may hope to learn something of its nature. On the other hand, the captain who in manoeuvring the vessel through a naval fight should think it necessary to bring the mouldy biscuit into his calculations would very

---

[14] *Ibid.*, 235.

[15] *Ibid.*, 218, 234. James accused the positivists of allowing no middle ground between the influence of outward environment and miracle (*ibid.*, 237), but he was guilty of the same charge in declaring that Spencer's approach "is of little more scientific value than the Oriental method of replying to whatever question arises by the unimpeachable truism, 'God is great.' " James made his task easy by focusing on "great men"—geniuses—and demonstrating not that their actions constituted a sufficient explanation of historical development, but only that they were not totally devoid of causal potency. The real issue—to what *extent* individuals *in general* can be regarded as causally potent—was evaded.

likely lose the battle by reason of the excessive "thoroughness" of his mind.

The causes which operate in these incommensurable cycles are connected with one another only *if we take the whole universe into account.* For all lesser points of view it is lawful—nay, more, it is for human wisdom necessary—to regard them as disconnected and irrelevant to one another.[16]

James was saying that the universe, after all, is not *totally* interdependent. It is not a seamless web. The very possibility of explanation hinges upon the legitimacy of regarding certain things as independent variables—"independent" meaning not that the thing is uncaused and miraculous, but that its causes lie outside the relevant "cycle of operation."[17] Although James did not in this essay develop the idea fully, he suggested in passing that the relevance of the cycle of operation was shaped by the purposes of the inquirer. Leaning on Darwin's idea of natural selection, James used his own concept of separate cycles of operation to argue that the true causes of genius were physiological (genetic) and that the social environment therefore did not *make* the great man, but had only the power to select him or reject him. Since the causes of his greatness lay in the physiological cycle, the great man could properly be regarded as "the cause" of whatever he accomplished in society, which was a cycle of operation quite distinct from the physiological.

James was correct, as far as he went, but he trivialized the views of his opponents in supposing that they were naive predestinarians. The issue was not whether there was *any* stopping place at all in the pursuit of causation, but *where* the stopping place was. When two men disagree about the location of the boundaries of a "cycle of operation," and therefore about the location of effective causation (as James and Spencer did), how is the dispute to be settled? James did not confront the question. The boundaries between cycles of operation are exceedingly slippery and unstable. To see how ephemeral they really are, one need only take James's own example and imagine that the crew of the man-of-war falls ill from bacteria hatched in those moldy biscuits. Now the ship captain would have to work within a broader cycle of opera-

---

[16.] *Ibid.,* 221. Italics in the original.

[17.] In the same spirit, Cushing Strout, protesting Charles Beard's characterization of causal judgments as subjective, arbitrary ruptures of the "seamless web" of history, says: "surely it is not so seamless that historians must follow Beard in believing that there is no more reason to explain American intervention in the First World War by reference to the Germany policy of unlimited submarine warfare than by reference to the Kaiser's moustaches." Lee Benson and Cushing Strout, "Causation and the American Civil War," in *The Craft of American History,* ed. A. S. Eisenstadt (New York: Harper and Row, 1966), 52–53. See also Bunge, *Causality,* 98–101.

tion, for a potential cause of his losing the battle does indeed lie in the storeroom. Cycles of operation serve to delimit the pursuit of causation only as long as there is no avenue of mutual dependence between them. In the long run the most profound consequence of social interdependence may lie in its tendency to undermine the boundaries between cycles of operation, making causal attribution very largely a matter of convention and fashion.[18]

James's triumph over positivism in "Great Men and Their Environment" was more literary than substantive, but in his *Principles of Psychology* (1890) he achieved a finer balance that acknowledged the strengths of positivism while preserving an essential core of voluntarism. In doing so he became one of the chief exemplars in America, England, and perhaps even the continent, of the intellectual ideals of the first generation of professional social scientists.[19] For the fundamental task facing that generation was to find a safe route between the extremes of positivism and voluntarism. On one side they faced the swirling Charybdis of radical positivism, with its tendency to divest human society of all centers of causal potency and thus of any basis for moral judgment. On the other they faced the Scylla of naive voluntarism, with its tendency to attribute causation superficially and to treat as autonomous the individual components of a society that was becoming increasingly interdependent.[20]

[18] Although this study stresses the role of communication-transportation technology, there is a broad sense in which all technological advances contribute to interdependence and erode the boundaries between cycles of operation. For example, it seems satisfactory to explain the havoc and destruction that took place in San Francisco on 18 April 1906 by reference to "an earthquake" as cause. This is true despite our awareness that a geologist could tell us the causes of the earthquake itself. But in the near future there probably will exist techniques for predicting earthquakes, diminishing their intensity, or preventing them altogether. If after the development of such technology another disastrous earthquake should occur, it would tend to appear as only the *proximate* cause of havoc and destruction, and the "real" cause would have to do with the errors and failings of the control apparatus and those responsible for operating it. What had been a one-way dependence of civil society upon geological factors would become a matter of mutual dependence between the two realms: the effect would be to drain the earthquake of causal potency, to some degree, and shift causation back into the antecedents of the event. This means of course that every technological advance potentially changes the conditions of adequate explanation. On the broadest consequences of technology, see Jacques Ellul's very murky but suggestive book, *The Technological Society*, trans. J. Wilkinson (New York: Random House, 1964). Regarding cycles of operation, see Bunge's discussion of the necessity of artificially isolating cause and effect sequences in order to arrive at a causal perception; *Causality*, 125–140.

[19] James's wide influence is confirmed by Hughes, *Consciousness and Society*, 111–113; Martin J. Wiener, *Between Two Worlds: The Political Thought of Graham Wallas* (Oxford: Clarendon Press, 1971), 86–91 and ch. VII; and Reba N. Soffer, "The Revolution in English Social Thought, 1800–1914," *American Historical Review*, 75 (Dec. 1970), 1938–64.

[20] Max Weber in *The Protestant Ethic and the Spirit of Capitalism* played much the same role vis-à-vis Marx that James played vis-à-vis Spencer. Weber's classic work was explicitly

Looking only at the polemic posture of the generation of the 1890's, one might very well conclude that its mission was not to steer a middle course between positivism and voluntarism, but to launch a "revolt against positivism."[21] But if one asks in what direction intelligent opinion about man and society moved in these years, it becomes clear that James, Dewey, and others of their generation faced a double task.

Intellectually the most pressing of the two dangers they saw facing them was positivism. Spencer and Marx became the negative focus for early professional social science because the positivist-materialist position was a plausible one; it was a live option in the minds of those who opposed it, and that is precisely why they devoted their energies to attacks upon it. They did not need to mount an elaborate attack on naive voluntarism, such as the ASSA represented, because it had been rendered implausible by a transformation in social circumstances. After the turn of the century it was no longer a live option for most serious students of man and society, because the conditions of day-to-day life in an interdependent society made it increasingly difficult for sensitive men to cling to habits of superficial causal attribution. The task of fending off the potentially nihilistic implications of positivism is amply recorded in the literature of the early social sciences, because the battle could never quite be won in a society whose multi-layered structure would always tempt men to go one level deeper in their search for causation. The other task left little record because the enemy was vanquished effortlessly. He simply could be ignored. Men who in the twentieth century located effective causation as proximately as Sanborn and White had done were dismissed as stupid, or immoral, or both.

Our understanding of the origins of modern social science has been distorted by the failure to accord a legitimate place to the voluntaristic tradition represented by the ASSA—a tradition which before the 1890's had at least as strong and perhaps a stronger hold on the mind of the educated public than the more memorable, more systematic work of Spencer and his disciples. In any broad view the transitional generation of newly professional social inquirers of the 1890's and

---

intended to illustrate the erroneous explanation of the rise of capitalism that would result from treating ideas (like the Protestant Ethic) as merely epiphenomenal reflexes of the economic substructure. Weber did not argue that the Protestant Ethic was an autonomous cause which sufficed in itself to explain the rise of capitalism (much less the *origins* of capitalism); on the contrary, he admitted that the ethic was itself caused, and that its causes were in part material. But he insisted that even though the ethic can be seen as the effect of more remote causes, it must also be admitted to possess a degree of independent causal potency if the rise of capitalism is to be properly understood. See p. 183 of the Parsons translation (New York: Scribner's, 1958).

[21.] Hughes, *Consciousness and Society*, ch. II. See also ch. I, above.

early 1900's did not *reject* positivism in the sense of remote causal attribution; on the contrary, they accepted all but the most extreme versions of it.[22]

Social scientists in the 1890's and in the Progressive era located the effective causation of human affairs sometimes in heredity, sometimes in environment; sometimes in the play of economic interests, sometimes in particular historical conditions, such as the frontier experience; sometimes in stimulus-response associations, sometimes in instinct, sometimes in culture, sometimes in that alien area within the person but remote from his conscious mind, the subconscious. They seldom located causation close to the surface of events or in the conscious, willing minds of individuals.[23] They often were individualists in the sense that they wished to enhance individuality and defend the individual, as George Simmel said, against "overwhelming social forces, of historical heritage, of external culture, and of the technique of life."[24] But they were not individualists in the same sense as Sanborn and most of his ASSA contemporaries, who saw the social universe as actually composed of autonomous individuals who normally were masters of their own fates. The difference is between an individualistic perception of social reality and individualism as a defensive assertion of what does not exist but is desirable. For the same reasons that they

---

[22.] The distortion inherent in sociology's foreshortened and selective view of its own past is evident in the terms chosen by Roscoe C. Hinckle and Gisela J. Hinckle to characterize the fundamental bias of American sociologists. Assuming in practice that the origins of sociology in America lie no farther back than 1905, when the American Sociological Society was founded, or 1896, when Albion W. Small founded the *American Journal of Sociology,* and assuming also that the origins of sociology are quite distinct from the origins of the other social sciences, the Hinckles conclude that "perhaps the outstandingly persistent feature of American sociology is its *voluntaristic nominalism.*" By that term they mean the assumption that the structure of any social group is "the consequence of the aggregate of its separate, component individuals and that social phenomena ultimately derive from the motivations of these knowing, feeling and willing individuals." *The Development of Modern Sociology: Its Nature and Growth in the United States* (New York: Random House, 1954), v. Italics in the original. If Franklin H. Giddings, Albion W. Small, Talcott Parsons, and Robert K. Merton—and everyone in between—are to be called "voluntaristic nominalists," what shall we call Frank Sanborn and Andrew D. White, or Louis Agassiz, or the moral philosophers before them? American sociologists have been voluntaristic only in contrast to the contemporary school of thought that has always seemed to them a live option and therefore in need of refutation—positivism, or its twentieth-century variant, behaviorism. From Frank Sanborn's point of view, Parsons's "voluntaristic" theory is a species of positivism.

[23.] The various disciplines differ in this respect, of course. History has always been a stronghold for particularists and voluntarists; anthropology and psychology have perhaps the strongest inclination toward remote causation and determinism. Within each discipline there is a considerable range of viewpoints as well, and the "new social sciences" of the 1960's may represent a shift toward voluntarism within the established social science disciplines.

[24.] Quoted by Robert A. Nisbet, *The Sociological Tradition* (New York: Basic Books, 1966), 305.

refused to locate effective causation in society's smallest components, the first generation of professional social scientists also refused to regard the surface of public life as entirely real. The Progressive generation's view of reality was crisply articulated by Richard Hofstadter: "It was rough and sordid; it was hidden, neglected, and, so to speak, off-stage; and it was essentially a stream of external and material events, of which psychic events were a kind of pale reflex."[25]

The decisive experience of the first generation of professional social inquirers, the experience which alienated them from the ASSA, was intense social interdependence and the habits of remote causal attribution encouraged by it. The men and women who founded the ASSA and manned it in its early years were in fact responding to the increasing interdependence of society, but they never captured it in thought. The young scholars of the 1880's and 1890's, however, conceptualized interdependence, detached it from the organicist metaphor with which it has always tended to merge, and put it to use in an energetic if unsystematic way as an analytic tool, perhaps even a paradigm in Thomas Kuhn's sense.[26]

The British social scientist Graham Wallas gave the concept of interdependence classic formulation in his book *The Great Society* (1914):

> During the last hundred years the external conditions of civilised life have been transformed by a series of inventions which have abolished old limits to the creation of mechanical force, carriage of men and goods, and communication by written and spoken words. One effect of this transformation is a general change of social scale. Men find themselves working and thinking in relation to an environment which, both in its worldwide extension and its intimate connection with all sides of human existence, is without precedent in the history of the world.[27]

Interdependence has been a staple of modern American liberal reform thought, but it has been an assumed condition, a backdrop rather than a focus for analysis. The concept was a fundamental premise of Woodrow Wilson's *The New Freedom* (1913). It was a recurring theme in the philosophy of John Dewey, and he gave it close attention in *The Public and Its Problems* (1927):

> The new technology applied in production and commerce resulted in a social revolution. The local communities without intent or forecast found their affairs conditioned by remote and invisible organizations. The scope of the latter's activities was so vast and their impact upon face-to-face

---

[25.] Hofstadter, "The Progessive View of Reality," in *Charles Beard*, ed. Howard K. Beale (Lexington: University of Kentucky Press, 1954), 87.

[26.] Kuhn, *Structure of Scientific Revolutions*. See ch. I, above.

[27.] Graham Wallas, *The Great Society: A Psychological Analysis* (New York: Macmillan, 1914), 3.

associations so pervasive and unremitting that it is no exaggeration to speak [as Woodrow Wilson did] of "a new age of human relations." ... The invasion of the community by the new and relatively impersonal and mechanical modes of combined human behavior is the outstanding fact of modern life.[28]

Richard T. Ely, aided perhaps by hindsight, portrayed an awareness of interdependence as the common denominator of the new economists who rejected laissez-faire and organized the American Economic Association: "We saw a multiplication of social and economic relations among men which was changing the character of economic society. More and more we were becoming increasingly dependent upon others, and more and more this dependence was becoming interdependence. The forces of life were getting beyond the control of individuals."[29] Albion W. Small thought the justification of sociology lay in its sensitivity to what he called "vortex causation":

... Important enough in itself to mark an epoch in the growth of thought, [is the fact] that in the generation since the sociological movement began, the presumption that linear causation is the main connection of human events has given place to the presumption which we may call vortex causation.... [Today] we are convinced that every actual social situation ... is a resultant of causal factors which run in on that center from every point of the compass—to speak in a figure of only two dimensions.... The best picture we have of social causation more nearly resembles a chemical reaction than a cable transmitting an electric current straight down from the beginning of the world. If this were all that had come from the sociological movement, it would have been worth more than it has cost.[30]

John Dewey, the preeminent philosophical spokesman for modern social science, gave the root notion classic expression: "The fundamental postulate of the discussion is that isolation of any one factor, no matter how strong its workings at a given time, is fatal to understanding and to intelligent action."[31]

Social inquirers so impressed with the interdependence of human affairs and so eager to avoid the parochialism of proximate causal attribution would naturally lean toward functionalism as a style of

---

[28] John Dewey, *The Public and Its Problems* (Denver: Alan Swallow, 1954), 98.
[29] Ely, *Ground Under Our Feet*, 152.
[30] Small, *Origins of Sociology*, 332.
[31] John Dewey, *Freedom and Culture* (New York: Putnams, 1939), 23. Similar statements abound in Dewey's work. For example, "there is no such thing as sheer self-activity possible—because all activity takes place in a medium, in a situation, and with reference to its conditions": from "The Child and the Curriculum" in *Dewey on Education*, ed. M. S. Dworkin (New York: Bureau of Publications, Teachers College, Columbia University, 1959), 110.

explanation—for the meaning of a part lies in its relation to the whole. They would be relativists—for the individual mind no less than the individual himself is caught in a web of external influences that shape and distort its perception of reality. To phrase it more generally, it is precisely the intense interrelatedness of everything in the universe of human affairs that makes it necessary to see things in a relative light. Inquirers eager to uncover hidden interdependencies would be antiformalists, suspicious of all attempts to abstract from the flux of reality—for abstraction is a process of isolation. They often would adopt an aggressively inductive stance, as if that were the only proper mode of thought—for fact-gathering is all that men can do when they find inherited paradigms and theoretic structures inadequate bases for deduction and prediction. In social policy their touchstone would be a quest for community—for interdependence devitalized the island community and simultaneously opened up the tantalizing promise of a grander community, one embracing the whole nation, or even all mankind.[32]

In terms of moral perspective they lived in a world practically unknown to Samuel Gridley Howe, to Frank Sanborn, to nineteenth-century humanitarians and pioneers in social work—unknown even to the men and women of that century who worked most closely with the human problems of poverty and degradation attending the urban-industrial transformation. Though there were hints of it earlier among a few radicals, some social gospel ministers, and some of the younger generation of academic social inquirers, it was not until the 1890's that the new moral perspective acquired a legitimacy, a self-confident aggressiveness and sense of momentum that made it a real contender for dominance in American culture. For those like Sanborn whose opinions had formed in the profoundly different society of only a few decades earlier, the new moral perspective would be inaccessible. For younger educated men and women who knew only the highly interdependent society of the nineties and beyond, it would seem incredible that sensitive people had ever believed anything else.

To Sanborn and his generation the individual had appeared to be a causally potent creature, normally master of his own fate and thus responsible for his own situation in life. Dependence, in the older view, was an exceptional and pathological condition: one referred to the

---

[32.] These capsule characterizations of social thought in the Progressive era are meant only to bring to mind an extensive body of literature. The major landmarks are White, *Social Thought in America;* May, *End of American Innocence;* Wiebe, *Search for Order;* Wilson, *In Quest of Community;* Herbst, *German Historical School in American Scholarship;* Buck, ed., *Social Sciences at Harvard;* Quandt, *From Small Town to Great Community;* Purcell, *Crisis of Democratic Theory.*

poor, the insane, and the crippled, all together, as the "dependent classes." So powerful was the presumption of individual autonomy that even dependence was often thought to be caused by the individual himself. The causes of dependence were as likely to be internal as external: poverty could be viewed as an *effect* of vice (and certainly of imprudence and lack of thrift), and so even could insanity or blindness or cholera.[33] Since each man was ordinarily master of his own fate, he himself—his will, character, and inner merit—was seen as the primary cause of his place in society. If a man was of high position, that in itself was evidence of merit; if he was lowly, that showed a lack of merit. The hierarchy of society, therefore, had an opaque quality of givenness to it; indeed, a quality of divinity, because the moral philosophers taught that God had so constructed the universe that merit was always ultimately rewarded and evil punished.[34]

To the new generation of professional social scientists the causally potent individual was an ideal perhaps, but not a plausible description of reality. Dependence in varying degrees was seen as the natural condition of mankind. As a consequence, merit was partially divorced from status: the powerful were seen as the beneficiaries of external circumstance, and the weak were seen as its victims. Neither were fully responsible for their places in life. Vice and internal corruption, far from being causes of poverty, were seen as its effects. Among the new breed of settlement house workers, exemplified by Jane Addams, aid for the poor was motivated not by religious duty or simple generosity, whereby the strong condescended to "lend a hand" to the weak; instead, aid for the poor came from a spirit of guilt, a sense that the reformer himself was causally implicated in poverty since the social arrangements that "made" him affluent "made" others poor. The technique of aiding the poor shifted away from the effort to inculcate thrift, prudence, and the other elements of self-mastery so essential to nineteenth-century humanitarianism. The new emphasis fell on redistribution of wealth or the design of institutional shelters where the poor might be shielded from life's worst hazards.[35]

---

[33] See Charles Rosenberg, *The Cholera Years: The United States in 1832, 1849 and 1866* (Chicago: University of Chicago Press, 1962), 229–230.

[34] A related point is discussed by Irving Kristol, " 'When Virtue Loses All Her Loveliness'—Some Reflections on Capitalism and 'the Free Society,' " *The Public Interest*, 21 (Fall 1970), 3–15.

[35] The literature on poverty, social work, and related topics at the turn of the century is abundant and growing rapidly. Much of it is flawed, I believe, by a parochial moralism that underestimates the magnitude of the gulf between the moral world of Sanborn's generation and that of the Progressives and ourselves. Historians can be forthright moral critics without merely carping at rival moral systems. All systems of morality set finite limits to moral responsibility: just where the limits are set is to a considerable extent an

The new generation did not deny human freedom—their assumptions included an important voluntaristic element, as Talcott Parsons said—but their view of man was a far cry from the transcendent individualism of earlier American theorists. Eager to relieve actual human degradation, they willingly adopted what seemed a means to that end: a view of man that was in itself degraded, comparatively speaking; a view which portrayed man as a frail and inelegant creature, always struggling to be something more than a dependent reflex of his interdependent surroundings.

---

arbitrary and conventional matter. To criticize one system from the vantage point of another is a very difficult task. Especially insightful are Lasch, *New Radicalism in America;* Hace Sorel Tishler, *Self-Reliance and Social Security, 1870–1917* (Port Washington, N.Y.: Kennikat Press, 1971), the first three chapters of which neatly document the recession of causation out of individuals; and Irwin Yellowitz, *The Position of the Worker in American Society, 1865–1896* (Englewood Cliffs, N.J.: Prentice-Hall, 1969), which traces the gradual abandonment of autonomy as a practical goal among workers. All three books place special stress on the 1890's. Thernstrom's *Poverty and Progress* is very useful, but I think he is wrong to construe the "promise of mobility" (in my terms, an aspect of the presumption of individual autonomy) as primarily an ideological mechanism. In the first place, the presumption existed not only at the popular, "Horatio Alger" level, but was deeply rooted in the basic possessive individualist assumptions of liberal thought going all the way back to the seventeenth century: see C. B. Macpherson, *Political Theory of Possessive Individualism*. In the second place, the presumption was sustained not only by class interest, but, more important, by the habits of proximate causal attribution appropriate to simple societies. The success ethic was finally killed not by the accumulation of evidence of immobility (there had been far more evidence of immobility and dependence in the seventeenth century, when individualist doctrines began to develop, than there was in the late nineteenth century, when they began to decline), but by a change in the habits of causal attribution that made the presumption of autonomy implausible. Other works I found useful were Nathan I. Huggins, *Protestants Against Poverty: Boston's Charities, 1870–1900* (Westport, Conn.: Greenwood, 1971); Allen F. Davis, *Spearheads for Reform: The Social Settlements and the Progressive Movement, 1890–1914* (New York: Oxford University Press, 1967); Robert H. Bremner, *From the Depths: The Discovery of Poverty in the United States* (New York: New York University Press, 1956) and " 'Scientific Philanthropy,' 1873–93," *The Social Service Review,* 30 (June 1956), 168–173; Bruno, *Trends in Social Work;* Gettleman, "Charity and Social Classes in the U.S."; Roy Lubove, *The Professional Altruist: The Emergence of Social Work as a Career, 1880–1930* (Cambridge: Harvard University Press, 1965); Rothman, *Discovery of the Asylum;* and Paul T. Ringenbach, *Tramps and Reformers: The Discovery of Unemployment in New York* (Westport, Conn.: Greenwood, 1973).

Appendix A

Chief Officers of the American Social Science Association

Presidents
  William Barton Rogers (1865–68)
  Samuel Eliot (1868–73)
  George William Curtis (1873–74)
  Theodore Dwight Woolsey (Winter 1874–75)
  David Ames Wells (1875–78)
  Benjamin Peirce (acting 1878)
  Daniel Coit Gilman (1879–80)
  Francis Wayland (1881–82)
  John Eaton (1883–84)
  Carroll D. Wright (1885–88)
  Andrew Dickson White (1888–91)
  Heman Lincoln Wayland (1891–93)
  Frederick J. Kingsbury (1893–96)
  James Burrill Angell (1896–97)
  Simeon Eben Baldwin (1898–99)
  Charles Dudley Warner (1899–1900)
  Oscar S. Straus (1901–1903)
  John Graham Brooks (1903–1905)
  John Huston Finley (1905–1909)

Secretaries
  Franklin B. Sanborn and various others (1865–68)
  Henry Villard (1868–70)
  no regular secretary (1870–73)
  Franklin B. Sanborn (1873–98)
  Rev. Frederick Stanley Root (1898–1905)
  Isaac Franklin Russell (1905–1909)

Appendix B

An Index by Author of All Articles Appearing in the Journal of Social Science, *Volume 1 (June 1869) through Volume 39 (November 1901)*

This index does not cover the last seven volumes of the *Journal.*

ADAMS, Charles Francis Jr. "The Election of Presidents," 2 (1870), 148–158.
———. "Protection of the Ballot," 1 (June 1869), 91–111.
ADAMS, Henry Carter (professor, Cornell University). "The Financial Standing of States," 19 (Dec. 1884), 27–46 (corrections and additions, vii–viii).
ADAMS, Herbert Baxter. "New Methods of Study in History," 18 (May 1884), 213–263.
ALDRICH, P. Emory (Worcester, Mass.). "Prohibitory Legislation," 14 (Nov. 1881), 71–89 (temperance).
ALLISON, H. E. (medical superintendent, Matteawan State Hospital, Fishkill-on-Hudson, N.Y.). "Methods of Securing Health of Insane Convicts," 35 (Dec. 1897), 155–166.
AMOS, Sheldon (professor, London). "Extradition," 11 (May 1880), 117–122 (read at 1877 meeting).
ANDERSON, Mr. (president, Rochester University). "Relations of Christianity to the Common Law," 10 (Dec. 1879), 55–74.
ANDERSON, Joseph (Rev., Waterbury, Conn.). "A Sketch of Recent Movements in the Educational Domain," 35 (Dec. 1897), 69–76.
———. "Remarks as Chairman of the Department of Education and Art," 36 (Dec. 1898), 64–67.
———. "Address by the Chairman [of the Education and Art Department]," 37 (Dec. 1899), 22–25.
ANDREWS, E. Benjamin (president, Brown University). "The Economic Law of Monopoly," 26 (Feb. 1890), 1–12.

———. "A Single Land Tax from the Point of View of Public Finance," 27 (Oct. 1890), 29–33.

ANGELL, George T. "The Protection of Animals," 6 (July 1874), 164–177.

ANGELL, James B. "The Progress of International Law," 8 (May 1876), 40–53.

——— (president, Michigan University). "The Diplomatic Relations between the United States and China," 17 (May 1883), 24–36.

ASHLEY, Clarence D. (dean, School of Law, New York University). "The Training of the Lawyer and Its Relation to General Education," 37 (Dec. 1899), 229–238.

ATKINSON, Edward. "Inefficiency of Economic Legislation," 4 (1871), 123–132.

———. "What Makes the Rate of Wages?" 19 (Dec. 1884), 47–116 (published later with two other essays by G. P. Putnam's Sons as *The Distribution of Products, or the Mechanism and the Metaphysics of Exchange*).

———. "Remarks of Edward Atkinson, Esq., [on the Single Tax]," 27 (Oct. 1890), 55–71.

———. "Final Remarks [on Single Tax]," 27 (Oct. 1890), 122–124.

ATWATER, W. O. (professor, Wesleyan University). "False and True Teaching in Our Schools Concerning Alcohol," 38 (Dec. 1900), 107–116 (debate, 125–126).

BACON, Francis. "Civilization and Health," 3 (1871), 58–77.

BACON, Francis, William A. HAMMOND, and David F. LINCOLN. "Vaccination," 2 (1870), 129–147.

BACON, Leonard Woolsey (Norwich, Conn.). "Considerations in Favor of License Laws for Restraining the Liquor Traffic," 14 (Nov. 1881), 118–128.

BACON, Theodore (Rochester, N.Y.). "Professional Ethics," 17 (May 1883), 37–46 (discussion, 46–48).

BAKER, Henry B. (doctor). "The Michigan Plan for Boards of Health [Abstract]," 16 (Dec. 1882), 24–27 (followed by debate, 27–29).

BALDWIN, Simeon E. "Recent Changes in Our State Constitutions," 10 (Dec. 1879), 136–151.

——— (professor, Yale University). "Graduate Courses, at Law Schools," 11 (May 1880), 123–137 (read at 1877 meeting).

———. "How to Deal with Habitual Criminals," 22 (June 1887), 162–171.

———. "Absolute Power an American Institution," 35 (Dec. 1897), 1–20 (presidential address).

———. "The History of American Morals," 36 (Dec. 1898), 1–55 (presidential address).

———. "The Natural Right to a Natural Death," 37 (Dec. 1899), 1–17.

BALDWIN, William H. Jr. (president, Long Island Railroad; trustee, Tuskegee Institute). "The Present Problem of Negro Education," 37 (Dec. 1899), 52–63 (debate, 64–68).

BARRETT, George C. "The Administration of Criminal Justice," 2 (1870), 167–175.

BARTLETT, George B. "The Recreations of the People," 12 (Dec. 1880), 135–146.

BARTON, Clara. "International and National Relief in War," 16 (Dec. 1882), 75–97.

BECKNER, W. M. (Winchester, Ky.). "City Schools as Compared with Country Schools," 21 (Sept. 1886), 224–231.

BEMIS, Edward W. (student, Johns Hopkins University; professor, Vanderbilt University). "Local Government in Michigan and the Northwest," 17 (May 1883), 49–69 (discussion by Goldwin Smith, 1870).

———. "Socialism and State Action," 21 (Sept. 1886), 33–68.

———. "Relation of Trades-Unions to Apprentices," 28 (Oct. 1891), 108–123 (discussion, 124–125).

BERNSTEIN, Charles (M.D.; assistant superintendent of Rome State Custodial Asylum, Rome, N.Y.). "The Relationship between the Physical and Mental Being in Connection with the Training of Idiots," 35 (Dec. 1897), 149–154.

BETTS, Frederic H. (New York City). "The Policy of Patent Laws," 10 (Dec. 1879), 151–180.

BIRD, F. W. (Walpole, Mass.). "The Province of Legislation in the Suppression of Intemperance," 14 (Nov. 1881), 90–117.

BLODGET, Lorin. "Waste of Existing Social Systems," 4 (1871), 8–18.

BLODGETT, Albert N. (M.D., Boston). "The Management of Chronic Inebriates and Insane Drunkards," 16 (Dec. 1882), 52–70.

BOLTON, Mrs. Sarah H. (Cleveland, Ohio). "The Relations between Employers and Employed," 18 (May 1884), 279–288.

BOWLES, Samuel (editor, Springfield *Republican*). "The Relations of State and Municipal Governments, and the Reform of the Latter," 9 (Jan. 1878), 140–146.

BRACE, Charles L. "What Is the Best Method for the Care of Poor and Vicious Children?" 11 (May 1880), 93–98 (debate, 98–103).

———. "Christianity and the Relations of Nations," 13 (Mar. 1881), 136–155.

BRACE, Charles Loring (secretary, New York Children's Aid Society). "Child-Helping as a Means of Preventing Crime in the City of New York," 18 (May 1884), 289–305.

BRADFORD, Gamaliel. "Financial Administration," 6 (July 1874), 46–59.

———. "The Financial Policy of England and the United States," 8 (May 1876), 128–139.

BRIMMER, Martin. "Some Relations of Art to the American People," 15 (Feb. 1882), 148–156.

BROCKWAY, Zebulon R. (superintendent, Elmira Reformatory, N.Y.). "Reformation of Prisoners," 6 (July 1874), 144–154.

———. "Indeterminate Sentences and Their Results in New York," 13 (Mar. 1881), 156–169.

———. "The Best Treatment of Criminals, Whether Felons or Misdemeanants," 39 (Nov. 1901), 196–216.

BROOKS, John Graham (Rev.; doctor, Brockton, Mass.; professor, Harvard University). "Labor Organizations: Their Political and Economic Service to Society," 23 (Nov. 1887), 68–75.

———. "Sweating in Germany," 30 (Oct. 1892), 59–64.

BROSIUS, Marriott (chairman, House Committee on Banking and Currency). "Progress toward an Ideal Currency," 38 (Dec. 1900), 212–217.

BROWN, Elizabeth Stow (M.D., N.Y.). "The Working-Women of New York: Their Health and Occupations," 25 (Dec. 1888), 78–92.

BROWN, Joseph T. (N.Y.). "The Progress of the Financial Credit of the Government of the United States, 1861–1890," 29 (Aug. 1892), 34–48.

BROWNE, Hugh M. (Washington, D.C.). "Debate on Higher Education for Negroes," 34 (Nov. 1896), 89–97.

BRYAN, Mrs. R. S. "Physical Education in Women's Colleges," 20 (June 1885), 45–48.

BRYSON, Louise Fiske (M.D., N.Y.). "The Education of Epileptics," 31 (Jan. 1894), 100–106.

BURK, Addison B. (Philadelphia). "The City of Homes and Its Building Societies," 15 (Feb. 1882), 121–134.

CARRET, James R. (Boston). "Remarks of James R. Carret [on Single Tax]," 27 (Oct. 1890), 99–112 (single tax debate).

CARROL, Alfred L. "Hygiene in Schools and Colleges," 7 (Sept. 1874), 266–269.

CARSON, J. C. (M.D.; superintendent of Syracuse State Institution for Feeble-Minded Children). "Feeble-Minded Children," 35 (Dec. 1897), 142–148.

CHAMBERLAIN, D. H. (N.Y.). "The American System of Trial by Jury," 23 (Nov. 1887), 85–124.

CHANNING, Walter (M.D., Boston). "The Treatment of Insanity in Its Economic Aspects," 13 (Mar. 1881), 89–93.

———. "Address of the Chairman [of the Health Department]," 14 (Nov. 1881), 164–177 (insanity).

———. "Address of the Chairman [of the Health Dept.]: Social Questions for Health Boards," 16 (Dec. 1882), 17–23.

———. "A Consideration of the Causes of Insanity," 18 (May 1884), 68–92.
CHAPIN, Henry Dwight (M.D.; professor of Diseases of Children at New York Post Graduate Medical School). "The Struggle for Subsistence: How Can It Be Most Efficiently Aided?" 25 (Dec. 1888), 93–96.
CHUBB, Percival (London). "Practical Measures of Socialism in England," 26 (Feb. 1890), 126–144.
CLARK, John Bates (professor, Smith College). "The Moral Basis of Property in Land," 27 (Oct. 1890), 21–28 (single tax debate).
CLARKE, Samuel B. (N.Y.). "What the Single Tax of Henry George Is," 27 (Oct. 1890), 1–7 (single tax debate).
COAN, Titus Munson (M.D., N.Y.). "Mineral Waters at Home and Abroad," 22 (June 1887), 50–62.
COHEN, Miss Mary M. (Philadelphia). "Hebrew Charities," 19 (Dec. 1884), 168–176.
COLBY, James F. "Disfranchisement for Crime," 17 (May 1883), 71–98.
COLLINS, Charles A. (professor, Cornell University). "Modern Methods of Treating Criminals," 31 (Jan. 1894), 93–99 (not read at ASSA meeting).
COMMITTEE ON PROVIDENT INSTITUTIONS. "Life Insurance," 25 (Dec. 1888), 158–163.
COOKE, Alice R. (Sandwich, Mass.). "Family Care for the Insane," 26 (Feb. 1890), 97–102 (discussions, 95–96, 102–106).
COXE, Brinton. "Von Mohl on Different Modes of Filling Offices in the Civil Service," 4 (1871), 74–94.
COXE, Eckley B. "Mining Legislation," 4 (1871), 19–32.
CROTHERS, T. D. (M.D., Hartford). "The Disease of Inebriety and Its Social Science Relations," 18 (May 1884), 100–111.
CURTIS, George William. "Opening Address," 6 (July 1874), 33–35.
———. "Civil Service Reform," 14 (Nov. 1881), 36–51.
CURTIS, H. Holbrook (M.D., N.Y.). "A Paper Entitled Noses," 22 (June 1887), 75–84.
———. "The Necessity of Additional Requirements for Obtaining a Medical Degree," 25 (Dec. 1888), 15–20 (discussion, 20–27).
DAINGERFIELD, Miss Henderson (Lexington, Ky.). "Social Settlement and Educational Work in the Kentucky Mountains," 39 (Nov. 1901), 176–189.
DANA, C. L. (M.D.; president, New York Neurological Society; professor of Nervous Mental Diseases, New York Post Graduate Medical School). "Immigration and Nervous Diseases," 24 (Apr. 1888), 43–54 (discussion, 55–56).
DANIEL, Anna S. (M.D.; outdoor physician to the New York Infirmary for Women and Children). "Conditions of the Labor of Women and

Children: Observed by a Dispensary Physician of New York in 1892," 30 (Oct. 1892), 73–85.

DAVIDSON, Thomas (Orange, N.J.). "The Place of Art in Education," 21 (Sept. 1886), 159–187.

———. "Property," 22 (June 1887), 107–112.

———. "The Single Tax," 27 (Oct. 1890), 8–14 (single tax debate).

DAVIS, R. T. "Pauperism in the City of New York," 6 (July 1874), 74–83 (report of Department of Social Economy).

DAWES, Henry L. "The Mode of Procedure in Cases of Contested Elections," 2 (1870), 56–68.

DEPARTMENT OF FINANCE. "Abstract of the Financial Discussion [21 May 1874]," 6 (July 1874), 20–29.

DEPARTMENT OF SOCIAL ECONOMY. "[Report of Committee on Provident Institutions] Life Insurance," 25 (Dec. 1888), 158–163.

———. "Report on the Sweating System in the Year 1893," 31 (Jan. 1894), 63–69.

———. "The Relief of the Unemployed in the United States during the Winter of 1893–1894," 32 (Nov. 1894), 1–51 (including reports from Charity Organization Societies and other groups of New Haven, Baltimore, New York City, Buffalo, N.Y., Cleveland, Denver, Pittsburgh, Waterbury, Conn., Louisville, Chicago, Philadelphia, St. Louis, and Syracuse).

DEWEY, Melvil (director, New York State Library). "The Future of the Library Movement in the United States in the Light of Andrew Carnegie's Recent Gift," 39 (Nov. 1901), 139–147 (discussion, 147–157).

DEXTER, Seymour (Hon., Elmira, N.Y.). "Co-operative Building and Loan Associations in the State of New York," 25 (Dec. 1888), 139–148.

———. "Compulsory Arbitration," 28 (Oct. 1891), 86–100.

DIKE, Samuel W. (Rev., Royalton, Vt.). "The Effect of Lax Divorce Legislation upon the Stability of American Institutions," 14 (Nov. 1881), 152–163.

DOUGLASS, Frederick. "The Negro Exodus from the Gulf States," 11 (May 1880), 1–21.

DREHER, Julius D. (president, Roanoke College, Salem, Va.). "Education in the South: Some Difficulties and Encouragements," 33 (Nov. 1895), 185–207.

DUNCAN, Samuel Augustus. "The American System of Patents," 3 (1871), 78–96.

DUTTON, Samuel T. (New York City). "Educational Resources of the Community," 38 (Dec. 1900), 117–124.

——— (superintendent of Public Schools, New Haven). "Education as a Cure for Crime," 26 (Feb. 1890), 55–65.

——— (Brookline, Mass.). "The Relation of Education to Vocation," 34 (Nov. 1896), 52–62.
DWIGHT, Theodore W. "The Public Charities of the State of New York," 2 (1870), 69–91.
EARLE, Pliny (doctor, Northampton, Mass.). "Popular Fallacies in Regard to Insanity and the Insane," 26 (Feb. 1890), 107–117.
EATON, Dorman B. "Municipal Government," 5 (1873), 1–35.
———. "The Experiment of Civil Service Reform in the United States," 8 (May 1876), 54–78.
EATON, John. "A Word on the Scientific Method in the Common Affairs of Life," 21 (Sept. 1886), ix–xxiii (presidential address).
ELIOT, Samuel. "Civil Service Reform," 1 (June 1869), 112–119.
———. "Relief of Labor," 4 (1871), 133–149.
ELIOT, W. G. (D.D., St. Louis). "Treatment of the Guilty," 8 (May 1876), 79–84.
EMERSON, George B. "Houses in the Country for Working Men and Working Women," 3 (1871), 169–184.
FARNAM, Henry W. (professor, Yale University). "The German Socialist Law of October 21, 1878," 13 (Mar. 1881), 36–53.
FIELD, David Dudley. "An International Code," 2 (1870), 188–199.
———. "Representation of Minorities," 3 (1871), 133–147.
FIELD, Matthew D. (M.D.). "Examination and Commitment of the Public Insane in New York City," 30 (Oct. 1892), 19–28.
FLAGG, Willard C. "The Farmer's Movement in the Western States," 6 (July 1874), 100–113 (discussion, 113–115).
FLOOD, Everett (M.D.; superintendent of the Hospital Cottages for Children, Baldwinsville, Mass.). "The Home Care of Epileptic Children," 35 (Dec. 1897), 132–138.
FLOYD, Silas X. (professor, Atlanta University). "Debate on Higher Education of Negroes," 34 (Nov. 1896), 84–86.
FOLKMAR, Daniel (professor; Paris, France). "Sociology as Based upon Anthropology," 36 (Dec. 1898), 260–263.
FOLKMAR, Mrs. Daniel (Paris, France). "The Short Duration of School Attendance: Its Causes and Remedies," 36 (Dec. 1898), 68–81.
FOSTER, Frank P. "On Public Vaccination," 5 (1873), 98–108.
FOSTER, John W. (Hon., Washington, D.C.). "The Latin-American Constitutions and Revolutions," 39 (Nov. 1901), 39–48.
GALLAUDET, Edward M. "The Deaf-Mute College at Washington," 6 (July 1874), 160–163.
———. "International Ethics," 18 (May 1884), 151–161.
GARDINER, C. A. (professor, Brooklyn, N.Y.). "The Race Problem in the United States," 18 (May 1884), 266–275.

GARDINER, Charles A. (Ph.D. "of the N.Y. Bar"). "The Proposed Anglo-American Alliance," 36 (Dec. 1898), 148–159.

GARFIELD, James A. "The American Census," 2 (1870), 31–55.

GARRISON, William Lloyd. "Remarks [on the Single Tax]," 27 (Oct. 1890), 15–20 (single tax debate).

GATES, ——— (president, Rutgers). "Land and Law as Agents in Educating Indians," 21 (Sept. 1886), 113–146.

GEORGE, Henry. "Remarks of Henry George [on the Single Tax]," 27 (Oct. 1890), 73–86 (single tax debate).

GERHARD, William Paul. "A Plea for Rain Baths in the Public Schools," 38 (Dec. 1900), 30–49 (rain baths = showers, or "inclined overhead douches").

GIDDINGS, Franklin H. (professor, Columbia College). "The Relation of Sociology to Other Scientific Studies," 32 (Nov. 1894), 144–150.

GILMAN, Arthur. "Thoughts on the Collegiate Instruction of Women," 24 (Apr. 1888), 68–86.

GILMAN, Daniel C. "American Education, 1869–1879," 10 (Dec. 1879), 1–27 (presidential address).

———. "The Purposes of the American Social Science Association, and the Means That May be Employed to Promote These Ends," 12 (Dec. 1880), xxii–xxiv (abstract furnished by Gilman—presidential address, 1880 meeting).

GLADDEN, Washington (Rev., Columbus, Ohio). "Arbitration of Labor Disputes," 21 (Sept. 1886), 147–158.

GODKIN, E. L. "Legislation and Social Science," 3 (1871), 115–132.

———. "Libel and Its Legal Remedy," 12 (Dec. 1880), 69–73.

GOMPERS, Samuel (president, American Federation of Labor). "Trade-Unions: Their Achievements, Methods, and Aims," 28 (Oct. 1891), 40–47.

GOODRICH, Caspar F. (commander, U.S. Navy). "Naval Education," 33 (Nov. 1895), 29–48.

GOULD, John Stanton. "Texas Cattle Disease," 1 (June 1896), 56–71.

GRAY, John H. (professor, Northwestern University). "Some Preliminary Municipal Problems and Their Solution," 34 (Nov. 1896), 174–181.

GREELEY, Horace. "Method of Diffusing Knowledge," 1 (June 1869), 88–90.

GREEN, S. S. (Worcester, Mass.). "The Relation of the Public Library to the Public Schools," 12 (Dec. 1880), 13–28.

GREENE, Jacob L. (president, Connecticut Mutual Life Insurance Company, Hartford, Conn.). "Bimetallism, or the Double Standard," 31 (Jan. 1894), 9–26.

———. "An Ideal Currency," 38 (Dec. 1900), 196–205.

GREENE, J. Warren (Brooklyn, N.Y.). "Legislation in Its Relation to Jurisprudence," 34 (Nov. 1896), 117–125.

GREENER, R. T. (professor, Howard University). "The Emigration of Colored Citizens from the Southern States," 9 (May 1880), 22–35.

GREENOUGH, William W. "Some Conclusions Relative to Public Libraries," 7 (Sept. 1874), 323–333.

GREGORY, John M. (U.S. Civil Service Commission). "The American Civil Service System," 18 (May 1883), 178–193.

GREGORY, J. M. (doctor, Chicago). "The American Newspaper and American Education," 12 (Dec. 1880), 61–68.

GUNTON, George (editor, *Social Economist*). "Social Influence of Labor Organizations," 28 (Oct. 1891), 101–107.

HALL, G. Stanley. "The Moral and Religious Training of Children," 15 (Feb. 1882), 56–76.

HALL, Lucy M. (M.D., Sherborn Reformatory). "Inebriety in Women: Its Causes and Results," 18 (May 1884), 93–99.

——— (M.D., Brooklyn, N.Y.). "Tenement-Houses and Their Population," 20 (June 1885), 91–97.

———. "The Physical Training of Women," 21 (Sept. 1886), 100–104.

———. "Unsanitary Conditions in Country Homes," 25 (Dec. 1888), 59–74 (discussion, 74–77).

HALLOWELL, Miss Anna (Philadelphia). "The Care and Saving of Young Children," 12 (Dec. 1880), 117–124.

HAMMOND, W. G. (professor, University of Iowa). "Legal Education and the Study of Jurisprudence in the West and North-West," 8 (May 1876), 165–176.

HARE, Thomas. "Minority Representation in Europe," 3 (1871), 185–191 (see also "Application of Mr. Hare's System of Voting," 192–198).

HARRIS, Daniel L. "Municipal Economy," 9 (Jan. 1878), 147–163 (includes valuable tables on municipal expenses).

HARRIS, Elisha. "Health Laws and Their Administration," 2 (1870), 176–187.

———. "Vital Registration—Public Uses of Vital Statistics," 7 (Sept. 1874), 221–228.

HARRIS, William Torrey. "The Method of Study in Social Science," 10 (Dec. 1879), 28–34.

———. "The Education of the Family, and the Education of the School," 15 (Feb. 1882), 1–5.

———. "Address of the Chairman [of the Department of Education]," 1882, 17 (May 1883), 133–155.

———. "Moral Education in the Common Schools," 18 (May 1884), 122–134.

———. "On the Function of the Study of Latin and Greek in Education," 20 (June 1885), 1–13.

———. "The Definition of Social Science and the Classification of the Topics Belonging to Its Several Provinces," 22 (June 1887), 3–6.

———. "Right of Property and the Ownership of Land," 22 (June 1887), 116–155.

——— (U.S. Commissioner of Education). "The Single Tax," 27 (Oct. 1890), 113–121 (single tax debate).

——— (U.S. Commissioner of Education). "The Uses of Higher Education," 36 (Dec. 1898), 93–104.

——— (U.S. Commissioner of Education). "A Year's Progress in Education," 38 (Dec. 1900), 69–78.

——— (U.S. Commissioner of Education). "Higher Education in the South," 39 (Nov. 1901), 123–131.

HARRIS, William Torrey, Henry BARNARD, and Emily TALBOT. "Report from an Education Department Sub-Committee on Kindergartens," 12 (Dec. 1880), 8–12.

HARTER, Michael D. (Hon., Ohio). "Free Silver Debate," 1895, 33 (Nov. 1895), 110–113 (letter from non-participant).

HARTWELL, Edward Mussey (M.A.; fellow of Johns Hopkins University). "The Study of Anatomy, Historically and Legally Considered," 13 (Mar. 1881), 54–88.

HENRY, Guy V. (general, Military Government of Puerto Rico). "Remarks on the Financial Administration of Colonial Dependencies," 37 (Dec. 1899), 153–163.

HERRICK, Mary E. (M.D., N.Y.). "The Tenement House: Its Influence upon the Child," 29 (Aug. 1892), 25–33.

HEWITT, Charles N. "The Minnesota Board of Health," 7 (Sept. 1874), 84–386.

HICKS, William L. (chief inspector, Boston Board of Health). "[Abstract of] Tenement House Work-Rooms in Boston," 30 (Oct. 1892), 103–104.

HIGGINS, ——— (senator, Del.). "Free Silver Debate, 1895," 33 (Nov. 1895), 114–123.

HIGGINSON, Thomas Wentworth. "Higher Education of Women," 5 (1873), 36–45.

——— (chairman, Department of Education). "Abstract of Address as Chairman of Department of Education," 24 (Apr. 1888), 63–65.

HILL, Alfred (Birmingham, England). "A Scheme of the Late Edwin Hill for Extinguishing Crime," 17 (May 1883), 99–108.

HILL, David Jayne (president, University of Rochester). "International Justice," 34 (Nov. 1896), 98–116.

HILL, Hamilton Andrews (Boston). "The Relation of the Business Men of the United States to the National Legislation," 3 (1871), 148–168.
———. "The Navigation Laws of Great Britain and the United States," 9 (Jan. 1878), 101–116.
———. "The Place of the Practical Man in American Public Affairs," 10 (Dec. 1879), 75–90.
———. "Penalties for Crimes against Property," 17 (May 1883), 109–116.
HILL, Robert T. (professor, Austin, Tex.). "Notes on Provident Institutions in Arkansas, Tennessee, and Texas," 25 (Dec. 1888), 152–157.
HINMAN, Miss Mary W. (Havana, N.Y.). "Home Life in Some of Its Relations to Schools," 12 (Dec. 1880), 44–60.
HITCHCOCK, Edward (professor, Amherst). "Athletics in American Colleges," 20 (June 1885), 27–44.
HITCHCOCK, Henry (LL.D., St. Louis; professor). "Modern Legislation Touching Marital Property Rights," 13 (Mar. 1881), 12–35.
HITZ, John (Washington, D.C.). "Homes for the People in the City of Washington," 15 (Feb. 1882), 135–145.
HOADLY, George (N.Y.). "The Constitutional Guarantees of the Right of Property as Affected by Recent Decisions," 26 (Feb. 1890), 13–54.
HOLLAND, W. J. (chancellor, Western University of Pennsylvania). "Remarks upon the Filtration of Municipal Water Supplies," 36 (Dec. 1898), 246–255.
HOLMAN, D. Emery (M.D., N.Y.). "The Function of the Lungs," 24 (Apr. 1888), 12–23.
HOLT, George C. "Lynching and Mobs," 32 (Nov. 1894), 67–81.
HOMANS, Sheppard. "Life Insurance," 2 (1870), 159–166.
———. "Suggestions for a Plan of Life Insurance without Large Accumulations, or Reserves," 8 (May 1876), 157–164.
HORR, Roswell G. (Hon., N.J.). "Free Silver Debate, 1895," 33 (Nov. 1895), 76–85, 104–108.
HOTCHKISS, Samuel M. (commissioner, Connecticut Bureau of Labor Statistics, Hartford, Conn.). "Practical Suggestions on Labor Organization," 28 (Oct. 1891), 28–39.
———. "Mutual Benefit Societies in Connecticut," 31 (Jan. 1894), 54–62.
HOVEY, William A. (editor, Boston *Transcript*). "Cooperative Distribution," 11 (May 1880), 105–113.
HOWE, Julia Ward. "Changes in American Society," 13 (Mar. 1881), 170–188.
HOWE, William Wirt (Hon.; LL.D., New Orleans). "The Law of Our New Possessions," 38 (Dec. 1900), 256–263.

HUBBARD, Gardiner G. "American Railroads," 6 (July 1874), 134–143.

HUBBELL, Charles Buckley (Hon., New York City). "Obligations of the State to Public Education," 36 (Dec. 1898), 212–219.

HUNT, Ezra M. (M.D.; Secretary of New Jersey Board of Health). "The Health Care of Households with Special Reference to House Drainage," 16 (Dec. 1882), 30–40.

——— (chairman of New Jersey Department of Health). "Health and Social Science," 18 (May 1884), 29–43.

HYDE, Henry D. "Customs Laws and Their Administration," 9 (Jan. 1878), 132–139.

IRELAND, Alleyne (London). "Financial Administration of Colonial Dependencies," 37 (Dec. 1899), 136–154 (debate, 155–157).

IVES, Charles Acton (Newport, R.I.). "A New Plan for the Labor of Prisoners," 18 (May 1884), 306–314 (debate follows, 315–320).

JACOBI, Mary Putnam (M.D., N.Y.). "Some Considerations on the Moral, and on the Non-Asylum Treatment of Insanity," 15 (Feb. 1882), 77–96 (followed by debate, 96–103).

JAMES, Edmund J. (professor, University of Pennsylvania). "Pedagogy in American Colleges," 24 (Apr. 1888), 66–67 (abstract only).

JAMES, H. A. (N.Y.). "Private Corporations and the State," 23 (Nov. 1887), 145–166.

JARVIS, Edward. "Vital Statistics of Different Races," 7 (Sept. 1874), 229–234.

JENKINS, J. Foster. "Tent Hospitals," 7 (Sept. 1874), 270–290 (discussion, 291–293).

JENKS, Jeremiah W. (Galesburg, Ill.). "Report of Savings Banks and Building Associations of Illinois," 25 (Dec. 1888), 125–138.

——— (professor, Cornell University). "Trades-Unions and Wages," 28 (Oct. 1891), 48–58.

——— (secretary, Finance Department). "The Present Aspect of the Silver Problem," 32 (Nov. 1894), xxiii–xxxix.

——— (professor, Cornell University). "The Causes of the Fall in Prices since 1872," 35 (Dec. 1897), 31–49.

———. "The George Junior Republic," 35 (Dec. 1897), 65–68.

JEVONS, W. Stanley. "The Silver Question," 9 (Jan. 1878), 14–20.

JOHNSON, S. W. (professor, Yale University). "Adulterations of Food," 13 (Mar. 1881), 99–123 (debate, 123–135).

JONES, J. H. (Rev., North Abington, Mass.). "Ten Hours," 16 (Dec. 1882), 149–164.

KAPP, Friedrich. "Immigration," 2 (1870), 1–30.

KEDZIE, R. C. "Ventilation of the Dwellings of the Poor," 7 (Sept. 1874), 244–247 (previously published: *Transactions,* State Medical Society of Michigan, 1874).

KELLEY, Mrs. Florence (Hull House, Chicago; chief factory inspector of Illinois). "The Working-Boy," 34 (Nov. 1896), 43–51.

———. "Principles and Aims of the Consumers League," 37 (Dec. 1899), 111–122.

KELLOGG, Charles D. (N.Y.). "Child Life in City and Country," 21 (Sept. 1886), 207–223.

KELLOGG, D. O. (Rev., Philadelphia). "The Principle and Advantage of Association in Charities," 12 (Dec. 1880), 84–90.

KIMBALL, Arthur Reed (of the *Waterbury American*, Waterbury, Conn.). "Education by Newspaper," 37 (Dec. 1899), 26–43.

KINGSBURY, Frederick J. (Waterbury, Conn.). "Pensions in a Republic," 13 (Mar. 1881), 1–11.

———. "Profit Sharing as a Method of Remunerating Labor—Some Limitations to be Considered," 23 (Nov. 1887), 25–36.

———. "The Development of an Organized Industry," 28 (Oct. 1891), 59–65.

———. "The Three Factors That Produce Wealth—Capital, Labor, and Management," 31 (Jan. 1894), 1–8.

——— (LL.D.). "The Reign of Law," 32 (Nov. 1894), ix–xxii (presidential address, 1894).

———. "The Tendency of Men to Live in Cities," 33 (Nov. 1895), 1–19 (presidential address).

———. "A Sociological Retrospect," 34 (Nov. 1896), 1–14 (presidential address).

LAFLEUR, Eugene (lawyer, Montreal). "American Marriages and Divorces before Canadian Tribunals," 36 (Dec. 1898), 196–211.

LANGHORNE, Mrs. Orra (Lynchburg, Va.). "Colored Schools in Virginia," 11 (May 1880).

———. "Changes of a Half-Century in Virginia," 38 (Dec. 1900), 168–176.

———. "Domestic Service in the South," 39 (Nov. 1901), 169–175.

LARCOM, Miss Lucy (Beverly, Mass.). "American Factory Life—Past, Present and Future," 16 (Dec. 1882), 141–146.

LARNED, J. N. (Buffalo, N.Y.). "Public Libraries and Public Education," 18 (May 1884), 196–208.

LEE, Elmer (doctor, N.Y.). "Health in [Army] Camps," 36 (Dec. 1898), 238–244.

——— (M.D., New York City). "The Genesis of Disease," 38 (Dec. 1900), 57–65.

LEE, Joseph (Brookline, Mass.). "The Sweating System," 30 (Oct. 1892), 105–137, appendix, 138–146.

———. "The Argument for Trade Schools," 33 (Nov. 1895), 208–212.

———. "The Textile School at Lowell," 35 (Dec. 1897), 56–64.

LEIPZIGER, Henry M. (Ph.D., N.Y.). "The Educational Value of the Popular Lecture," 36 (Dec. 1898), 82–89.

LEVASSEUR, Emil. "Malthus and the Laws of Increasing Population," 28 (Oct. 1891), ix–xxiv (from 3d vol. of Levasseur's *La Population Française*).

LINCOLN, D. F. (M.D.). "A Report on School Hygiene," 7 (Sept. 1874), 261–265.

——. "The Nervous System as Affected by School-Life," 8 (May 1876), 87–110.

——. "The Health of Boy's Boarding Schools [Abstract]," 16 (Dec. 1882), 40–46.

LINDSAY, Samuel M. (professor, University of Pennsylvania). "Growth and Significance of Municipal Enterprises for Profit," 34 (Nov. 1896), 154–161.

LOCKWOOD, Mrs. Florence Bayard. "The Principle of Volunteer Service," 12 (Dec. 1880), 125–134.

LOEW, Miss Rosalie (N.Y.). "Bar Lawyers' Work among the Poor," 39 (Nov. 1901), 17–23 (Legal Aid Society).

LOGAN, T. M. (general). "The Opposition in the South to the Free-School System," 9 (Jan. 1878), 92–100.

LOGAN, Walter S. (N.Y.). "A Mexican Lawsuit," 33 (Nov. 1895), 134–162.

LOWELL, Mrs. Charles Russell (N.Y.). "Wisdom Is Better Than Weapons of War *Ecclesiastes*," 28 (Oct. 1891), 66–85.

LUMSDEN, Miss Louisa Innes. "On the Higher Education of Women in Great Britain and Ireland," 20 (June 1885), 49–60.

McCOOK, Jno J. (professor, Hartford, Conn.). "Possibilities of Social Amelioration," 32 (Nov. 1894), 160–175.

McCRACKAN, W. D. (N.Y.). "A Trio of Sub-Alpine Scholars: Alessandro Manzoni, Antonio Rosmini, and Antonio Stoppani," 35 (Dec. 1897), 112–119.

McCULLOCH, Oscar C. (Rev., Indianapolis). "General and Special Methods of Operation in the Association of Charities," 12 (Dec. 1880), 91–100.

M'KEEN, James (N.Y.). "A Reply to Mr. Eugene Smith's Paper on 'Retribution in Its Relations to Crime,'" 31 (Jan. 1894), 83–92.

M'KELWAY, St. Clair (of the Brooklyn *Eagle*). "Modern Municipal Reform," 34 (Nov. 1896), 126–139.

——. "Medical and Other Experts," 36 (Dec. 1898), 222–237.

MACY, Jesse (professor, Iowa College). "Practical Instruction in Civics," 32 (Nov. 1894), 151–159.

MARKOE, George F. H. "Legislation in Relation to Pharmacy," 5 (1873), 122–135.

MARTIN, Henry A. "Animal Vaccination," 7 (Sept. 1874), 251–259.

MARTIN, Miss Myra B. (N.Y.). "Art Education in American Life," 30 (Oct. 1892), 12–18.

MAYO, A. D. (Rev., Boston). "National Aid to Education," 17 (May 1883), 3–22.

———. "The Third Estate of the South," 27 (Oct. 1890), xxi–xlii.

MEADE, Edward Sherwood (doctor, University of Pennsylvania). "The Stability of the Gold Standard," 38 (Dec. 1900), 206–211.

MEANS, D. McGregor (N.Y.). "Labor Unions under Democratic Government," 21 (Sept. 1886), 69–78.

MERRILL, Edward B. (N.Y.). "County Jails as Reformatory Institutions," 30 (Oct. 1892), 29–43.

———. "George William Curtis: A Tribute to His Life and Public Service," 31 (Jan. 1894), ix–xxxix.

MILLER, Kelly (professor, Harvard University). "The Education of the Negro," 39 (Nov. 1901), 117–122.

MINOT, William Jr. "Local Taxation and Public Extravagance," 9 (Jan. 1878), 67–77.

MORGAN, T. J. (general; ex-Indian commissioner). "Debate on Higher Education for Negroes," 34 (Nov. 1896), 79–84.

MOSHER, Eliza M. (doctor, Sherborn Reformatory). "The Health of Criminal Women," 16 (Dec. 1882), 46–51, 72.

MOTT, Valentine (M.D., N.Y.). "Rabies and How to Prevent It," 22 (June 1887), 63–74.

MOZOOMDAR, P. C. (Calcutta). "The Religion of India," 18 (May 1884), 209–212.

MUNSON, C. La Rue (Williamsport, Pa.). "The Great Coal Combination and the Reading Leases," 30 (Oct. 1892), 147–163.

MURRAY, Grace Peckham (M.D., N.Y.). "Health Fads of Today," 39 (Nov. 1901), 70–78 (discussion, 79–83).

MYGIND, Holder (M.D., Denmark). "The Care of Deaf-Mutes in Denmark," 36 (Dec. 1898), 257–259.

NELSON, N. O. (St. Louis). "Letter on Profit Sharing," 23 (Nov. 1887), 61–67.

NORTH, S.N.D. (secretary, National Association of Wool Manufacturers, Boston). "Industrial Education in Old and New Boston," 34 (Nov. 1896), 29–42.

———. "Some Fallacies of Industrial Statistics," 34 (Nov. 1896), 140–153.

NOURSE, B. F. "The Silver Question," 9 (Jan. 1878), 21–43.

NOYES, William (M.D.; assistant physician, Bloomingdale Asylum, N.Y.). "The Criminal Type," 24 (Apr. 1888), 31–42.

OLMSTED, Frederick Law. "Public Parks and the Enlargement of Towns," 3 (1871), 1–36.

———. "The Justifying Value of a Public Park," 12 (Dec. 1880), 147–164.
OSBORNE, Thomas M. "The George Junior Republic," 36 (Dec. 1898), 134–138.
PAGE, Charles E. (M.D., Boston). "Are Bacilli the Cause of Disease or a Natural Aid to Its Cure?" 38 (Dec. 1900), 25–29.
PAINE, Robert Treat Jr. (president, Associated Charities of Boston). "The Work of Volunteer Visitors of the Associated Charities among the Poor," 12 (Dec. 1880), 101–116.
———. "Homes for the People," 15 (Feb. 1882), 105–120.
PARKMAN, Mrs. S. "National Education in England," 4 (1871), 150–159.
PARRISH, Samuel L. (N.Y.). "American Expansion Considered as an Historical Evolution," 37 (Dec. 1899), 99–110.
PARTRIDGE, William Ordway. "The Relation of the Drama to Education," 21 (Sept. 1886), 188–206.
PATTERSON, Josiah (Hon., Memphis, Tenn.). "Free Silver Debate," 1895, 33 (Nov. 1895), 54–61, 95–99.
PEABODY, A. P. (professor, Cambridge, Mass.). "The Voting of Women in School Elections," 10 (Dec. 1879), 42–54.
PEARSON, R. A. (Washington, D.C.). "A Proposed Plan to Bring about the Improvement of City Milk Supply," 38 (Dec. 1900), 19–24.
PECKHAM, Grace (M.D., N.Y.). "Influence of City Life on Health and Development," 21 (Sept. 1886), 79–89.
———. "The Nervousness of Americans," 22 (June 1887), 37–49.
———. "Relation of the Physician to the Community, and of the Community to the Physician," 24 (Apr. 1888), 1–11.
PEIRCE, Benjamin. "Ocean Lanes for Steamships," 6 (July 1874), 116–119.
——— (acting president, 1878 meeting). "The National Importance of Social Science in the United States," 12 (Dec. 1880), xii–xxi (read at 1878 meeting).
PERKINS, Charles C. "Art Education in America," 3 (1871), 37–57.
———. "Art Schools," 4 (1871), 95–104.
PETERSON, Frederick (M.D., New York Hospital for Nervous and Epileptic). "Outline of a Plan for an Epileptic Colony," 29 (Aug. 1892), 57–61.
——— (M.D., N.Y.; instructor in Nervous and Mental Diseases, College of Physicians & Surgeons, N.Y.; professor of Nervous Diseases, University of Vermont). "Recent Progress in Surgery and Medicine," 31 (Jan. 1894), l–lx.
PETTUS, Isabella Mary (lecturer, New York University). "The Legal Education of Women," 38 (Dec. 1900), 234–244.
PHILBRICK, John D. "Inspection of Country Schools," 1 (June 1869), 11–23.

PLATT, Johnson T. (professor; New Haven). "The Assertion of Rights," 18 (May 1884), 138–150.

PLATT, Walter B. (M.D., Baltimore). "Certain Injurious Influences of City Life and Their Removal," 24 (Apr. 1888), 24–30.

POPE, Emily F. (M.D.). "The Practice of Medicine by Women in the United States," 14 (Mar. 1881), 173–187.

PORTER, D. G. (Waterbury, Conn.). "English as a Universal Tongue," 32 (Nov. 1894), 117–130.

———. "The Perversion of Funds in the Land Grant Colleges," 35 (Dec. 1897), 77–98.

PORTER, Robert P. (Chicago). "Recent Changes in the West," 9 (May 1880), 46–66.

POST, Louis F. (N.Y.). "Address of Louis F. Post [on the Single Tax]," 27 (Oct. 1890), 48–54 (single tax debate).

POTTER, Edward T. (Newport, R.I.). "Systematized Alternate Ventilation as Applicable to Russian *Étapes* and New York Tenements," 38 (Dec. 1900), 15–18.

POTTS, Joseph D. "The Science of Transportation," 2 (1870), 115–128.

POWELL, George May (Philadelphia). "Profit Sharing, Historically and Theoretically Considered," 23 (Nov. 1887), 47–57.

PURRINGTON, W. A. (New York City). "How Far Can Legislation Aid in Maintaining a Proper Standard of Medical Education?" 25 (Dec. 1888), 28–41 (discussion, 42).

PUTNAM, James J. (doctor). "Gymnastics for Schools," 8 (May 1876), 110–124.

QUINCY, Josiah (mayor of Boston). "Playgrounds, Baths, and Gymnasia," 36 (Dec. 1898), 139–145 (debate, 145–147).

QUINN, Daniel (professor, Catholic University). "The Higher Education in Greece," 32 (Nov. 1894), 131–138.

———. "The Duty of Higher Education in Our Times," 34 (Nov. 1896), 15–28.

RAYMOND, George L. (professor, Princeton University). "The Influence of Art upon Education," 36 (Dec. 1898), 105–127.

———. "The Artistic Versus the Scientific Conception in Educational Methods," 38 (Dec. 1900), 92–106.

RAYNOLDS, Edward V. (Grand Rapids, Mich.). "The Constitution in Its Relation to National Development," 21 (Sept. 1886), 105–112.

——— (New Haven). "The Referendum and Other Forms of Direct Democracy in Switzerland," 33 (Nov. 1895), 213–226.

RICHARD, Ernst (principal of Public Schools of Dolgeville, N.Y.). "Report on Profit-Sharing Plan of Mr. Alfred Dolge," 23 (Nov. 1887), 37–46.

ROBINSON, Harriet Hanson (Malden, Mass.). "The Life of the Early Mill-Girls," 16 (Dec. 1882), 127–140.

ROBINSON, William C. (professor, Yale Law School). "The Diagnostics of Divorce," 14 (Nov. 1881), 136–151.

ROGERS, Henry Wade (president, Northwestern University). "The Acquisition and Government of Territory," 37 (Dec. 1899), 173–192.

ROMERO, ———— (minister of Mexico at Washington). "Address on Mexican Jurisprudence," 33 (Nov. 1895), 163–184.

ROOT, Frederick Stanley (Rev.). "The Educational Features of the Drama," 35 (Dec. 1897), 99–111.

————. "Annual Report of the General Secretary," 36 (Dec. 1898), 56–63.

————. "Annual Report of the General Secretary [1899]," 37 (Dec. 1899), 18–21.

ROPES, Joseph S. "Restoration and Reform of the Currency," 5 (1873), 46–70.

ROSENGARTEN, J. G. "Civil Service Reform," 4 (1871), 33–48.

ROSSE, Irving D. (M.D., Washington, D.C.). "Brief Mention of a Few Ethnic Features of Nervous Disease," 37 (Dec. 1899), 239–244.

ROUND, W.M.F. (secretary, New York Prison Association). "Immigration and Crime," 26 (Feb. 1890), 66–78.

————. "How Far May We Abolish the Prisons?" 35 (Dec. 1897), 197–206.

RUSSELL, Isaac Franklin (professor, New York University Law School). "Can International Disputes Be Judicially Determined?" 36 (Dec. 1898), 187–195.

————. "Why Law Schools Are Crowded," 37 (Dec. 1899), 164–172.

————. "The Domain of the Written Law," 38 (Dec. 1900), 219–226 (discussion, 227–233).

RUSSELL, James E. (dean, Teacher's College, Columbia University). "The Advanced Professional Training of Teachers," 38 (Dec. 1900), 79–91.

SAMUELSON, James (Liverpool, England). "Cooperation in England," 9 (May 1880), 113–116.

SANBORN, Edwin W. (New York City; remote cousin of F. B. Sanborn). "Social Changes in New England in the Past Fifty Years," 38 (Dec. 1900), 147–167.

SANBORN, Franklin Benjamin. "Supervision of Public Charities," 1 (June 1869), 72–87.

————. "The Work of Social Science in the United States," 6 (July 1874), 36–45.

————. "Training-Schools for Nurses," 7 (Sept. 1874), 294–298.

————. "The Work of Social Science, Past and Present," 8 (May 1876), 23–39 (secretary's report, May 1875).

————. "Social Science in Theory and in Practice," 9 (Jan. 1878), 1–13.

———. "[Extracts from] the Report of the General Secretary," 11 (May 1880), vi–x (1879 meeting).

———. "The General Secretary's Report," 11 (May 1880), xi–xiii (annual meeting, 14 Jan. 1880).

———. "Report of the Social Economy Department, for the Year 1879," 11 (May 1880), xxiv–xxvii (read at annual meeting, in Boston, Jan. 1880).

———. "Report of the Department of Social Economy," 11 (May 1880), 86–92 (Sept. 1879 general meeting).

———. "Extended Obituary on Benjamin Peirce," 12 (Dec. 1880), ix–xi.

———. "Report of the General Secretary [1880]," 12 (Dec. 1880), 1–7.

———. "The Three-Fold Aspect of Social Science in America," 14 (Nov. 1881), 26–35.

———. "Report of the General Secretary [1882]," 16 (Dec. 1882), 6–16.

———. "Opening Address of the Chairman of the Department of Social Economy," 16 (Dec. 1882), 98–100.

———. "Annual Report of the General Secretary [1883]," 18 (May 1884), 19–28.

———. "The Commonwealth of Social Science," 19 (Dec. 1884), 1–10.

———. "The Social Sciences: Their Growth and Future," 21 (Sept. 1886), 1–12.

———. "Address of the Chairman [of the Social Economy Department, 1886]," 22 (June 1887), 98–106.

———. "Address of the Chairman of the Department [of Social Economy]," 23 (Nov. 1887), 21–24.

———. "The Opportunities of America [1887 Report of the General Secretary]," 24 (Apr. 1888), 57–62.

———. "Co-operative Building Associations," 25 (Dec. 1888), 112–124.

———. "Report [of the Social Economy Department] on Co-operative Building and Loan Associations," 26 (Feb. 1890), 118–125.

———. "The Work of Twenty-Five Years," 27 (Oct. 1890), xlii–xlix.

———. "Annual Report on Co-operative Building and Loan Associations," 27 (Oct. 1890), liii–lviii.

———. "Syllabus of Four Lectures on the Duties and Opportunities of the Medical Profession," 28 (Oct. 1891), 20–22.

———. "Aids in the Study of Social Science," 29 (Aug. 1892), 49–56 (report of general secretary, 1891).

———. "Social Science in the Nineteenth Century," 30 (Oct. 1892), 1–11.

———. "Socialism and Social Science," 31 (Jan. 1894), xl–xlix.

———. "Phases of Social Economy," 31 (Jan. 1894), 44–52.

———. "Society and Socialism," 33 (Nov. 1895), 20–28.

———. "Report of the General Secretary [1897]," 35 (Dec. 1897), 21–30.

———. "Our Progress in Social Economy since 1874," 35 (Dec. 1897), 50–55.

———. "Co-operative Banking in the United States (1873–1898)," 36 (Dec. 1898), 128–132.

———. "Social Relations in the United States," 37 (Dec. 1899), 69–74 (debate, 75–77).

———. "Past and Present Requirements of Prison Science," 37 (Dec. 1899), 123–131.

———. "Social Changes in the United States in the Half-Century, 1850–1900," 38 (Dec. 1900), 134–146.

———. "Land Owning and Home-Building by the Industrious Classes," 39 (Nov. 1901), 158–168.

SANBORN, Franklin B., and John AYRES. "A Preliminary Report by a Sub-Committee of the Department of Social Economy [on Prison Reform: Advocating Irish System—Date 4 Sept. 1874]," 7 (Sept. 1874), 357–374.

SARGENT, D. A. (professor, Harvard University). "Physical Training in Homes and Training Schools," 18 (May 1884), 44–52.

———. "The Evils of the Professional Tendency of Modern Athletics," 20 (June 1885), 87–90 (abstract only).

SCHLOSS, David F. (London). "The 'Sweating System' in the United Kingdom," 30 (Oct. 1892), 65–72.

SCOVEL, Sylvester F. (president of Wooster University, Wooster, Ohio). "The Value of a Liberal Education Antecedent to the Study of Medicine," 25 (Dec. 1888), 43–53.

SELIGMAN, Edwin R. A. (professor, Columbia College). "Address of Professor Edwin R. A. Seligman [on the Single Tax]," 27 (Oct. 1890), 34–44 (single tax debate).

———. "Remarks of Professor Seligman [on the Single Tax]," 27 (Oct. 1890), 87–98.

SHELDON, Joseph (New Haven). "Free Silver Debate, 1895," 33 (Nov. 1895), 61–76, 99–104.

SHEPARD, Edward M. (Brooklyn, N.Y.). "The Extension of Reform Methods to the Civil Service of States and Cities," 20 (June 1885), 98–131.

SIMCOX, Miss Edith J. (London). "School Board Educational Progress in England," 12 (Dec. 1880), 29–44.

SLOANE, William M. (professor, Columbia University). "The Life and Public Services of the Late President of the Association, Charles Dudley Warner," 39 (Nov. 1901), 1–15.

APPENDIX B

SMITH, Eugene (New York City). "Conflict of State Laws—The Evil and the Remedy," 19 (Dec. 1884), 132–144.

———. "Reformation or Retribution," 31 (Jan. 1894), 71–82 ("A Reply to Mr. Eugene Smith's Paper" by James M'Keen, 83–92).

———. "Edward Livingston and His Criminal Code," 39 (Nov. 1901), 27–38.

SMITH, Goldwin. "University Education," 1 (June 1869), 24–55.

SMITH, Stephen (M.D., N.Y.). "The International Sanitary Conference of Paris," 32 (Nov. 1894), 90–109.

SOUTHARD, C. F. (N.Y.). "The Dangerous Side of Building Associations," 25 (Dec. 1888), 149–151.

SPAHR, Charles B. (doctor, N.Y.). "The Present Status of Silver," 31 (Jan. 1894), 27–40.

SPOFFORD, Ainsworth R. "The Public Libraries of the United States," 2 (1870), 92–114.

SPRATLING, W. P. (M.D.; superintendent of the Craig Colony, Sonyea, N.Y.). "The Epileptic," 35 (Dec. 1897), 125–131.

STERNBERG, George M. (doctor; surgeon general of the United States). "Army Transmission of Yellow Fever by Mosquito," 39 (Nov. 1901), 84–96.

STETSON, George R. (Washington, D.C.). "The Racial Problem," 39 (Nov. 1901), 100–116.

STIMSON, F. J. (Boston). "Democracy and the Laboring Man," 35 (Dec. 1897), 167–196.

STOKES, Anson Phelps. "Free Silver Debate, 1895," 33 (Nov. 1895), 124–126.

STRAUS, Oscar S. "The United States Doctrine of Citizenship and Expatriation," 39 (Nov. 1901), 49–64.

STRONG, William. "The Study of Social Science," 4 (1871), 1–7.

SUMNER, William Graham. "American Finance," 6 (July 1874), 181–189.

TALBOT, Mrs. Emily (Boston). "Report of the Department of Education for the Year 1880–1881," 15 (Feb. 1882), 53–55.

———. "Report of the Secretary [of the Education Department]," 20 (June 1885), 14–26.

———. "Social Science Instruction in Colleges," 22 (June 1887), 7–27.

TALBOT, Mrs. I. T. (secretary of Department of Education). "Report of the Department of Education," 10 (Dec. 1879), 34–42.

TALBOT, W. T. (M.D., Boston). "Summer Camp for Boys," 29 (Aug. 1892), 1–8.

TAYLOR, Henry Ling (M.D.). "American Childhood from a Medical Standpoint," 30 (Oct. 1892), 44–55.

THACHER, Thomas (N.Y.). "The Law's Uncertainty," 23 (Nov. 1887), 125–139.

THIRY, J. H. (Long Island City, N.Y.). "The Early History of School Savings Banks in the United States," 25 (Dec. 1888), 165–177.

THOMAS, W. Cave (London). "The True 'Higher Culture,' or Proportioned Culture," 18 (May 1884), 53–67.

THOMPSON, C. O. "Industrial Drawing," 4 (1871), 105–112.

THURBER, Francis B. (lawyer, N.Y.). "The Right to Combine," 37 (Dec. 1899), 215–226 (debate, 227–228).

TIEDEMAN, Christopher G. (LL.D., New York City). "Suppression of Vice: How Far a Proper and Efficient Function of Popular Government," 38 (Dec. 1900), 245–255.

TOPPAN, Robert Noxon. "International Coinage—A Unit of Eight Grammes," 11 (May 1880), 82–85.

TORREY, Henry W. "Some Topics in Criminal Law," 1 (June 1869), 120–126.

TOURGUENEFF, N. "Economic Results of the Emancipation of Serfs in Russia," 1 (June 1869), 141–149 (letter).

TOWNSEND, John P. (N.Y.). "Savings Banks," 9 (Jan. 1878), 44–66.

———. "Savings Banks in the United States," 25 (Dec. 1888), 100–111.

TRENHOLM, W. L. "The Southern States: Their Social and Industrial History, Condition, and Needs," 9 (Jan. 1878), 78–91.

TYLER, M. F. (New Haven). "The Legal History of the Telephone," 18 (May 1884), 163–177.

UPHAM, J. Baxter. "Music in Public Schools," 4 (1871), 113–122.

VANDERLIP, Frank A. (assistant secretary of the Treasury). "War Financiering," 36 (Dec. 1898), 164–177.

VILLARD, Henry. "Historical Sketch of Social Science," 1 (June 1869), 5–10.

———. "People's Banks of Germany," 1 (June 1869), 127–136.

WADLIN, Horace G. (chief, Massachusetts Bureau of Statistics of Labor). "The Sweating System in Massachusetts," 30 (Oct. 1892), 86–102.

WALKER, Francis A. "Some Results of the Census," 5 (1873), 71–97.

——— (president, MIT). "Industrial Education," 19 (Dec. 1884), 117–131.

WARING, George E. Jr. "The Sewerage of the Small Towns," 10 (Dec. 1879), 180–194.

———. "House-Building with Reference to Plumbing and House Drainage," 18 (May 1884), 112–121.

WARNER, A. J. (Hon., Marietta, Ohio). "Free Silver Debate, 1895," 33 (Nov. 1895), 51–53, 92–95, 108–109.

WARNER, Charles Dudley. "The American Newspaper," 14 (Nov. 1881), 52–70.

———. "The Elmira System," 32 (Nov. 1894), 52–66.

———. "The Education of the Negro," 38 (Dec. 1900), 1–14 (presidential address—read for Warner, then on deathbed).

WASHBURN, Emory (professor of law, Harvard University). "Limitations of Judicial Power," 8 (May 1876), 140–146.

WASHINGTON, Booker T. "Debate on Higher Education for Negroes," 34 (Nov. 1896), 86–88.

WASSON, D. A. "The International," 5 (1873), 109–121.

WATSON, William (professor, Boston). "A Report from the Committee on the Protection of Life from Casualties in the Use of Machinery," 11 (May 1880), 67–81.

WAYLAND, Francis (professor, Yale Law School). "On Certain Defects in Our Method of Making Laws," 14 (Nov. 1881), 1–25 (presidential address).

———. "Opening Address," 16 (Dec. 1882), 1–5 (presidential address, 1882).

———. "Opening Address [1883]," 18 (May 1884), 1–18 (presidential address, on abolition of capital punishment, debate, 135–136).

———. "The Pardoning Power: Where Should It Be Lodged and How Should It Be Exercised?" 19 (Dec. 1884), 145–155.

———. "Incorrigible Criminals," 23 (Nov. 1887), 140–144.

WAYLAND, Heman Lincoln (Rev., Philadelphia; brother of Francis Wayland). "The Progressive Spelling," 17 (May 1883), 117–132.

———. "The Unnamed Third Party," 21 (Sept. 1886), 25–32.

———. "The State and the Savings of the People," 22 (June 1887), 156–161.

———. "Social Science in the Law of Moses," 23 (Nov. 1887), 167–177.

———. "The Dead Hand," 26 (Feb. 1890), 79–88 (discussion, 88–90).

———. "Has the State Abdicated?" 30 (Oct. 1892), v–xviii.

———. "Compulsory Arbitration," 31 (Jan. 1894), lxiii–lxxii.

———. "State Surgery," 32 (Nov. 1894), 82–89.

———. "The Higher Education of the Colored People of the South," 34 (Nov. 1896), 68–78.

WEEKS, Joseph D. (Pittsburg). "Industrial Arbitration and Conciliation," 10 (Dec. 1879), 194–203.

WELCH, Mrs. Margaret H. "Is Newspaper Work Healthful for Women?" 32 (Nov. 1894), 110–116.

WELLS, David A. "Rational Principles of Taxation," 6 (July 1874), 120–133.

———. "Influence of the Production and Distribution of Wealth on Social Development," 8 (May 1876), 1–22 (presidential address, 1875).

WHARTON, Joseph. "International Industrial Competition," 4 (1871), 49–73.

WHITE, Andrew Dickson. "The Relation of National and State Governments to Advanced Education," 7 (Sept. 1874), 299–322.

———. "Instruction in Social Science," 28 (Oct. 1891), 1–22 (presidential address, 1890).

———. "Opening Remarks at the General Meeting of 1891," 28 (Oct. 1891), 23–27 (presidential address, 1891).

WHITE, Miss Flora J. (Concord, Mass.). "The Boers of South Africa in Their Social Relations," 38 (Dec. 1900), 177–188.

WHITE, Horace. "The Tariff Question and Its Relations to the Present Commercial Crisis," 9 (Jan. 1878), 117–131.

WILLARD, Frances E. (Chicago). "Woman and the Temperance Question," 23 (Nov. 1887), 76–84.

WILLCOX, W. F. (professor, Cornell University). "Methods of Determining the Economic Production of Municipal Enterprises," 34 (Nov. 1896), 162–173.

——— (professor, Cornell University; chief statistician in the Census Office). "Negro Criminality," 37 (Dec. 1899), 78–98.

WILLIAMS, S. Wells (professor, New Haven). "Chinese Immigration," 10 (Dec. 1879), 90–123.

WILSON, Charles G. (New York City Health Department). "The New York City Health Department," 29 (Aug. 1892), 9–24.

WILSON, George G. (professor, Brown University). "The Place of Social Philosophy," 32 (Nov. 1894), 139–143.

WINES, F. H. (Rev., Springfield, Ill.). "The Threefold Basis of the Criminal Law," 19 (Dec. 1884), 156–167.

——— (secretary, Illinois Board of Public Charities). "The Law for the Commitment of Lunatics," 20 (June 1885), 61–77.

WINGATE, Charles F. (N.Y.). "The Health of American Cities," 21 (Sept. 1886), 90–99.

WISE, P. M. (president of New York Lunacy Commission). "Physical Health of the Insane," 35 (Dec. 1897), 120–124.

WOOD, Wallace (M.D., N.Y.). "The Science of Dietetics," 22 (June 1887), 85–97.

WOODFORD, A. B. (N.Y.). "Free Silver Debate, 1895," 33 (Nov. 1895), 86–91.

WOODFORD, Arthur Burnham (Ph.D., New Haven). "Twentieth-Century Education," 37 (Dec. 1899), 44–51.

WOODWARD, John (Hon.; Justice, New York Supreme Court). "The Tendency of the Courts to Sustain Special Legislation," 37 (Dec. 1899), 193–208.

WOOLSEY, Theodore Dwight (president, Yale University). "Nature and Sphere of Police Power," 3 (1871), 97–114.

———. "The Exemption of Private Property upon the Sea from Capture," 7 (Sept. 1874), 191–209.

———. "The United States and the Declaration of Paris," 10 (Dec. 1879), 124–135.

———. "The Moral Statistics of the United States," 14 (Nov. 1881), 129–135.

WRIGHT, A. O. "Lunacy Legislation in the Northwest," 20 (June 1885), 78–86.

WRIGHT, Carroll D. "The Factory System as an Element in Civilization," 16 (Dec. 1882), 101–126.

——— (chief, Massachusetts Bureau of Statistics of Labor). "The Scientific Basis of Tariff Legislation," 19 (Dec. 1884), 11–26.

———. "Popular Instruction in Social Science," 22 (June 1887), 28–36 (presidential address).

———. "Problems of the Census," 23 (Nov. 1887), 1–20.

———. "The Growth and Purposes of Bureaus of Statistics of Labor," 25 (Dec. 1888), 1–14 (presidential address, 1888).

WRIGHT, Elizur. "Life Insurance for the Poor," 8 (May 1876), 147–157.

WYMAN, Walter (M.D.; Surgeon General, U.S. Marine Hospital Service). "Suppression of Epidemics," 38 (Dec. 1900), 50–56.

Bibliographical Note

In view of the somewhat Teutonic character of my footnotes, I have not thought it necessary to supply a bibliography. In lieu of a bibliography, I have prepared an author-title index. The reader will find it easy, I believe, to look up either the author or the title of a work in this index, turn to the first footnote specified, and find there the full citation for the work in question. Later footnotes referring to the same work are also listed, if important, so that the reader can trace my references to frequently mentioned sources.

The most important primary sources for this study were the hundreds of articles published in the *Journal of Social Science* from 1869 to 1909, and the personal correspondence of Frank Sanborn and other leading members of the Association. Anyone interested in using the *Journal* will find a useful index-by-author of every article published between 1869 and 1901 in Appendix B of my "Safe Havens for Sound Opinion: The American Social Science Association and the Professionalization of Social Thought in the United States, 1865–1909" (Ph.D. dissertation, Stanford University, 1973; University Microfilm no. 73-14,903).

To the best of my knowledge, after extensive correspondence with scholars and archivists all over the country, there is no major collection of ASSA papers. When Sanborn died his sons burned much of his correspondence, and whatever looked valuable was sold at auction. According to an auction catalogue in the possession of the Concord Free Public Library, some of the items sold were official ASSA documents.

The major manuscript collections consulted by me are listed below. My research was greatly simplified by the kindness of John W. Clarkson, who made available to me his extensive collection of copies of letters written by Sanborn, the originals of which are scattered all over the country. In my footnotes I have referred to this as the Clarkson Collection, and I have indicated in each case where the original letter is kept.

Manuscript Collections

Louis Agassiz Papers, Houghton Library, Harvard University.
Simeon Eben Baldwin Papers, Yale University Library.
George Gordon Battle Papers, Virginia Historical Society.

Caroline Healey Dall Papers, Houghton Library, Harvard University, and Massachusetts Historical Society.
John Huston Finley Papers, Manuscript Division, New York Public Library.
Daniel Coit Gilman Papers, Frieda C. Thies Manuscript Room, M. S. Eisenhower Library, Johns Hopkins University.
Edmund Janes James Papers, University Archives, University of Illinois Library.
Benjamin Peirce Papers, Houghton Library, Harvard University.
Franklin Benjamin Sanborn Papers, Concord Free Public Library; Concord Antiquarian Society (correspondence with Frederick J. Kingsbury); Houghton Library, Harvard University; and American Antiquarian Society, Worcester, Mass.
Henry Villard Papers, Houghton Library, Harvard University.
Charles Dudley Warner Papers, Watkinson Library, Trinity College, Hartford, Conn.
Andrew Dickson White Papers, Collection of Regional History & University Archives, Cornell University.
Carroll D. Wright Papers, Goddard Library, Clark University.
Theodore Dwight Woolsey Papers, Yale University Library.

Author-Title Index

Full bibliographical information for every work cited in the text can be found by turning to the first footnote listed in this index. (E.g., footnote IV-18 is the eighteenth footnote in Chapter IV.) Important footnote references subsequent to the initial one are also listed for the benefit of readers who wish to trace the use of a frequently mentioned work.

AARON, Daniel: *America in Crisis,* I-5; *Men of Good Hope,* IX-25; *Unwritten War,* VI-3
ABRAMS, Philip: *Origins of British Sociology,* I-1; V-17, 29; IX-57; XI-9
ADAMS, Herbert Baxter: *Papers of the American Historical Association,* VIII-10
ADDAMS, Jane: *Second Twenty Years at Hull House,* VI-52
ALLEN, Gay Wilson: *William James,* II-18
American Academy of Political and Social Science: *Handbook,* X-9
American Political Science Association, Sub-committee on Personnel of the Committee on Policy: "The Teaching Personnel in American Political Science Departments," X-57
American Social Science Association: *Address by Samuel Eliot,* V-48; *Constitution, Address, and List of Members,* V-16, 31, 40; VIII-20; *Documents Published by the Association,* V-48; *Handbook for Immigrants,* V-51
ARNOLD, Matthew: *Culture and Anarchy,* IV-3
ASCHER, Abraham: "Professors as Propagandists," VIII-30

BALDWIN, Simeon Eben: "Absolute Power an American Institution," X-18; "Corporal Punishments for Crime," X-20; "Flog the Kidnappers," X-20; "Founding of the American Bar Association," X-24; "Restoration of Whipping as a Punishment for Crime," X-20; "Whipping and Castration as Punishments for Crime," X-20
BARKER, Charles A.: *Henry George,* IX-25, 30
BARNES, Harry Elmer: *Introduction to the History of Sociology,* IX-1, 45
BASTIAT, Frederick: *Harmonies of Political Economy,* XI-8
BEACH, Mark: "Andrew Dickson White as Ex-President," VI-5; "Was There a Scientific Lazzaroni?" IV-15
BEALE, Howard K.: *Charles Beard,* XI-25
BEARD, Charles A.: *Discussion of Human Affairs,* I-27
BECKER, Carl L.: *Cornell University,* IX-5
BEECHER, Henry Ward: "Study of Human Nature," IV-47, 48
BELL, Daniel: "Theory of Mass Society," II-7
BENSON, Lee: *Merchants, Farmers and Railroads,* II-11; V-39; *Turner and Beard,* II-11
BERGER, Peter, Brigitte BERGER, and Hansfried KELLNER: *Homeless Mind,* I-31
BERKHOFER, Robert F., Jr.: *Behavioral Approach to Historical Analysis,* I-26
BERNARD, L. L., and Jessie BERNARD: *Origins of American Sociology,* IV-1; VI-38; VII-8; VIII-19
BERTHOFF, Rowland: "American Social

Order," II-12; *Unsettled People,* II-12
BLASSINGAME, John W.: *Slave Community,* III-23
BLEDSTEIN, Burton J.: "Cultivation and Custom," I-1; III-28; IX-42; *Culture of Professionalism,* IV-3
BLODGETT, Geoffrey: *Gentle Reformers,* IV-2
BLOOMFIELD, Maxwell: "Lawyers and Public Criticism," IV-35
BOLLER, Paul F., Jr.: *American Thought in Transition,* I-1; XI-12
BORDIN, Ruth: *Andrew Dickson White,* IX-4
BOURKE, Paul F.: "Social Critics and the End of American Innocence," I-11
BOUWSMA, William J.: "Lawyers and Early Modern Culture," VII-33
BREMNER, Robert H.: *From the Depths,* XI-35; "'Scientific Philanthropy' 1873–93," XI-35
BROADHURST, Betty P.: "Social Thought, Social Practice and Social Work Education," VII-8
BROOKS, John Graham: "Labor Organizations," X-53
BROWN, Richard: "Knowledge Is Power," II-11
BROWN, Richard M.: "Legal and Behavioral Perspectives on American Vigilantism," X-20
BRUCE, Robert V.: "Democracy and American Scientific Organizations," IV-12; "Universities and the Rise of the Professions," IV-16
BRUNO, Frank: *Trends in Social Work,* V-9; XI-35
BRYSON, Gladys: "Comparable Interests of Old Moral Philosophy and Modern Social Sciences," V-35; "Emergence of Social Sciences from Moral Philosophy," V-35; *Man and Society,* II-5; V-35; "Sociology Considered as Moral Philosophy," V-35
BUCK, Paul: *Social Sciences at Harvard,* II-1; XI-32
BUNGE, Mario: *Causality,* II-15; XI-17, 18
BURGESS, John W.: *Reminiscences of an American Scholar,* II-21
BURROW, J. W.: *Evolution and Society,* II-17, 18
BUTTERFIELD, Herbert: *Origins of Modern Science,* V-33

CALHOUN, Daniel H.: *Professional Lives,* IV-36
CARTER, Paul A.: *Spiritual Crisis of the Gilded Age,* II-17

CHANNING, Walter: "Consideration of the Causes of Insanity," IV-46
CHROUST, Anton-Hermann: *Rise of the Legal Profession in America,* IV-36
CHURCH, Robert L.: "Development of the Social Sciences as Academic Disciplines at Harvard," II-1; VIII-17; "Economists Study Society: Sociology at Harvard," X-53
CLARK, Terry Nichols: *Prophets and Patrons,* II-1
COATS, Alfred W.: "First Two Decades of the AEA," II-1; VIII-25; "Political Economy Club," II-1; VIII-28
COBEN, Stanley, and Lorman RATNER: *Development of an American Culture,* II-12
COHEN, Morris R.: *American Thought,* VI-59; *Chance, Love and Logic,* XI-3
COMMAGER, Henry Steele: *American Mind,* I-3
CONRAD, Robert: *Destruction of Brazilian Slavery,* III-23
CORDASCO, Francesco: *Daniel Coit Gilman and the Protean Ph.D.,* VII-1
CRAVENS, Hamilton: "American Scientists and the Heredity-Environment Controversy," II-1
CRICK, Bernard: *American Science of Politics,* II-1; V-64

DANIELS, George H.: *Darwinism Comes to America,* V-21; "Process of Professionalization in American Science," IV-17; *Nineteenth Century Men of Science,* IV-15
DARWIN, Charles: *Origin of Species,* XI-8
DAVIS, Allen F.: *Spearheads for Reform,* XI-35
DEGLER, Carl N.: *Neither Black nor White,* III-23
DEWEY, John: "Child and the Curriculum," XI-31; *Freedom and Culture,* XI-31; *Public and Its Problems,* XI-28
DONNAN, Elizabeth, and Leo F. STOCK: *Historian's World,* VIII-6
DORFMAN, Joseph: *Economic Mind in American Civilization,* II-1; IV-25
DOSTOEVSKY, Fyodor: *Notes from Underground,* XI-11
DUPREE, A. Hunter: *Science in the Federal Government,* IV-10
DURKHEIM, Emile: *Division of Labor,* II-12; XI-5
DWORKIN, M. S.: *Dewey on Education,* XI-31

EISENSTADT, A. S.: *Craft of American History,* XI-17

ELKINS, Stanley: *Slavery,* II-12; III-23, 24
ELLUL, Jacques: *Technological Society,* XI-18
ELMORE, Andrew E.: "President's Address [National Conference of Charities and Correction]," VI-48
ELY, Richard T.: "American Economic Association," VIII-38; *Ground Under Our Feet,* VIII-24, 32; X-54; XI-1, 29; *Report of the Organization of the American Economic Association,* VIII-25, 40; *Science Economic Discussion,* VIII-38

FAIRCHILD, H. L.: "History of the American Association for the Advancement of Science," IV-10
FLEXNER, Abraham: *Daniel Coit Gilman,* VII-1
FONER, Eric: *Free Soil, Free Labor, Free Men,* II-12
FONER, Laura, and Eugene D. GENOVESE: *Slavery in the New World,* III-23
FOX, Daniel M.: *Discovery of Abundance,* VIII-29, 32
FRANKLIN, Fabian: *Life of Daniel Coit Gilman,* IV-26; VII-4
FREDRICKSON, George M.: *Inner Civil War,* III-18, 26; V-6
FURNER, Mary O.: *Advocacy and Objectivity,* II-1; VII-31; VIII-45, 48; IX-22

GALAMBOS, Louis: "Emerging Organizational Synthesis," II-12
GAWALT, Gerard W.: "Massachusetts Legal Education in Transition," IV-39; "Sources of Anti-Lawyer Sentiment in Massachusetts," IV-2
GEIGER, Roger Lewis: "Development of French Sociology," II-1
GERTH, H. H., and C. W. MILLS: *From Max Weber,* II-18
GETTLEMAN, Marvin E.: "Charity and Social Classes in the U.S.," X-55; XI-35; "John H. Finley and the Academic Origins of Social Work," X-55; "John H. Finley at CCNY," X-55
GIDDINGS, Franklin H.: "Relation of Sociology to Other Scientific Studies," IX-47, 49; "Theory of Sociology," IX-53
GILMAN, Daniel Coit: "Alliance of Universities and Learned Societies," VII-29
GLAZER, Nathan: "Rise of Social Research in Europe," II-17
GOODE, William J.: "Encroachment, Charlatanism, and the Emerging Profession," II-2

GOULDNER, Alvin W.: "Reciprocity and Autonomy in Functional Theory," II-6
GRAY, John H.: "German Economic Association," VIII-30
GREEN, Nicholas St. John: "Proximate and Remote Cause," IV-50
GROB, Gerald N.: *Mental Institutions,* V-1, 45
GROSS, L.: *Symposium on Sociological Theory,* II-6

HADDOW, Anna: *Political Science in American Colleges and Universities,* IV-27
HALLER, John S.: "Civil War Anthropometry," V-5
HANSON, Norwood Russell: *Patterns of Discovery,* I-30
HARDING, Walter: "Franklin B. Sanborn and Thoreau's Letters," III-16
HARRIS, William Torrey: "Method of Study in Social Science," VI-61
HART, Albert Bushnell: "Standardization of the Accounts of Learned Societies," X-8
HASKELL, Thomas L.: "Safe Havens for Sound Opinion," IX-12
HAWKINS, Hugh: *Pioneer,* VII-1; VIII-3
HAYS, Samuel P.: "New Organizational Society," I-9, II-12; *Response to Industrialism,* I-9, II-12
HEIDER, Fritz: "Social Perception and Phenomenal Causality," II-15
HERBST, Jurgen: *German Historical School in American Scholarship,* II-1; V-35; VIII-30; XI-32
HICKOK, Benjamin B.: "Political and Literary Careers of F. B. Sanborn," III-2
HIGHAM, John: *History,* IV-3; VIII-1; *From Boundlessness to Consolidation,* III-28; IV-3, 51; *Writing American History,* I-5; "Reorientation of American Culture in the 1890's," I-5; IV-3
HINCKLE, Roscoe C., and Gisela J. HINCKLE: *Development of Modern Sociology,* XI-22
HOFSTADTER, Richard: *Anti-Intellectualism in American Life,* IV-2; "Manifest Destiny and the Philippines," I-6; "Progressive View of Reality," XI-25; *Social Darwinism,* I-1; II-19; VI-53; XI-12
HOFSTADTER, Richard, and Walter P. METZGER: *Development of Academic Freedom in the United States,* II-1
HOFSTADTER, Richard, and Wilson SMITH: *American Higher Education,* IV-31

HOLLINGER, David: "T. S. Kuhn's Theory of Science," I-32
HOLT, W. Stull: *Historical Scholarship in the United States*, VIII-3; "Idea of Scientific History in America," VIII-21
HOOGENBOOM, Ari: *Outlawing the Spoils*, V-54; VI-15
HOPKINS, Charles H.: *Rise of the Social Gospel in American Protestantism*, VIII-25
HOXIE, R. Gordon: *History of the Faculty of Political Science, Columbia University*, II-1
HUGGINS, Nathan I.: *Protestants Against Poverty*, XI-35
HUGHES, Everett C.: "Professions," II-3
HUGHES, H. Stuart: *Consciousness and Society*, I-15; XI-7, 19
HYMAN, Harold M.: *More Perfect Union*, V-28

ISAJIW, Wsevolod W.: *Causation and Functionalism in Sociology*, II-6
ISRAEL, Jerry: *Building the Organizational Society*, I-9

JACKSON, Frederick H.: *Simeon Eben Baldwin*, X-17
JAMES, Edmund J.: "Anniversary Meeting [of the AEA]," VIII-27
JAMES, William: "Great Men and Their Environment," XI-12; *Will to Believe and Other Essays*, XI-12
JAMESON, John Franklin: "American Historical Association, 1884–1904," VIII-7
JOHNSON, Alexander: *Adventures in Social Welfare*, VI-49; "Appreciation of Frank B. Sanborn," III-14
JOHNSON, Robert Underwood: *Remembered Yesterdays*, X-48
JONES, Robert A.: "John Bascom," VI-62
JOYNER, Fred Bunyan: *David Ames Wells*, VII-2

KAUFMANN, Walter: *Existentialism*, XI-11
KELSO, Robert W.: *History of Public Poor Relief*, V-3
KETT, Joseph F.: *Formation of the American Medical Profession*, IV-36
KINGSBURY, Frederick J.: "Sociological Retrospect," X-14
KIRKLAND, E. C.: *Dream and Thought in the Business Community*, V-39
KNAPP, Martin A.: "Social Effects of Transportation," II-11; "Transportation and Combination," II-11
KOHLSTEDT, Sally Gregory: "Step toward Scientific Self-Identity," IV-9; *Formation of the American Scientific Community*, IV-10

KORNHAUSER, William: *Politics of Mass Society*, II-7
KRISTOL, Irving: "When Virtue Loses All Her Loveliness," XI-34
KUHN, Thomas S.: *Structure of Scientific Revolutions*, I-30; XI-26

LAKATOS, Imre, and Alan MUSGRAVE: *Criticism and the Growth of Knowledge*, I-32
LANE, Ann J.: *Debate over "Slavery,"* III-23
LASCH, Christopher: *New Radicalism in America*, I-7; XI-35
LEE, Joseph: "Argument for Trade Schools," IX-24
LEIDECKER, Kurt F.: *Yankee Teacher*, VI-59; X-1
LERNER, Daniel: *Human Meaning of the Social Sciences*, II-17
LEVERETTE, William E., Jr.: "E. L. Youmans' Crusade for Scientific Autonomy and Respectability," VI-53
LIEBY, James: *Carroll Wright and Labor Reform*, IX-22
LOEB, Jacques: *Mechanistic Conception of Life*, I-19
LUBOVE, Roy: *Professional Altruist*, XI-35
LURIE, Edward: *Louis Agassiz*, IV-10; VI-1; *Nature and the American Mind*, VI-1
LYNN, Kenneth: *Professions in America*, II-3

MACIVER, R. M.: *Social Causation*, II-15
MCKELVEY, Blake: *American Prisons*, V-44; X-15
MACPHERSON, C. B.: *Political Theory of Possessive Individualism*, II-16
MALIN, James C.: *John Brown and the Legend of Fifty-six*, III-5
MANDELBAUM, Maurice: *History, Man and Reason*, I-1
MANN, Arthur: *Yankee Reformers in the Urban Age*, III-4
MANNING, D. J.: *Mind of Jeremy Bentham*, V-38
Massachusetts: *Annual Reports of the Board of State Charities of Massachusetts*, III-21; V-8, 30
MAY, Henry: *End of American Innocence*, I-11; IV-53; X-50; XI-32
MEYER, Donald Harvey: "American Moralists," II-10; IV-31; V-35; XI-8; *Instructed Conscience*, IV-31
MEYERS, Marvin: *Jacksonian Persuasion*, II-12
MICHOTTE, A.: *Perception of Causality*, II-15
MILL, John Stuart: "On the Definition of Political Economy," VI-42
MILLER, Howard S.: *Dollars for Research*,

IV-6; "Science and Private Agencies," IV-5
MILLER, Perry: *American Thought*, II-20; *Errand into the Wilderness*, II-16; *Jonathan Edwards*, II-16; *Life of the Mind in America*, IV-36; "Marrow of Puritan Divinity," II-16
MILLS, C. Wright: "Professional Ideology of Social Pathologists," IX-1; *Sociological Imagination*, II-5; V-84; *Sociology and Pragmatism*, IV-7; XI-3
MILNE, Gordon: *George William Curtis*, III-31
MOONEY, James E.: *John Graham Brooks*, IX-27; X-53
MOORE, Wilbert E.: *Professions: Roles and Rules*, II-4
MURPHEY, Murray G.: *Development of Peirce's Philosophy*, IV-7

NETTL, J. P.: "Ideas, Intellectuals, and Structures of Dissent," VII-31
NEWCOMB, Simon: "Review of R. T. Ely, *Outlines of Economics*," VIII-46
NISBET, Robert A.: *Sociological Tradition*, XI-24
NOBLE, David W.: *Paradox of Progressive Thought*, I-10; *Progressive Mind*, I-10
NORTH, Douglass: *Growth and Welfare in the American Past*, II-11
NORTHCOTT, Clarence H.: "Sociological Theories of Franklin Henry Giddings," IX-46

OATES, Stephen B.: *To Purge This Land with Blood*, III-6
OBERSCHALL, Anthony: *Empirical Social Research in Germany*, II-1; VII-28; VIII-30; IX-39

PARSONS, Talcott: "Professions," II-3; *Structure of Social Action*, I-12; V-29
PATTEN, Simon Nelson: "Facts about the Accounts of Learned Societies," X-8
PEEL, J. D. Y.: *Herbert Spencer*, I-1; VI-57; XI-7
PEIRCE, Benjamin: "National Importance of Social Science in the United States," VII-3
PEIRCE, Charles S.: "Fixation of Belief," XI-3; "How to Make Our Ideas Clear," XI-3
PERRY, Ralph Barton: *Thought and Character of William James*, III-7
PERSONS, Stow: *American Minds*, I-1; IV-31; *Decline of American Gentility*, IV-2; *Free Religion*, IV-4; XI-4
PESSEN, Edward: *Jacksonian America*, IV-34

PFAUTZ, Harold W.: *Charles Booth*, V-29
PHILLIPOVICH, Eugen von: "Verein für Sozialpolitik," VIII-30
PLATT, Anthony M.: *Child Savers*, X-15
POCHMANN, Henry A.: *New England Transcendentalism and St. Louis Hegelianism*, X-1
POLANYI, Karl: *Great Transformation*, II-11
PRED, Allen: *Urban Growth and the Circulation of Information*, II-11
PURCELL, Edward A., Jr.: *Crisis of Democratic Theory*, I-1; XI-32

QUANDT, Jean B.: *From Small Town to Great Community*, I-10; II-11; XI-32

RADER, Benjamin G.: *Academic Mind and Reform*, VIII-24, 38
RAWLE, Francis: "How the [American Bar] Association Was Organized," X-24
RICHARDS, Laura E.: *Letters and Journals of Samuel Gridley Howe*, V-4
RIEFF, P.: *On Intellectuals: Theoretical Studies, Case Studies*, VII-31
REINGOLD, Nathan: *Science in Nineteenth-Century America*, IV-13
RINGENBACH, Paul T.: *Tramps and Reformers*, XI-35
RINGER, Fritz: *Decline of the German Mandarins*, VIII-30, 44
RITT, Lawrence: "Victorian Conscience in Action," V-17, 38
ROGERS, Walter P.: *Andrew D. White and the Modern University*, V-37; IX-6
ROSENBERG, Charles: *Cholera Years*, XI-33
ROSENGARTEN, Joseph G.: "Work of the Philadelphia Social Science Association," V-60
ROSS, Dorothy: *G. Stanley Hall*, I-1
ROSS, Earle Dudley: *Liberal Republican Movement*, VI-15
ROTHMAN, David J.: *Discovery of the Asylum*, III-22; V-13
ROTHSTEIN, William G.: *American Physicians*, IV-36

SANBORN, Franklin B.: "Aids in the Study of Social Science," V-28; "Cooperative Banking in the U.S.," IX-41; "Our Progress in Social Economy Since 1874," IX-41; *Recollections of Seventy Years*, III-3; *A. Bronson Alcott*, III-16; *Dr. S. G. Howe*, III-16; *Hawthorne and His Friends*, III-16; *Henry D. Thoreau*, III-16; *Life and Letters of John Brown*, III-16; *Memoirs of Pliny Earle, M.D.*, III-16; *Ralph Waldo Emerson*, III-16; "Social Relations in the United States," X-13; "So-

cial Sciences: Their Growth and Future," VIII-26; IX-8; "Work of Twenty-five Years," V-16; IX-17
SCHIRMER, Daniel B.: *Republic or Empire: American Resistance to the Philippine War*, X-13
SCHLESINGER, Arthur M., Jr.: *Age of Jackson*, IV-34
SCHLESINGER, Arthur M., Jr., and M. WHITE: *Paths of American Thought*, II-5
SCHNEIDER, Louis: *Scottish Moralists on Human Nature and Society*, II-9; XI-8
SCHNEIDER, Louis, and C. BONJEAN: *Idea of Culture in the Social Sciences*, II-9
SCHUMPETER, J. A.: *History of Economic Analysis*, VIII-30, 44
SCHWARTZ, Harold: *Samuel Gridley Howe*, V-6, 13
SCOVEL, Sylvester F.: "Value of Liberal Education Antecedent to Study of Medicine," X-34
SHILS, Edward: "Contemplation of Society in America," II-5
SILLS, David L.: *International Encyclopedia of the Social Sciences*, II-3
SILVERMAN, Robert, and Mark BEACH: "National University for Upstate New York," IV-23
SMALL, Albion: "Era of Sociology," IX-54; "Fifty Years of Sociology," IX-43; *Origins of Sociology*, II-22; V-35; VIII-30, 44; XI-6, 30
SMITH, Adam: *Inquiry into the Nature and Causes of the Wealth of Nations*, II-8
SMITH, Wilson: *Professors and Public Ethics*, IV-31; V-35
SOFFER, Reba N.: "Revolution in English Social Thought," XI-19
SOMIT, Albert, and Joseph TANENHAUS: *Development of Political Science*, II-1; X-57
SPENCER, Herbert: *Study of Sociology*, IX-57
SPROAT, John G.: *The "Best Men,"* III-32; IV-2
STANWOOD, Edward: "Memoir of Franklin Benjamin Sanborn," III-19
STARK, Werner: *Fundamental Forms of Social Thought*, I-13
STARR, Harris E.: *William Graham Sumner*, II-19
STEIN, Maurice R.: *Eclipse of Community*, II-7
STEVENS, Rosemary: *American Medicine and the Public Interest*, IV-36
STOCKING, George W., Jr.: "Franz Boas and the Founding of the American Anthropological Association," II-1
STORR, Richard J.: *Beginnings of Graduate Education in America*, IV-16, 31

STROUT, Cushing: "Causation and the American Civil War," XI-17
STRUIK, Dirk J.: *Origins of American Science (New England)*, IV-10; V-18
SWANSON, Richard A.: "Edmund J. James," IX-19

TAYLOR, George Rogers: *Transportation Revolution*, II-11
THELEN, David P.: "Rutherford B. Hayes and the Reform Tradition," I-34
THERNSTROM, Stephan: *Poverty and Progress*, V-4; XI-35
THOMAS, John L.: "Romantic Reform in America," III-27
TISHLER, Hace Sorel: *Self Reliance and Social Security*, XI-35
TOCQUEVILLE, Alexis de: *Democracy in America*, IV-28, 29
TOMSICH, John: *Genteel Endeavor*, IV-2
TOPLIN, Robert B.: *Abolition of Slavery in Brazil*, III-23
TYACK, David B.: *George Ticknor*, IV-30

VAN TASSEL, David D.: *Recording America's Past*, VIII-16
VAN TASSEL, D., and M. G. HALL: *Science and Society in the United States*, IV-5
VEBLEN, Thorstein: *Higher Learning in America*, X-58
VEYSEY, Laurence R.: *Emergence of the American University*, II-1; V-64
VIDICH, Arthur J., and Joseph BENSMAN: *Small Town in Mass Society*, II-7

WALLAS, Graham: *Great Society*, XI-27
WARD, John William: "Jacksonian Democratic Thought," II-12; *Red, White and Blue*, 1-28; II-12;
WARD, Lester F.: "Place of Sociology," VI-42; "Sociology and Cosmology," IX-51
WARNER, Amos: "Philanthropology in Educational Institutions," IX-44; *Sociology in Institutions of Learning*, IX-44
WARNER, Charles Dudley: "Elmira System," X-16
WARREN, Austin: "Concord School of Philosophy," X-1
WARREN, Charles: "Inquiry Concerning Vital Statistics of College Graduates," V-37
WATSON, John B.: "Psychology as the Behaviorist Views It," I-19
WATSON, Frank D.: *Charity Organization Movement*, X-55
WAYLAND, Francis: *Thoughts on the Present Collegiate System in the U.S.*, IV-32

WEBER, Max: *Protestant Ethic and Spirit of Capitalism,* XI-20

WEINSTEIN, Allen, and Frank Otto GATELL: *American Negro Slavery,* III-23

WHITE, Andrew D.: *Autobiography,* VIII-17; IX-8; *History of the Warfare of Science with Theology,* VI-6; "Instruction in Social Science," V-39; IX-7, 21; "Opening Remarks," IX-23

WHITE, Andrew Strong: "Institutions Visited by Professor Sanborn and His Class in Social Science," IX-11

WHITE, Morton: *Origin of Dewey's Instrumentalism,* I-21; *Social Thought in America,* I-22; VI-42; XI-32

WHITEHEAD, Alfred North: *Science and the Modern World,* I-35

WIEBE, Robert H.: *Search for Order,* I-9; II-12; V-63; XI-32

WIENER, Martin J.: *Between Two Worlds,* XI-19

WIENER, Philip P.: *Evolution and the Founders of Pragmatism,* IV-4, 50; IX-4

WILSON, R. Jackson: *In Quest of Community,* I-1; III-25, 28; IV-4; VI-7; XI-5, 10, 32

WRIGHT, Carroll D.: "Address of C. D. Wright, President of the American Statistical Association," V-18; "Popular Instruction in Social Science," IX-22; "Study of Statistics in Colleges," IX-22

WYATT-BROWN, Bertram: "Stanley Elkins' *Slavery,*" III-24

YELLOWITZ, Irwin: *Position of the Worker in American Society,* XI-35

YOUMANS, E. L.: "Editor's Table," VI-53; "Harris on Social Science," VI-61

ZIFF, Larzer: *The American 1890's,* X-50

Index

Abbot, Francis E., 66n, 238
Abbott, Lyman, 178–179, 188n
Academic freedom: foundation stone of AEA, 183. *See also* Community of the competent
Academy of Letters: Agassiz proposes, 124
Academy of Moral and Social Sciences: Agassiz proposes, 124
Academy of Sciences (Paris), 135
Activist social sciences. *See* Investigation vs. agitation
Adams, Charles Francis, 124, 128
Adams, Charles Kendall, 170, 171, 172, 174n
Adams, Henry, 157, 227, 237n
Adams, Henry Carter, 157, 177, 182, 188n, 202
Adams, Herbert Baxter: organizes AHA, 169–177; collaborates with Ely in organizing AEA, 182, 185; mentioned, 25, 145
Addams, Jane, 191n, 203
Advocacy and objectivity. *See* Investigation vs. agitation
Agassiz, Louis: and founding of AAAS, 69–74; plan for reform of Harvard, 123–124; proposes Academies of Letters and Moral and Social Sciences, 124–125; and founding of Cornell, 125; idealist predilections of, 125–126; revives ASSA, 129; on ASSA's lack of authority, 134–135; social attitudes of, 147–148; mentioned, 45, 75, 99, 246, 251n
Alcott, Bronson, 53, 212
Aldrich, Thomas Bailey, 227
American Academy of Arts and Letters: sponsored by NIAL, 227; mentioned, 86
American Academy of Arts and Sciences, 131
American Academy of Political and Social Science: absorbs Philadelphia branch of ASSA at founding, 215; beats ASSA at own game, 215; merger with ASSA considered, 215–216; finances challenged, 216n; mentioned, 118, 182. *See also* Edmund J. James
American Association for the Advancement of Science (AAAS): founding of, 68–72; merger with Johns Hopkins proposed by B. Peirce, 149–150; social science in, 214; mentioned, 111n, 123, 131, 134, 152, 170–171
American Association for the Promotion of Social Science (ASSA), 100
American Bar Association: founded by Baldwin, 221; mentioned, 86
American Economic Association (AEA): organizes at joint meeting of ASSA and AHA, 177; shares much common ground with ASSA, 179, 182–183; Ely's platform contrasted with James's and Patten's, 181; Ely's platform not representative of founders' opinions, 181–182, 185–188; professional motives of founders, 183–189; modified platform adopted, 185n; platform formally repudiated, 186–187; superficiality of commitment to reform, 187n; first officers of, 188n; mentioned, 54, 99, 145, 200, 231, 236n
American Historical Association (AHA): founding of at Gilman's suggestion, 145, 168–177; organized under auspices of ASSA, 172; little overlap of membership with ASSA, 173; contrasted to ASSA, 173–177; first officers, 174n; compared to AEA, 178; mentioned, 226
American Institute of Christian Sociology, 187
American Journal of Science, 70
American Journal of Sociology, 203, 239, 253n
American Oriental Society, 170, 171n

American Philological Association, 152, 170

American Political Science Association: founding of is death blow to ASSA, 230–231; mentioned, 145, 201

American Public Health Association, 86, 221

American Social Science Association (ASSA): idealist bias of, 2–3, 206; forum for anti-Spencerian social science, 7; style of inquiry, 24, 110–115; publishes annual *Journal of Social Science*, 54, 116, 127, 131, 141, 150n, 158, 159, 228, 234; dominated by small circle, 56; reputation as center of humanitarian reform sentiment misleading, 63; "headquarters" of movement to establish authority, 63–64, 86; conception of social science, 78, 86–88, 100–110, 164–167, 175, 204–206; departmental structure, 87, 100, 104–110, 167, 176n, 195–197, 240; founding, 97–100, 110–111; founders' assumptions, 100–110; membership, 100, 109, 118, 133–134, 173, 225, 229; publishes *Handbook for Immigrants*, 115–116; and civil service reform, 115–121; finances, 118, 130–132, 155n, 214; branch associations, 118, 132–134; disrupted by failure of Liberal Republican movement, 126–130; dissolution proposed, 127–130; maintains office in Boston, 127, 130, 158, 164; Sanborn appointed permanent secretary, 129–130; women on governing board become issue, 129; Agassiz questions credibility of, 134–135; sponsors formation of National Conference of Charities and Correction, 136–138; merger with Johns Hopkins proposed by B. Peirce, 144, 149–150; merger negotiations between Peirce, Gilman, and Sanborn, 151–156; merger rejected, 156; constitutional reforms adopted under leadership of Gilman and Wayland, 158; significance of Gilman's rejection of merger, 160–167; sponsors organization of AHA, 172; indirectly sponsors organization of AEA, 177; "mother of associations," 177, 221, 232; threatened by rise of academic sociology, 190; model curriculum project based on Sanborn's Cornell course, 195–197; drifts toward popularization, 198–199; Single Tax Debate (1890), 199–200; Free Silver Debate (1895), 200–201; cooperative banking a major concern of, 202; criticized by A. W. Small and A. G. Warner, 203; F. H. Giddings declares social science obsolete, 203–206; Sanborn admits limitations of, 213; merger with AAAS proposed, 214; merger with American Academy of Political and Social Science considered, 215–216; growing conservatism of, 216–220; professionalism discussed at 1888 meeting, 222–223; sponsors organization of NIAL, 224–228; death blow is founding of American Political Science Association, 230–231; sponsors organization of National Institute of Social Sciences, 232–233; failure virtually inescapable, 234

—Department of Education, 104–110, 114, 126, 127, 196, 223n
—Department of Education and Art, 224, 226
—Department of Finance, 104–110, 114, 127, 136, 182, 200, 229
—Department of Health, 82, 104–110, 114, 126, 127, 221, 223, 229
—Department of History (proposed), 168–169
—Department of Jurisprudence, 104–110, 114, 221
—Department of Social Economy, 105, 106n, 108, 136–137, 179, 182, 196, 197n, 201, 217, 229

American Sociological Association. *See* American Sociological Society

American Sociological Society: founding of, 231; mentioned, 145

American Statistical Association, 111n

American Statistical Society, 98

Amherst College, 152

Anderson, Joseph, 216, 223, 226

Andrew, John Albion, 52, 92, 93, 96, 98, 123

Andrews, E. Benjamin, 185, 200

Angell, James Burrill, 76, 148, 232

Annals of the American Academy of Political and Social Science, 215

Anti-formalism, 9–14, 254

Anti-institutionalism, 56–62, 85

Association for the Protection of the Insane and the Prevention of Insanity, 221n

Association of American Geologists, 68

Association of American Geologists and Naturalists, 68

Atkinson, Edward, 99, 114, 163n, 177n, 184, 200, 214n

Atkinson, William P., 111, 112

Atlantic Monthly, 129n, 212

INDEX

Auburn State Asylum for Insane Criminals, 194
Auburn State Prison, 194
Austin, John, 10
Authority. *See* Movement to establish authority
Autonomy. *See* Interdependence; Causal attribution; Individualism; Voluntarism

Bache, Alexander Dallas, 70–72, 93
Bacon, Francis, 106
Bacon, Robert, 226
Bacon, Theodore, 222
Bailey, J. Warren, 202
Bain, Alexander, 83
Baker, William Emerson, 100
Baldwin, Simeon Eben: ASSA's most conservative president, 216, 219–220; organizes NIAL, 222–228; writes flexible federal charter for ASSA, 226–227, 232; recommends that political scientists organize within ASSA, 230; mentioned, 86, 232
Baltimore, 149, 152, 157
Bancroft, George, 156
Barnard, Henry, 114
Barnard, James M., 126–127
Barnes, Harry Elmer, 191n
Barzun, Jacques, 227n
Bascom, John, 114
Bastiat, Frederick, 244n
Beard, Charles: on history as seamless web, 248; mentioned, 10
Beecher, Henry Ward, 82–84, 198n
Bellamy, Edward, 199
Belmont, August, 226
Bemis, Edward W., 177, 179, 201
Bentham, Jeremy, 10
Bird, Francis W., 52, 129n
Bird Club, 52, 54, 72, 92, 129n
Blatchford, J. S., 127
"Block universe," 14, 141
Blodgett, Lorin, 118
Board of Charities. *See* Massachusetts Board of State Charities
Board of Health, New York City, 114
Booth, Charles, 103n
Borman, Frank, 232
Boston, 92, 103, 111, 115, 129
Boston *Advertiser*, 212
Boston University, 126n, 222
Boutwell, George S., 99, 104n
Bowen, Clarence W., 173n, 174n
Bowen, Francis, 32
Bowker, Richard R., 226
Bowles, Samuel, 54, 129
Brace, Charles Loring, 112–113

Bradford, Gamaliel, 149, 153
Bridgman, Laura, 96
British Association for the Advancement of Science, 69
Brockway, Zebulon: defended against charges of cruelty, 218–219; mentioned, 194
Brooks, John Graham: organizes ASSA Single Tax Debate, 200; mentioned, 203, 228–229
Brougham, Lord Henry, 98
Brown, John, 49–50, 53, 195n, 212
Browne, Hugh M., 201n
Brown University, 76, 185
Bruguiere, Emile, 226
Burgess, John W.: on research, 46–47; mentioned, 171, 173n, 195n
Burlingame, Anson, 52
Businessmen, merchants: role of in early social science, 109–111

Cambridge, Mass., 152, 171
Carey, Henry C., 112
Carpenter, Mary, 112
Carrett, James R., 200
Causal attribution: influenced by interdependence, 13, 17, 29n, 39–47, 208, 241–249; and common sense, 41–42; central to professional role, 81; superficiality of, on part of Charity Board members, 95–97; ASSA's presumption of individual causal potency obscures environmental causes, 165; sociology identified with remote causal attribution by Giddings, 205–208; changing habits of underlie new humanitarianism of 1890's, 241–242, 250–256; positivists carry to radical extreme, 242–245; conventional character of, 243, 248–249, 255n; based on "cycles of operation" (W. James), 247; A. W. Small on new style of, 253; Dewey on new style of, 253. *See also* Recession of causation; Interdependence
Chandler, Mrs. Norman, 232
Channing, Ellery, 53
Channing, Walter, 82, 84
Charities Review: A Journal of Practical Sociology, 229
Charity. *See* Massachusetts Board of State Charities
Charity Organization movement, 138n
Chautauqua meetings, 187
Chicago, 116
Child study, 201–202
Cincinnati Prison Congress (1870), 129
City College of New York, 229

Civil Service Commission, 128
Civil service reform: aspect of movement to establish authority, 91, 119–121; launched by ASSA, 115–121; mentioned 86, 112
Clark, Henry G., 111
Clark, John Bates, 188n, 200
Clarke, James Freeman, 52
Clarke, Samuel B., 200
Clemens, Samuel, 219, 227
Columbia University, 104, 200, 221
Community of inquiry. *See* Community of the competent
Community of the competent: essential functions of, 65–68, 73; AAAS and NAS as prototypical approximations to, 68–74; relation to tyranny of the majority, 75n; neutrality of, 88–90; Gilman recognized necessity for, 162; presupposes consensus about the criteria of competence, 163; Winsor proposes for historians, 176; underlies Historical and Economic Associations but not ASSA, 176, 178; academic freedom expresses demand for autonomy of, 183; epistemological necessity of discussed by C. S. Peirce, 237–239; mentioned, 80, 235–240
Comte, Auguste, 195, 204
Concord, Mass., 52–59 *passim*
Concord Summer School of Philosophy: organized by Sanborn, 212; mentioned, 196
Congress of Provident Institutions (Paris), 202
Conrad, Johannes, 181
"Consensus of the competent," 238. *See also* Community of the competent
Convergence of idealism and positivism. *See* Idealism; Positivism
Conway, Moncure, 128
Cooley, Thomas M., 156, 157
Copyright laws, 226
Cornell, Ezra, 192
Cornell University: Agassiz compares to Harvard, 125; Sanborn teaches course in social science 1885–88, 193–195; mentioned, 86, 148, 172, 204
Crawford, Marion F., 225n
Crisis of professional authority, 77–85. *See also* Movement to establish authority
Cultural organicism, 10–12
Culture: concept of, 143
Curtis, George William: on organization, 60; mentioned, 117, 119, 128, 146n
Curtis, H. Holbrook: organizes, 1888 ASSA session on professionalism, 222; organizes NIAL, 224–228; organizes National Institute of Social Sciences, 232
Custodians of culture. *See* Gentry class
"Cycles of operation," 247–249

Dall, Caroline Healey, 99, 100, 113, 231
Dalton, Edward B., 114
Damrosch, Walter, 225n
Dana, James Dwight, 70
Dartmouth College, 152
Darwin, Charles: theory of evolution encourages positivistic style of explanation, 243–244; mentioned, 8, 44, 201, 248
Davidson, Thomas, 200
Davison, Henry P., 232n
Deane, Charles, 173, 174n
De Koven, Reginald, 225n
Delmar, Alexander, 114
Devine, E. T., 191n
Dewey, John: and anti-formalism, 9–10, 12; and interdependence, 14–16, 253; on implausibility of autonomy, 253; mentioned, 86, 250
Division of labor in social science. *See* American Social Science Association—departmental structure
Dostoevsky, Fyodor, 245
Dugdale, Richard L., 114
Dunbar, Charles, 184, 187
Durkheim, Emile: on interdependence, 36n, 38n; on death of dilettantism, 238n; mentioned, 4, 16n
Dwight, T. W., 221

Eaton, John: urges historians to organize within ASSA, 172, 175; mentioned, 86, 198n
Edwards, Jonathan, 42
Eliot, Charles William, 73, 86, 125, 148, 192n, 232
Eliot, Samuel, 115, 116, 118n
Elkins, Stanley: on "transcendentalist as abolitionist," 56–62
Elliott, A. Marshall, 171
Elmira State Reformatory (N.Y.), 194, 218
Elmore, Andrew E., 137–138
Ely, Richard T.: organizes AEA, 178–189; outmaneuvers James and Patten, 181n; on interdependence of society, 253; mentioned, 25, 136, 145, 157, 161, 163n, 229, 236n, 237n
Emerson, George B., 100
Emerson, Ralph Waldo: invites Sanborn to teach Concord school, 52; mentioned, 49, 52, 58, 59, 124, 126, 171n, 212
Emerton, Ephraim, 171, 174n

Falkner, Roland P., 215
Farnam, Henry W., 155
Finley, John Huston, 228-232
Floyd, Silas X., 201n
Forbes, John M., 52
Formalism: signifies failure to recognize interdependence, 12; mentioned, 9-14, 43, 254
Frazer, John F., 70
Free Religious Association, 213
Functionalism, 253-254

Gallaudet, Edward M., 113
Garland, Hamlin, 226
Garrison, Wendell Phillips, 118
Garrison, William Lloyd, 99, 115
Garrison, William Lloyd II, 200
"Genteel tradition," 52
Gentry class: social base of movement to establish authority, 63; composed largely of professional men, 63n; loses touch with movement to establish authority after Gilman's abandonment of ASSA, 165-167; mentioned, 90, 91, 108, 162, 174, 211-212, 228
George, Henry: and ASSA Single Tax Debate, 199-201
German Historical School of Economics, 107
Gheel, Belgium, 97n
Gibbs, Oliver Wolcott, 70, 93
Giddings, Franklin H.: announces displacement of social science by sociology, 203-208; mentioned, 215, 251n
Gillin, J. L., 191n
Gilman, Daniel Coit: and founding of ASSA, 99, on reform of ASSA, 133; career of, 148; responds to Peirce's merger proposal, 152-160; distinguishes between agitation and investigation, 154-156, 158, 160-167, 191; tries to invigorate ASSA, 157-159; reasons for rejecting merger of ASSA and Johns Hopkins, 160-167; recommends that ASSA form History Department, 168-169; encourages professionalizing activities at Hopkins, 168-172, 182; mentioned, 74, 75, 86, 122, 123, 184n, 192
Gladden, Washington, 179, 188n
Godkin, E. L., 86, 114, 117, 128, 129
Gompers, Samuel, 201
Gould, Benjamin A., 70, 72, 93
Grant, Ulysses S., 128, 132
Greeley, Horace, 128-129, 198
Green, Nicholas St. John, 84-85
Gumplowicz, Ludwig, 177

Hadley, Arthur Twining, 184, 186
Hall, G. Stanley, 157
Hammond, W. G., 221
Hampton Falls, N.H., 51
Harpers Ferry raid, 50
Harris, William Torrey: attacked positivism, 141-143; and ASSA's model social science curriculum, 196; participates in ASSA Free Silver Debate, 200; child study project, 201-202; mentioned, 86, 164, 212
Hart, Albert Bushnell, 216n
Hartley, Robert M., 113
Harvard College and University: Ticknor proposes reforms, 75-76, 86; criticized by Agassiz, 123-124; Agassiz's plans for reform of, 124; mentioned, 49, 51, 73-75, 84, 104, 121n, 125, 126, 132, 148, 152, 171, 187, 192n, 216n, 221
Hawthorne, Julian, 53
Hay, John, 227
Haymarket incident: Sanborn's reaction to, 217
Hecker, Isaac, 60
Henry, Joseph, 66, 68-74, 77, 176
Hesburgh, Theodore M., 233n
Higginson, Thomas Wentworth, 49, 52, 99, 100, 231
Hill, Thomas. 99, 104, 111-112, 126
Historicism, 10-12
Hoar, Ebenezer, 53
Hobbes, Thomas, 41
Holmes, Oliver Wendell, Sr., 75, 124
Holmes, Oliver Wendell, Jr., 10
Howe, Samuel Gridley: and Massachusetts Board of Charities, 92-97; and founding of ASSA, 99, 104, 110; mentioned, 45, 49, 52, 55, 56, 113, 114, 124, 164, 203, 212, 254
Howells, William Dean, 129n, 226
Hoyt, John Gibson, 51
Hoyt, John W., 116, 132-133
Hughes, H. Stuart, 4-9
Humanitarianism: new style exemplified by attack on Brockway, 218-219; new style reflects remote causal attribution, 241-242, 250-256
Hume, David, 10

Idealism: of ASSA, 2-3, 140-143; converges with positivism in 1890's, 4-8, 140-143, 207-208, 249-251; growing implausibility of in social thought, 8; Concord Summer School of Philosophy, 212; mentioned, 10, 126, 206, 207, 242. *See also* Positivism; Voluntarism

Independent, 172
Individualism, 5, 13, 23, 42, 56–62, 85, 95–97, 164–165, 180–181, 235, 251, 255–256. *See also* Voluntarism; American Social Science Association—style of inquiry
Interdependence: presupposed by antiformalists, 12–14; influences causal attribution, 13, 17, 29n, 39–47, 243; recognition of by professional social scientists, 15, 252–253; serves purposes similar to Kuhnian paradigm, 20–21; defined, 28–29; and power, 29n; recognized by Adam Smith, 30–32; and market discipline, 30–33; and transportation revolution, 33–36; and specialization, 36–37; and declining sovereignty of "island communities," 37–38; causes recession of causation, 39–42; devitalizes personal milieux, 40; creates shortage of independent variables, 40–41; invites uniformitarian perception of society, 43; supplies paradigm of social action, 43; undermines traditional belief systems, 43–47; generates demand for expertise in human affairs, 43–47; disrupts authority of traditional professions, 81–85; merchants first to experience, 109–110; supplies common ground for convergence of idealism and positivism, 141–143; undermines ASSA style of inquiry, 167; rise of sociology signifies growing recognition of, 207–210; epitomized in Spencer's iron plate analogy, 209–210; alters conditions of adequate explanation, 235–249; contributes to breakdown of conventions of causal attribution, 243; limitations of stressed by W. James, 247–248; Wallas on, 252; Dewey on, 252; Ely on, 253; Small on, 253. *See also* Recession of causation; Causal attribution
Investigation *vs.* agitation: ASSA's founders do not distinguish, 100–101, 160–161; shortage of facts obscures distinction, 102–104; Gilman draws distinction in response to Peirce's merger proposal, 154; Sanborn plays down distinction, 155–156; increasingly sharp distinction in late 19th century attributable to altered conditions of explanation, 162–163; purpose of Gilman's distinction not to condemn agitation, 163; conservatives often endorsed advocative style, 163n; formation of AHA signals new division of labor geared to investigators, 175–176; relative priority controversial in early years of AEA, 178; AEA fundamentally oriented toward investigation, 187–188; early sociology nearly succeeds in combining the two functions, 191
Investigatory commission: idea of, 91, 103

Jackson, Barbara Ward, 233n
James, Edmund Janes: on ASSA, 179; plan for organization of economists, 180–181; and ASSA's model social science curriculum, 196; participates in ASSA Single Tax Debate, 200; organizes American Academy of Political and Social Science, 215; mentioned, 136, 177, 188, 202, 214n
James, Henry, Sr., 45, 53
James, Henry, Jr., 227
James, Robertson, 50
James, Wilkinson, 50
James, William: declines membership in American Academy of Arts and Letters, 228; defends "great man" theory of history against Spencer, 246–249; on "cycles of operation," 247–249; mentioned, 45, 50, 84, 140–141, 212, 229
Jameson, John Franklin, 25, 169, 170, 174n, 175n, 237n
Jarvis, Edward, 98, 99, 100, 104n, 111, 113
Jenckes, Thomas A., 116–117
Jenks, Jeremiah W., 25, 200, 202, 230
Johns Hopkins University: merger with ASSA contemplated, 144–167 *passim;* original staff weak in social science, 156–157; intense organizing fever in 1880's cultivated by Gilman, 169–172; mentioned, 74, 86, 192
Johnson, Alexander, 138n
Johnson, Emory R., 233n
Journal of Social Science. See American Social Science Association
Journal of Speculative Philosophy, 141, 196. *See also* Harris, W. T.
Juvenile delinquency, 113

Kelley, Florence, 201, 203
Kellogg, Paul V., 191n
Kingsbury, Frederick J., 216, 218
Knies, Karl, 177
Knox College, 229
Kuhn, Thomas S., 18–23, 252

Labor question, 201
Laissez-faire: crusade against by AEA, 178–185 *passim;* mentioned, 220, 223

Langdell, Christopher C., 221
Langhorne, Orra, 228
Laughlin, James Laurence, 163n, 184
Lawrence Scientific School (Harvard), 73, 74, 146, 246
Lazzaroni: vanguard of movement to establish authority in natural science, 71–74; control AAAS in early years, 71; prominent members in U.S. Sanitary Commission, 93; Agassiz and Peirce prominent in reinvigoration of ASSA, 129; mentioned, 90, 122, 125, 134, 146, 176
Lea, Henry C., 117–118
Lee, Joseph, 198
Liberal Republican movement: failure of disrupts ASSA, 126–130; mentioned, 54
Lieber, Francis, 70, 99, 104, 114, 192
Lincoln, David F., 126, 127
Lindsay, Samuel M., 191n
Loeb, Jacques, 8
Lounsbury, Thomas R., 227
Low, Seth, 232
Lowell Institute, 111, 116n

MacIver, Robert, 191n
McKelway, St. Clair, 226
McKim, Charles Follen, 225
McKim, James Miller, 118
Mandeville, Bernard, 243
Mann, Horace, 53, 96n
Market discipline, 30–33, 78–79
Marshall, Alfred, 4
Marx, Karl, 249n, 250
Massachusetts:
—Board of Alien Commissioners, 92
—Board of Commissioners of Prisons, 94
—Board of Education, 94
—Board of Health, 94
—Board of Health, Lunacy and Charity, 94
—Board of State Charities: origins, 91–97; model for other states, 92, 94; theoretical assumptions of founders, 95–97; sponsored founding of ASSA, 97–98; mentioned, 55, 212
Massachusetts Historical Society, 173
Massachusetts Institute of Technology, 99, 111, 172
Mass society theory, 30n
Menger, Carl, 187n
Merton, Robert K., 251n
Metaphysical Club, 66n, 84
Methodenstreit, 186n

Mill, John Stuart, 10, 136
Mitchell, Maria, 126n
Modern Language Association, 145, 171
Moral philosophy, 107, 124, 125
Moran, Charles, 114
Morgan, T. J., 201n
Morrill Tariff, 112
Movement to establish authority: named by John Higham, 63–64; Lazzaroni in vanguard of, 70; modernization of higher education an important phase of, 74–77; classic professions contribute driving force to, 77–78; premised on science of society, 87–88; transcends own narrow motives, 89; expressed in investigatory commissions, 91; expressed in civil service reform, 91, 119, 121; enters new phase with breakdown of traditional division of professional labor, 144, 166–167; founded on premise that sound opinion required institutional support, 162; divorced from gentry class that launched it, 166–167; mentioned, 147, 148, 164
Mugwump movement, 54, 219

Nation, 86, 114, 118, 129, 172
National Academy of Science (NAS), 71–72, 123, 152
National Association for Promotion of Social Science (Great Britain): criticized by A. G. Warner, 203; mentioned, 98, 106, 108–109, 136, 222
National Conference of Charities and Correction: ASSA sponsors founding of, 135–138; meets separately from ASSA after 1878, 138; mentioned, 55, 86, 212
National Consumer's League, 229
National Institute for the Promotion of Science, 68n
National Institute of Arts and Letters (NIAL): fulfills Agassiz's dream, 124; founding of, 224–228; modelled on Institute of France, 224; secret preparations for, 224; conflict with ASSA, 227; sponsors American Academy of Arts and Letters, 227–228; mentioned, 86, 211
National Institute of Social Sciences: founding sponsored by ASSA, 232–233; mentioned, 211
National League of Building and Loan Associations, 202
National Prison Association, 221n
National University, 133n
Newcomb, Simon, 184, 186, 187n

New Englander, 212
New York, 92, 117, 171, 173n
New York Board of Charities, 218
New York *Times,* 229
North American Review, 141
Norton, Charles Eliot, 128, 227

Organicism, 32. See also Cultural organicism
Oriental Society, 152

Pace, Frank, Jr., 233n
Palmer, A. B., 99, 100, 111, 112
Paradigm, 19–22
Pareto, Vilfredo, 4, 16n
Parker, Theodore, 3, 49, 52, 56
Parsons, Talcott, 4–9, 10, 15, 16n, 103n, 251n, 256
Patten, Simon Nelson: plan for organization of economists, 180–181; mentioned, 216
Pavlov, Ivan, 8
Pearson, Lester B., 233n
Peirce, Benjamin: and founding of AAAS, 69–74; and Lawrence Scientific School, 73; idealist predilections of, 125–126; revives ASSA, 129; on reform of ASSA, 134; address on social science, 146–147; social attitudes of, 147–148; proposes merger of ASSA with Johns Hopkins, 149–160; conceived of ASSA as university for the people, 151; and Concord School of Philosophy, 212; mentioned, 68, 74, 75, 122, 123
Peirce, Charles S.: communal theory of truth, 237–240; mentioned, 67–68, 72, 84, 246
Peirce, James Mills, 126, 149
Pendleton Civil Service Act (1883), 64
Penn Monthly Magazine, 118
Perkins, James A., 233n
Perry, Arthur L., 113, 244n
Philadelphia, 117
Philadelphia Social Science Association: founded, 118–119; absorbed by American Academy of Political and Social Science, 215
Philanthropology, 203n
Phillips, Wendell, 99, 100, 203
Phillips Academy, 51
Political economy: contrasted with social economy, 136–137
Political Economy Club, 179–180
Popular Science Monthly, 83, 139, 238n
Porter, Noah, 46, 148, 212
Positivism: role in late 19th century thought, 2n; converges with idealism in 1890's, 4–8, 140–143, 207–208, 249–251; revolt against, 6–8, 250; Youmans's criticism of ASSA, 139–140; strength rooted in social change rather than emulation of natural science, 242–243; carries remote causal attribution to radical extreme, 242–245; unpalatable moral implications, 245; challenged by W. James, 246–249; all but most extreme versions accepted by generation of 1890's, 250–251; mentioned, 10, 13, 126, 206, 207–208. See also Idealism; Causal attribution
Potter, Alonzo, 32
Princeton University, 229
Prison reform, 112, 218–219, 220
Professional: defined, 27–28
Professionalization: T. S. Kuhn's conception of, 18–19, 21–22; defined, 19; characterized, 24–27; not merely a status-seeking strategy in late 19th century, 65; comes to focus in Gilman's life 1878–84, 146; had not proceeded far enough to render ASSA obsolete in 1880, 160; mentioned, 220–224, 240
Professional role, 81, 88, 166–167, 223
Professions: constitute social base of gentry class and movement to establish authority, 63, 77–78; breakdown of classical division of labor, 110
Progressive party, 139n

Quincy, Josiah, 114, 202

Race relations, 201
Ray, Isaac, 112
Recession of causation: principal consequence of interdependence, 39–42; disrupted traditional professional role, 82; recognized by professionals Channing, Beecher, and Green, 82–85; epitomized by Spencer's denial of great men in history, 246; mentioned, 256n. See also Interdependence; Causal attribution
Redpath, James, 99
Reformist social science. See Investigation *vs.* agitation
Relativism, 254
Relief of unemployed, 201
Republican party, 92, 123, 127–129
Research: skeptical implications of, 45–47
Richmond, Mary E., 191n
Riis, Jacob, 3
Robinson, James Harvey, 10, 215
Robinson, William S., 52
Rogers, William Barton: and founding of AAAS, 68–73; first president of ASSA, 99; mentioned, 115

Romanticism, 52, 56, 85
Roosevelt, Theodore, 219, 227, 229
Rosengarten, Joseph G., 118
Ross, Edward A., 25, 214n
Royal Statistical Society (London), 98n
Ruggles, Samuel B., 71
Russell, Robert, 225

St.-Gaudens, Augustus, 225
Sanborn, Charles, 51
Sanborn, Franklin B.: and John Brown's raid, 49–50; and Genteel Tradition, 52; personality, 54; on teaching profession, 59; and founding of ASSA, 97–100; and Department of Social Economy, 106n; on prison reform, 112; becomes editor of Springfield *Republican,* 115; on civil service reform, 119; on Grant administration, 128–129, 132; made permanent secretary of ASSA, 129; salary, 131; on need for local branches, 133–134; links ASSA with National Conference of Charities and Correction, 135–138; on social economy, 137; wished to resign, 149; interprets Peirce's merger proposal democratically, 151; discusses Hopkins merger proposal with Gilman, 153–156; on AEA, 179; teaches social science at Cornell, 193–195; cooperative banking his pet reform, 202; organizes Concord School of Philosophy, 212; admits failure of ASSA, 213; favors merger with American Academy of Political and Social Science, 216; growing conservatism of, 217–219; on professional ethics, 222–223; retires from ASSA Department of Social Economy, 229; unable to attend last ASSA meeting, 231; mentioned, 45, 47, 85, 87, 92, 103n, 113, 171n, 181, 182, 198n, 200n, 225, 238, 251, 254
Sanitary science, 112
Santayana, George, 52
Saratoga Springs, N.Y., 88, 145, 160, 171, 172, 173, 177
Sargent, John S., 225
Saturday Club, 124
Schirmer, Rudolph, 225
Schmoller, Gustav von, 177, 187n
Schurz, Carl, 128
Schuyler, Eugene, 188n
Science, 184n, 187n
Scott, Austin, 156, 157, 171
Scovel, Sylvester F.: on professionalism, 223
Seligman, E. R. A., 185, 186, 188n, 200, 236n

Shattuck, George, 80–81
Shattuck, Lemuel, 98
Shaw, Albert, 229
Sheffield Scientific School (Yale), 74, 99, 148, 222
Sheldon, George, 226
Silliman, Benjamin, Jr., 70
Simmonds, Edward, 225
Skilton, James A., 214
Small, Albion W.: on origins of sociology, 47; opinion of ASSA, 203; regards interdependence as justification for professional sociology, 208–209; on "vortex causation," 253; mentioned, 25, 206n, 239–240, 251n
Smith College, 200
Smith, Adam: on interdependence, 30–32; mentioned, 10, 243
Smith, Gerrit, 49, 198n
Smith, Hopkinson, 226
Smithsonian Institution, 66, 214
"Socialists of the chair," 180
Social economy: distinguished from political economy by Mill, 136; contrasted to political economy by Sanborn, 137. *See also* American Social Science Association—Department of Social Economy
Social gospel, 178–179
Social science: ASSA conception of, 78, 86–88, 100–110, 140, 164; indispensable premise for movement to establish authority, 87–88; and civil service reform, 119; conceptions of, 192, 193, 194, 204–206
Society for the Study of National Economy: proposed by James and Patten, 180–181
Sociological Institution (proposed), 214
Sociology: aspired to synthesize all social science fields, 190–191; contrasted to social science by Sanborn, 195; distinguished from social science by Giddings, 204; rival conceptions of Giddings and Small, 206n
Sparks, Jared, 124
Specialization of function, 36–37, 238n. *See also* American Social Science Association—departmental structure
Spellman, Cardinal Francis, 232
Spencer, Herbert: influence of, 2, 7, 245; and causal attribution, 45; likens social reform to hammering warped iron plate, 209n; attacked by William James on role of "great men" in history, 246–250; mentioned, 8, 83, 126, 139, 140, 191, 195, 204, 206, 207, 244n
Springfield *Republican,* 54–55, 115, 122,

128, 129, 172, 204, 212
Stanford University, 203
State Charities Aid Association of New York City, 229
State Charities Record, 229
Straus, Oscar S., 216, 228, 232
Stearns, Frank Luther, 49, 53
Stedman, Edmund Clarence, 227
Steele, G. M., 188n
Steinbeck, John, 3
Stern, Simon, 114
Stokes, Anson Phelps, 216
Stowe, Harriet Beecher, 51
Strong, Oliver S., 100
Strong, William, 118
Sumner, Charles, 52, 96n, 99, 124, 128, 129
Sumner, William Graham: abandons pulpit for sociology, 45–46; mentioned, 2, 163n, 184, 187, 200

Talbot, Emily, 126n, 197
Taussig, Frank W., 184
Taylor, Graham, 191n
Taylor, Maxwell D., 232
Thayer, James Bradley, 221
Thoreau, Henry David, 49, 53
Ticknor, George, 75–76, 77, 124
Tocqueville, Alexis de, 75
Tompkins County Jail, 194
Tompkins County Poorhouse, 194
Torrey, H. W., 171, 192
Toynbee, Arnold, 200
Transcendentalism, 52, 56–61 *passim*
Transportation-communication revolution, 33–36
Tucker, Ellen, 54
Tyler, Moses Coit, 170, 171, 172, 174n

Unintended consequences, 31–32, 243–244
Union League Club (New York), 192n
U.S. Coast Survey, 72
U.S. Sanitary Commission, 93
University of California, Berkeley, 133, 148
University of Chicago, 203
University of Illinois, 196
University of Iowa, 221
University of Michigan, 76, 99, 112, 148, 172
University of Pennsylvania, 180
University of Wisconsin, 133n, 148
University of Wyoming, 133n
Utilitarian dilemma, 6n
Utilitarianism, 6n, 10, 98, 103n, 109n

Value-free social science. *See* Investigation *vs.* agitation

Veblen, Thorstein, 10, 163, 231
Verein für Sozialpolitik, 159, 180, 200
Vico, Giovanni Battista, 6
Villard, Henry: becomes secretary of ASSA, 115; and civil service reform, 116–121; resigns, 126; mentioned, 103n, 133, 213
Volkswirtschaftlicher Kongress, 180
Voluntarism: espoused by ASSA, 190, 206–207; mentioned, 191, 241–256 *passim. See also* Idealism; Individualism
"Voluntaristic theory of action," 5–7

Walker, Amasa, 99, 100, 113, 114
Walker, Ariana Smith, 53–54
Walker, Francis Amasa, 54, 99, 172–173, 177n, 188n, 200
Walker, George, 113, 114, 192n
Wallas, Graham: on interdependence of society, 252
Ward, Lester, 2, 136n, 206, 207, 214n
Warner, Amos G.: opinion of ASSA, 203
Warner, Charles Dudley: presides over first meeting of NIAL, 226; president of ASSA, 227; mentioned, 219
Warren, Charles, 108
Washburn, Emory, 104n, 113, 132
Washington, Booker T., 201
Watson, John B., 8
Wayland, Francis, Sr., 76–77
Wayland, Francis, Jr.: and ASSA's model social science curriculum, 196; dominates ASSA Jurisprudence Department, 221; mentioned, 76, 158, 216, 218, 219
Wayland, Heman Lincoln, 76, 201n
Weber, Max 4, 6, 45, 159, 202n, 249n
Weeden, William B., 174n
Wells, David Ames, 74, 99, 104n, 113, 114, 146, 163n, 164, 182, 184
Western House of Refuge (Rochester, N.Y.), 199
Western Social Science Association (Chicago), 116, 133
Wharton, Joseph, 118
Wharton School of Finance and Commerce (University of Pennsylvania), 118
Wheelwright, H. B., 116n
Whistler, James Abbott McNeill, 225
White, Andrew Dickson: reform motives of, 175; defends Ely's platform for AEA, 185; continuing high opinion of ASSA, 190, 192, 197; invites Sanborn to teach social science at Cornell, 193; mentioned, 86, 104n, 110n, 114, 125–126, 148, 160, 164, 170, 171, 172, 174n, 182, 190, 198, 214n, 230, 232, 237n, 251n
White, Horace, 128

White, James C., 99
White, Morton, 4, 9–16
Whitehead, Alfred N., 19
Whittier, John Greenleaf, 52
Wilder, David M., 113
Willards Asylum, 194
Willoughby, W. W., 230
Wilson, Henry, 52, 72, 114
Wilson, Woodrow, 170, 229, 252
Wines, Enoch C., 113, 114
Winsor, Justin: urges historians to organize, 176; mentioned, 171, 174n
Wisconsin Academy of Sciences, Arts and Letters, 133n
Wisconsin Board of Charities, 137

Woods, Robert, 191n
Woolsey, Theodore Dwight, 99, 104, 146n, 221
Wooster University (Ohio), 223
Wright, Carroll D.: urges ASSA to broaden its public appeal, 198; mentioned, 188n, 230

Yale College and University, 73–74, 83, 99, 104, 148, 152, 158, 186, 196, 216, 218n, 219, 221
Yale Law Journal, 220
Youmans, Edward L.: criticizes the ASSA, 139–143; mentioned, 83–84, 204

Ingram Content Group UK Ltd.
Milton Keynes UK
UKHW011808080623
423123UK00001B/84